Topics in Current Genetics

18

Series Editor: *Stefan Hohmann*

Jens Nielsen · Michael C. Jewett (Eds.)

Metabolomics

A Powerful Tool in Systems Biology

With 53 Figures, 13 in Color; and 14 Tables

Springer

Professor Dr. Jens Nielsen
Dr. Michael C. Jewett
Center for Microbial Biotechnology
BioCentrum-DTU
Technical University of Denmark
Søltofts Plads
DK-2800 Kgs. Lyngby
Denmark
e-mail: jn@biocentrum.dtu.dk
mcj@biocentrum.dtu.dk

The cover illustration depicts pseudohyphal filaments of the ascomycete *Saccharomyces cerevisiae* that enable this organism to forage for nutrients. Pseudohyphal filaments were induced here in a wild-type haploid MATa Σ1278b strain by an unknown readily diffusible factor provided by growth in confrontation with an isogenic petite yeast strain in a sealed petri dish for two weeks and photographed at 100X magnification (provided by Xuewen Pan and Joseph Heitman).

ISBN 978-3-642-09418-7 e-ISBN 978-3-540-74719-2

DOI 10.1007/978-3-540-74719-2

Topics in Current Genetics ISSN 1610-2096

Cover: WMXDesign GmbH, Heidelberg, Germany

Printed on acid-free paper

9 8 7 6 5 4 3 2 1

springer.com

Topics in Current Genetics publishes review articles of wide interest in volumes that center around specific topics in genetics, genomics as well as cell, molecular and developmental biology. Particular emphasis is placed on the comparison of several model organisms. Volume editors are invited by the series editor for specific topics, but further suggestions for volume topics are highly welcomed.

Each volume is edited by one or several acknowledged leaders in the field, who invite authors and ensure the highest standard of content and presentation. Only solicited manuscript will be considered. All contributions are peer-reviewed. All papers are published online prior to the print version. Individual DOIs (digital object identifiers) make each article fully citable from the moment of online publication.

All volumes of Topics in Current Genetics are part of the Springer eBook Collection. The collection includes online access to more than 3,000 newly released books, book series volumes and reference works each year. In addition to the traditional print version, this new, state-of-the-art format of book publications gives every book a global readership and a better visibility.

Editorial office:
Topics in Current Genetics
Series Editor: Stefan Hohmann
Cell and Molecular Biology
Göteborg University
Box 462
40530 Göteborg, Sweden
FAX: +46 31 7862599
E-mail: editor@topics-current-genetics.se

Table of contents

List of contributors

Aguilera-Vázquez, Luciano
Depto.de Biotecnología, Universidad Politécnica de Pachuca, Ex-Hacienda de Sta. Barbara, CP 43830, Municipio de Zenpoala, Hgo, Mexico

Andersen, Birgitte
Center for Microbial Biotechnology, BioCentrum-DTU, Technical University of Denmark, Søltofts Plads 221, DK-2800 Kgs. Lyngby, Denmark

Frisvad, Jens C.
Center for Microbial Biotechnology, BioCentrum-DTU, Technical University of Denmark, Søltofts Plads 221, DK-2800 Kgs. Lyngby, Denmark

Hansen, Michael A. E.
Center for Microbial Biotechnology, BioCentrum-DTU, Technical University of Denmark, Søltofts Plads, DK-2800 Kgs. Lyngby, Denmark

Hardy, Nigel
Department of Computer Science, University of Wales, Aberystwyth, Penglais, Aberystwyth SY23 3DB, United Kingdom
nwh@aber.ac.uk

Hummel, Jan
Max Planck Institute of Molecular Plant Physiology (MPI-MP), Am Muehlenberg 1, 14476 Potsdam-Golm, Germany

Jenkins, Helen
Department of Computer Science, University of Wales, Aberystwyth, Penglais, Aberystwyth SY23 3DB, United Kingdom

Jewett, Michael C.
Center for Microbial Biotechnology, BioCentrum-DTU, Technical University of Denmark, Søltofts Plads, DK-2800 Kgs. Lyngby, Denmark

Kopka, Joachim
Max Planck Institute of Molecular Plant Physiology (MPI-MP), Am Muehlenberg 1, 14476 Potsdam-Golm, Germany
Kopka@mpimp-golm.mpg.de

Koulman, Albert
AgResearch Limited, Grasslands Research Centre, Private Bag 11008, Palmerston North 4442, New Zealand.

Lane, Geoffrey A.
AgResearch Limited, Grasslands Research Centre, Private Bag 11008, Palmerston North 4442, New Zealand.

Mauch, Klaus
Insilico Biotechnology AG, Nobelstrasse. 15, D-70569 Stuttgart

Nielsen, Jens
Center for Microbial Biotechnology, BioCentrum-DTU, Technical University of Denmark, Søltofts Plads, DK-2800 Kgs. Lyngby, Denmark
jn@biocentrum.dtu.dk

Pettolino, Filomena
School of Botany, The University of Melbourne, 3010 Victoria, Australia

Reuss, Matthias
Institute of Biochemical Engineering and Centre of Systems Biology, University Stuttgart, Allmandring 31, D-70569 Stuttgart
reuss@ibvt.uni-stuttgart.de

Robert, Martin
Institute for Advanced Biosciences, Keio University, 403-1 Daihoji, Tsuruoka, Yamagata, 997-0017 Japan
mrobert@ttck.keio.ac.jp

Roessner, Ute
Australian Centre for Plant Functional Genomics, School of Botany, The University of Melbourne, 3010 Victoria, Australia
ute.roessner@acpfg.com.au

Selbig, Joachim
University of Potsdam, Institute of Biochemistry and Biology, c/o MPI-MP, Am Muehlenberg 1, 14476 Potsdam-Golm, Germany

Smedsgaard, Jørn
Center for Microbial Biotechnology, BioCentrum-DTU, Technical University of Denmark, Søltofts Plads 221, DK-2800 Kgs. Lyngby, Denmark

Soga, Tomoyoshi
Institute for Advanced Biosciences, Keio University, 403-1 Daihoji, Tsuruoka, Yamagata, 997-0017 Japan

Thrane, Ulf
Center for Microbial Biotechnology, BioCentrum-DTU, Technical University of Denmark, Søltofts Plads 221, DK-2800 Kgs. Lyngby, Denmark
ut@biocentrum.dtu.dk

Villas-Bôas, Silas G.
AgResearch Limited, Grasslands Research Centre, Private Bag 11008, Palmerston North 4442, New Zealand.
silas.villas-boas@agresearch.co.nz

Walther, Dirk
Max Planck Institute of Molecular Plant Physiology (MPI-MP), Am Muehlenberg 1, 14476 Potsdam-Golm, Germany

Zamboni, Nicola
Institute of Molecular Systems Biology, ETH Zurich, Wolfgang-Pauli Strasse 16, 8093 Zurich, Switzerland
zamboni@imsb.biol.ethz.ch

The role of metabolomics in systems biology

Jens Nielsen and Michael C. Jewett

Abstract

The metabolome comprises the complete set of metabolites, the non-genetically encoded substrates, intermediates, and products of metabolic pathways, associated to a cell. By representing integrative information across multiple functional levels and by linking DNA encoded processes with the environment, the metabolome offers a window to map core attributes responsible for different phenotypes. Given increasing demand to quantitatively identify the metabolome and understand how trafficking of metabolites through the metabolic network impact cellular behavior, metabolomics has emerged as an important complementary technology to the cell-wide measurements of mRNA, proteins, fluxes, and interactions (e.g. protein-DNA). Metabolomics is already a powerful tool in drug discovery and development and in metabolic engineering. While maintaining these strengths, the field promises to play a heightened role in systems biology research, which is transforming the practice of medicine and our ability to engineer living organisms.

1 Metabolomics

Metabolome analysis, originally proposed by Oliver et al. in 1998, seeks to identify and quantify the entire collection of intracellular and extracellular metabolites. Conceptually, there are two basic analytical methodologies used in metabolomics (Fig. 1) (Villas-Bôas et al. 2005a). Mainly exploited for classification, *metabolite profiling* strategies investigate qualitative scanning of a large number of metabolites (i.e. more than 100). Here, the pattern of metabolites (or even spectra from chromatography or mass spectrometry) is used to find discriminatory elements via high-throughput detection followed by data deconvolution methods (Kell 2004; Goodacre et al. 2004). Metabolite profiling comprises of metabolic fingerprinting, which covers the endometabolome (intracellular metabolites), and metabolic footprinting, which covers the exometabolome (metabolites in the growth media or extracellular fluid) (Fig.1). The other general method used in metabolomics is *target analysis*. Here, absolute, or at least semi-quantification and unambiguous detection of metabolites are achieved. While historically target analysis has been reserved for interrogating relatively small numbers of metabolites (e.g. less than 20), new developments enable quantitative analysis of more expanded metabolome coverage (e.g. Villas-Bôas et al. 2005b; Soga et al. 2003; Roessner et al. 2000).

Topics in Current Genetics, Vol. 18
J. Nielsen, M. C. Jewett (Eds.): Metabolomics
DOI 10.1007/4735_2007_0228 / Published online: 13 June 2007

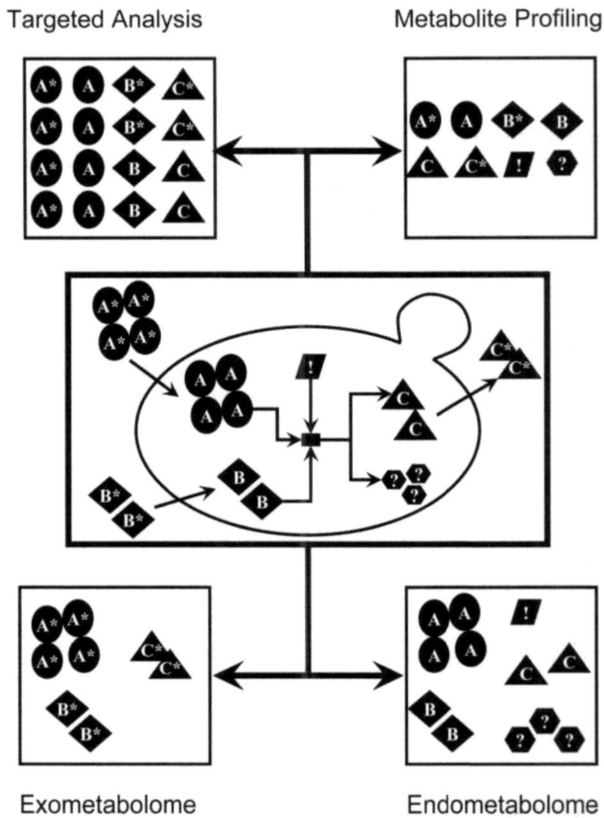

Fig. 1. Strategies for metabolome analysis. The metabolome is comprised of two parts, the endometabolome (intracellular metabolites) and the exometabolome (extracellular metabolites). Metabolome analysis seeks to identify cellular metabolites through targeted analysis (identification and quantification of pre-defined metabolites) or metabolite profiling (scanning of all metabolites identified by a specific analytical technique). Extracellular metabolites: A*, B*, and C*. Intracellular metabolites: A, B, C, !, ?. Note: ! and ? are unidentified metabolites.

A variety of analytical platforms have been utilized for metabolite detection (Villas-Bôas et al. 2005a). While most quantitative strategies couple a separation technique (e.g. capillary electrophoresis (CE), liquid chromatography (LC), and gas chromatography (GC)) with mass spectrometry (MS) or NMR based detection, it is not uncommon to only make use of direct infusion MS for metabolite profiling. From a practical standpoint, our inability to quantitatively extract and detect highly diverse families of metabolites in their original state over a large dynamic range with a single or even limited-set of analytical techniques makes analyzing the complete set of all metabolites associated to a cell impossible. Thus, metabolomics is more appropriately used to describe an "area of science rather than an analytical approach" (Villas-Bôas et al. 2005c).

2 Applications of metabolomics

The application of metabolomics has been widely pursued. Historically, metabolite profiling has been used for medical and diagnostic purposes (Horning and Horning 1971; Gates and Sweeley 1978) as well as strain classification and characterization (Frisvad and Filtenborg 1983). The latter has been particularly true for fungi and plants, which have extremely diverse metabolic landscapes. As an example, detection and quantification of mycotoxins from fungi has been a focal point for characterization studies (for overview of mycotoxins see Bennett and Klich 2003). The increase in public awareness over the safety of food and feed during the last several years has led to the establishment of many new laws and guidelines with respect to mycotoxins (FAO 2004). For the latest developments in analysis and detection methods, we refer the reader to a review detailing this topic (Krska et al. 2005). A key development is that unconventional biosensor methods, which typically rely on metabolite profiling; such as, electronic nose or tongue technology, have a strong potential to mature into key techniques for the detection of mycotoxins and toxigenic fungi (Logrieco et al. 2005).

Metabolome analysis is also an important tool in functional genomics, revealing the roles of genes from comprehensive analysis of the metabolome (Fiehn 2001; Trethewey 2001). For example, metabolite profiling and target analysis have been effectively used to classify molecular signatures responsible for the phenotype of silent and unknown mutations (Raamsdonk et al. 2001; Allen et al. 2003; Weckwerth et al. 2004). In one illustration, Weckwerth et al. (2004) demonstrated the application of target analysis, using GC-TOF-MS for quantification of more than 1,000 metabolites, to characterize the features responsible for a silent plant phenotype. Exploiting statistical tools, metabolic correlations were determined between identified metabolites (e.g. trehalose-erythritol) and used to reveal network maps which suggested hypotheses for the impact of an exact phenotype on carbohydrate and amino acid metabolism.

Hierarchical metabolomics, first reported in plants, is also well suited to guide targeted analysis of metabolism (Catchpole et al. 2005). Catchpole et al. (2005) used metabolome coverage of conventional and genetically modified (GM) potato crops to reveal that, apart from anticipated engineered differences, metabolic compositions were comparable among several types of cultivars. First, they applied metabolic fingerprinting of potato tuber extracts to classify several potato genotypes. Second, target analysis of defined and specific classes of metabolites using LC-MS and GC-TOF-MS was exploited to identify specific fructans responsible for the global classifications. Finally, data analysis tools were applied to remove the influence of anticipated differences in the GM crops and show that the GM and conventional crops were within the variation observed from investigating several unmodified metabolic phenotypes. Hierarchical analysis provides a rapid and relatively inexpensive screen for many functional genomics and screening applications.

3 The role of metabolomics in systems biology

In addition to finding utility in drug discovery, strain classification and functional genomics, metabolomics is emerging as a powerful tool in systems biology (Jewett et al. 2006; Wang et al. 2006). Systems biology is the quantitative study of an organism, viewed as a complex web of interacting and interchanging molecular participants (DNA, mRNA, proteins, and metabolites) and their environment. The overarching vision is that studying defined biological systems as a whole, through the combination of mathematical modeling and experimental biology, will provide insights into cellular behavior that are not apparent from investigating the components alone (Nielsen and Jewett, submitted). As a result of pioneering advances in genome sequencing, today's systems biology has dramatically enhanced our ability to study the relationships among active molecular players of the cell for describing and predicting cellular behavior. It promises to transform the practice of medicine and our ability to engineer living organisms by facilitating drug discovery, treating disease, and improving bioprocesses (Hood and Perlmutter 2004; Stephanopoulos et al. 2004; Weston and Hood 2004).

Realizing the promise of systems biology is hampered by the integrated and complex nature of cellular networks. First, there is not a one-to-one correlation between genes and metabolites (Nielsen and Oliver 2005). Not only can metabolites participate in many different biochemical reactions, but also multiple mRNAs can be formed from one gene, multiple proteins from one mRNA, and multiple metabolites from one enzyme. Second, interactions between proteins and small molecules, translational regulation, and other post-transcriptional mechanisms weaken the linkage between transcriptional state and metabolic phenotype. Third, the highly connected nature of cellular networks means that small perturbations rapidly traverse the cellular landscape; hence, impacting the overall functional operation of the network. Based on the yeast *Saccharomyces cerevisiae* genome-scale metabolic model containing about 800 metabolites and 1200 enzymatic reactions, for example, the average path length to get from any metabolite or enzyme to any other metabolite or enzyme is only about 5 (Patil and Nielsen 2005).

Given their central role as signals capturing information from all functional levels of the cell (Nielsen 2003) and also as nodes in dense metabolic networks (Jewett et al. 2006), determining concentrations of specifically identified metabolites is core to the systems biology agenda. We envision that the most powerful approach for using metabolomics data for systems biology is within the context of complex interactions, cellular pathways, molecular participants, and environmental stimuli that they connect. We believe that metabolic networks, which are well established, provide an effective and efficient scaffold for organizing systems biology data. Emerging computational strategies that exploit topological information from genome-scale metabolic models have already proven to be a compelling approach for inferring co-regulated cellular network structures (Patil and Nielsen 2005; Çakir et al. 2006).

Examples using metabolomics in systems biology mainly focus on quantifying metabolite levels and flows in primary metabolism. By relying on predefined connections between genetic sequences and metabolites, the information observed by acquiring a snapshot of the cellular metabolic composition is upgraded. Elucidating a metabolic image of the central carbon metabolism has provided insights for linking normal anabolic and catabolic trafficking with other branches of metabolism. The use of LC-MS, for example, has been exploited to map metabolic activity and flexibility through dynamic analysis of intracellular metabolites during the yeast cell-cycle (Wittmann et al. 2005) and the effect of culture age on metabolite pools (Mashego et al. 2005). Even though quantification of biomolecules involved in central metabolism offers many insights into key nodes of metabolism, other applications have also laid the foundation for target analysis of metabolic hubs that lay one step beyond the central metabolism. To quantify metabolites containing an amino or carboxylic acid group, Villas-Bôas et al. (2005b) applied a sensitive GC-MS method coupled to a statistical data-mining strategy for the integrated analysis of clearly identified and quantified intra- and extracellular metabolites in *S. cerevisiae* (approximately 60). By isolating statistically significant differences among metabolite levels from four biological conditions, they observed discriminatory metabolic features which hinted at the potential for future integration with comparative omic analyses. Highlighting the generality of this method, Panagiotou et al. (2005) have utilized this analytical approach to determine the influence of aerobic and anaerobic cultivation conditions on the metabolic state of *Fusarium oxysporum*.

Equally important in guiding a systems-level understanding of the overlapping layers of global regulation and network flexibility are efforts to experimentally measure the flow of material through central metabolism. Characterization of metabolic operation is achieved by using ^{13}C-labeled substrates followed by determination of characteristic metabolite patterns which can indicate directional flow (Sauer 2006). The most general approach uses proteinogenic amino acid analysis to infer labeling patterns and flux distributions. However, the application of rapid sampling and quenching has recently been applied to directly analyze intracellular metabolites from *S. cerevisiae* without being impeded by the high metabolic turnover rates (van Winden et al. 2005). This approach generates direct data without inference; however, caution must be exercised due to the rapid dynamics of exchange between metabolites and amino acids incorporated into cellular proteins (Grotkjær et al. 2004).

High-throughput efforts for comparative flux analysis offer an unprecedented view of the rigidity, flexibility, and performance of metabolic networks. In one illustration, Blank et al. considered flux data from over 30 mutants in *S. cerevisiae* to investigate potentially flexible fitness reactions during growth on glucose (Blank et al. 2005). Combination of measurements with mathematical modeling revealed that metabolic network robustness to single gene knockouts was principally a result of genetic redundancy, duplicate genes, with alternative pathways, redirection of carbon flow, having less importance. This approach was taken further in a larger scale, systematic flux analysis of 137 null mutants of *Bacillus subtilis* (selected from all major functional categories) on its preferred substrate (Fischer and Sauer 2005). As in the previous illustration, this strategy enabled identification of fundamental design principles of *in vivo* network operation. A key feature, which is likely to be universal, is the manifestation of rigid distribution patterns which are "largely independent of the rate and yield of biomass formation." The above cases represent powerful strategies to uncover the structure and function of the interplay between genetic regulatory networks and phenotype.

4 Outline of this book

Despite holding significant promise to elucidate key features of cellular behavior, there are several challenges to be addressed in the field of metabolomics. These include: interpreting metabolomic data, measuring concentrations of more and more metabolites using standardized and efficient methods, shifting towards more quantitative measurements, standardizing reporting practices, organizing data into user-friendly libraries and databases, identifying statistically relevant and discriminatory features in the data, and developing appropriate frameworks to integrate and map metabolite data with other X-omic data. For example, ensuring unbiased and robust quantification of a large number of metabolites is still a major concern. This issue is exacerbated relative to measurements of other participants in the cell (e.g. genes, mRNAs, and proteins) because metabolites exist on a considerably shorter time-scale (by more than an order of magnitude). In addition to short time-scales, the chemical diversity of metabolite classes and the physical barriers of the cell (e.g. cell structure and compartments) make metabolome coverage, particularly for the endometabolome, an issue. On top of sample preparation and chemical analysis are challenges of analyzing the growing volumes of data being generated. Although standard multivariate statistical methods can be applied, there are still a lot of difficult problems to be dealt with. Besides improving existing pre-processing methods, intelligent methods for finding patterns in (extremely) high-dimensional data related to prior functionality are currently, and will continue to be, on top of the agenda. These methods will have to detect complex patterns in ill-posed problems (many metabolites in relatively few samples). Beyond chemical analysis and data analysis, new data integration techniques are also required. We believe that the limiting step for utilization of metabolome data will be in improving our ability to develop appropriate frameworks to integrate and map data from multiple cellular levels.

This book brings together the latest in the field of metabolomics. We comprehensively present the current state of the metabolomics field by underscoring experimental methods, analysis techniques, standardization practices, and advances in specific model systems (e.g. *Escherichia coli*). In Chapter 2, issues of standardization are discussed. Standardization is very important for generating high quality data that are reproducible and can be consistently compared between different laboratories. Pre-analytical, intra-analytical, and post-analytical sources of variability are highlighted. Not only is standardization of analytical methods important, but we must also consider how the metabolomic data we generate are reported. In Chapter 3, Hardy and Jenkins focus on the importance of and recent developments in reporting standardization. In Chapter 4, another important issue for obtaining high-quality metabolome data is discussed; namely, the use of annotated databases of metabolites and mass spectrometry profiles. There are several initiatives to generate databases. Here, the group of Kopka describes the Golm Metabolome Database, which is specifically developed for handling data from gas-chromatography mass-spectrometry (GC-MS).

As discussed in Section 3, metabolome data play a very important role in improving our understanding of how metabolism operates. Chapters 5 and 6 discuss this objective from two levels. In Chapter 5, Reuss and co-workers describe how detailed kinetic models for individual enzymes can be put together to give a quantitative description of whole biochemical pathways. In particular they show how these models can describe the dynamic operation of such pathways, and how different reactions interplay through the common use of different co-factors. As the authors emphasize, the availability of high-quality metabolome data is central for determination of kinetic parameters of these models. In Chapter 6, Zamboni underscores a more global approach. Here, the objective is to map all fluxes in a metabolic network by quantifying the incorporation of ^{13}C-labelled substrates into different intracellular metabolites. This strategy enables quantitative information describing carbon flow throughout all of metabolism and is well suited for mapping a large number of mutants.

Data analysis is an integral part of metabolome analysis. There are many different ways to handle the often very large and high dimensional data-sets that are obtained. One way is to analyze the data in the context of either detailed kinetic models that can simulate dynamic operation of metabolic pathways (Chapter 5), or as a metabolic network model with the objective to get information about the fluxes (Chapter 6). However, often the objective is simply to identify specific biomarkers, i.e. significantly changed metabolites in one biological sample as compared to another, or groups of metabolites that have changed levels in one or more growth conditions. Chapter 7 describes a variety of statistical methods available for identifying key features in metabolome data.

Model organisms play a very important role in biology. In terms of metabolomics, *S. cerevisiae* and *E. coli* are two of the most important model microorganisms. Together with data analysis, metabolome analysis of the yeast *S. cerevisiae* is covered in Chapter 7. Chapter 8 is devoted to metabolome analysis of the bacterium *E. coli*. Here the group of Tomita discusses both efforts to measure a very large fraction of the metabolome, and also how metabolome data can be used to infer insights about the metabolic function of this bacterium.

The last two chapters of the book are devoted to metabolome analysis of filamentous fungi (Chapter 9) and plant cells (Chapter 10). Filamentous fungi play an important role as both pathogens and industrial cell factories. Understanding their secondary metabolism is very important for identification of new natural products, as well as for differentiating mycotoxin producing fungi from other fungi. Similarly, plant metabolomics plays an important role for rapid phenotypic characterization of food and feedstocks. Due to the extreme metabolic diversity of plant cells, mapping the complete plant metabolome is a significant challenge. However, many studies continue to expand our ability towards this ultimate goal.

Chapter 10 gives an overview of this field. In particular, the authors discuss the importance of the anatomy and physiology of plant cells in metabolomics studies.

References

Allen J, Davey HM, Broadhurst D, Heald JK, Rowland JJ, Oliver SG, Kell DB (2003) High-throughput classification of yeast mutants for functional genomics using metabolic footprinting. Nat Biotechnol 21:692-696

Bennett JW, Klich M (2003) Mycotoxins. Clin Microbiol Rev 16:497-516

Blank LM, Kuepfer L, Sauer U (2005) Large-scale 13C-flux analysis reveals mechanistic principles of metabolic network robustness to null mutations in yeast. Genome Biol 6:R49

Catchpole GS, Beckmann M, Enot DP, Mondhe M, Zywicki B, Taylor J, Hardy N, Smith A, King RD, Kell DB, Fiehn O, Draper J (2005) Hierarchical metabolomics demonstrates substantial compositional similarity between genetically modified and conventional potato crops. Proc Natl Acad Sci USA 102:14458-14462

Çakir T, Patil KR, Onsan ZI, Ulgen KO, Kirdar B, Nielsen J (2006) Integration of metabolome data with metabolic networks reveals reporter reactions. Mol Sys Biol 2:0050 doi:10.1038/msb4100085

Food and Agriculture Organization (2004) Worldwide regulations for mycotoxins in food and feed in 2003. FAO Food Nutrition paper 81

Fiehn O (2001) Combining genomics, metabolome analysis, and biochemical modelling to understand metabolic networks. Comp Funct Genomics 2:155-168

Fischer E, Sauer U (2005) Large-scale *in vivo* flux analysis shows rigidity and suboptimal performance of *Bacillus subtilis* metabolism. Nat Genet 37:636-640

Frisvad JC, Filtenborg O (1983) Classification of terverticillate *penicillia* based on profiles of mycotoxins and other secondary metabolites. Appl Environ Microbiol 46: 1301-1310

Gates SC, Sweeley CC (1978) Quantitative metabolic profiling based on gas chromatography. Clin Chem 24:1663-1673

Goodacre R, Vaidyanathan S, Dunn WB, Harrigan GG, Kell DB (2004) Metabolomics by numbers: acquiring and understanding global metabolite data. Trends Biotechnol 22:245-252

Grotkjaer T, Akesson M, Christensen B, Gombert AK, Nielsen J (2004) Impact of transamination reactions and protein turnover on labeling dynamics in C-13-labeling experiments. Biotechnol Bioeng 86:209-216

Hood L, Perlmutter RM (2004) The impact of systems approaches on biological problems in drug discovery. Nat Biotechnol 22:1215-1217

Horning EC, Horning MG (1971) Human metabolic profiles obtained by GC and GC/MS. J Chromatogr Sci 9:129–140

Jewett MC, Hofmann G, Nielsen J (2006) Fungal metabolite analysis in genomics and phenomics. Curr Opin Biotechnol 17:191-197

Kell DB (2004) Metabolomics and systems biology: making sense of the soup. Curr Opin Microbiol 7:296-307

Krska R, Welzig E, Berthiller F, Molinelli A, Mizaikoff B (2005) Advances in the analysis of mycotoxins and its quality assurance. Food Addit Contam 22:345-353

Logrieco A, Arrigan DW, Brengel-Pesce K, Siciliano P, Tothill I (2005) DNA arrays, electronic noses and tongues, biosensors and receptors for rapid detection of toxigenic fungi and mycotoxins: a review. Food Addit Contam 22:335-344

Mashego MR, Jansen ML, Vinke JL, van Gulik WM, Heijnen JJ (2005) Changes in the metabolome of *Saccharomyces cerevisiae* associated with evolution in aerobic glucose-limited chemostats. FEMS Yeast Res 5:419-430

Nielsen J (2003) It is all about metabolic fluxes. J Bacteriol 185:7031-7035

Nielsen J, Oliver S (2005) The next wave in metabolome analysis. Trends Biotechnol 23:544-546

Oliver SG, Winson MK, Kell DB, Baganz F (1998) Systematic functional analysis of the yeast genome. Trends Biotechnol 16:373-378

Patil KR, Nielsen J (2005) Uncovering transcriptional regulation of metabolism by using metabolic network topology. Proc Natl Acad Sci USA 102:2685-2689

Panagiotou G, Villas-Boas SG, Christakopoulos P, Nielsen J, Olsson L (2005) Intracellular metabolite profiling of *Fusarium oxysporum* converting glucose to ethanol. J Biotechnol 115:425-434

Raamsdonk LM, Teusink B, Broadhurst D, Zhang N, Hayes A, Walsh MC, Berden JA, Brindle KM, Kell DB, Rowland JJ, Westerhoff HV, van Dam K, Oliver SG (2001) A functional genomics strategy that uses metabolome data to reveal the phenotype of silent mutations. Nat Biotechnol 19:45-50

Roessner U, Wagner C, Kopka J, Trethewey RN, Willmitzer L (2000) Technical advance: simultaneous analysis of metabolites in potato tuber by gas chromatography-mass spectrometry. Plant J 23:131-142

Sauer U (2006) Metabolic networks in motion: 13C-based flux analysis. Mol Syst Biol 2:0062 doi:10.1038/msb4100109

Soga T, Ohashi Y, Ueno Y, Naraoka H, Tomita M, Nishioka T (2003) Quantitative metabolome analysis using capillary electrophoresis mass spectrometry. J Proteome Res 2:488-494

Stephanopoulos G, Alper H, Moxley J (2004) Exploiting biological complexity for strain improvement through systems biology. Nat Biotechnol 22:1261-1267

Trethewey RN (2001) Gene discovery via metabolic profiling. Curr Opin Biotechnol 12:135-138

van Winden WA, van Dam JC, Ras C, Kleijn RJ, Vinke JL, van Gulik WM, Heijnen JJ (2005) Metabolic-flux analysis of *Saccharomyces cerevisiae* CEN.PK113-7D based on mass isotopomer measurements of (13)C-labeled primary metabolites. FEMS Yeast Res 5:559-568

Villas-Boas SG, Mas S, Akesson M, Smedsgaard J, Nielsen J (2005a) Mass spectrometry in metabolome analysis. Mass Spectrom Rev 24:613-646

Villas-Boas SG, Moxley JF, Akesson M, Stephanopoulos G, Nielsen J (2005b) High-throughput metabolic state analysis: the missing link in integrated functional genomics of yeasts. Biochem J 388:669-677

Villas-Boas SG, Rasmussen S, Lane GA (2005c) Metabolomics or metabolite profiles? Trends Biotechnol 23:385-386

Wang QZ, Wu CY, Chen T, Chen X, Zhao XM (2006) Integrating metabolomics into a systems biology framework to exploit metabolic complexity: strategies and applications in microorganisms. Appl Microbiol Biotechnol 70:151-161

Weckwerth W, Loureiro ME, Wenzel K, Fiehn O (2004) Differential metabolic networks unravel the effects of silent plant phenotypes. Proc Natl Acad Sci USA 101:7809-7814

Weston AD, Hood L (2004) Systems biology, proteomics, and the future of health care: toward predictive, preventative, and personalized medicine. J Proteome Res 3:179-196

Wittmann C, Hans M, van Winden WA, Ras C, Heijnen JJ (2005) Dynamics of intracellular metabolites of glycolysis and TCA cycle during cell-cycle-related oscillation in *Saccharomyces cerevisiae*. Biotechnol Bioeng 89:839-847

Jewett, Michael C.
Center for Microbial Biotechnology, BioCentrum-DTU, Technical University of Denmark, Søltofts Plads, DK-2800 Kgs. Lyngby, Denmark

Nielsen, Jens
Center for Microbial Biotechnology, BioCentrum-DTU, Technical University of Denmark, Søltofts Plads, DK-2800 Kgs. Lyngby, Denmark
jn@biocentrum.dtu.dk

Analytical methods from the perspective of method standardization

Silas G. Villas-Bôas, Albert Koulman, and Geoffrey A. Lane

Abstract

Variability between laboratories, between instruments and between analysts within the same laboratory is an important issue of practical concern for metabolomics. Method standardization is essential for comparability of metabolomics data between experiments and laboratories in multi-disciplinary studies. However agreed standard requirements to extract metabolites, to concentrate cell extracts and to detect low molecular weight molecules in biological samples are lacking, and this significantly limits data comparison. This chapter reviews the sources of variation in analytical methods in current use and outlines possible quality specifications for global metabolite analysis. We categorize the sources of variability as pre-analytical (sampling and sample preparation), intra-analytical (instrumentation) and post-analytical (data mining and handling). The broad range of applicability of metabolomics precludes a generalised uniform approach. However by analyzing the factors influencing metabolite measurements, we aim to highlight areas for developing recommendations for method standardization that minimize analytical variation and specifications of performance standards including quality control procedures and measures of data quality in order to improve laboratory performance and to enable scientist to compare data across studies.

1 Introduction

Metabolomics or metabolome analysis is a rapidly evolving field that has gained increased popularity in recent years. Metabolomics is well-recognized for its potential as a functional genomics tool (Oliver et al. 1998; Fiehn et al. 2000; Raamsdonk et al. 2001; Trethewey 2001; Fiehn 2002; Sumner et al. 2003; Bino et al. 2004; and others) and by its broad range of applicability, mainly due to the development of powerful analytical methods capable of screening a large number of chemical compounds in biological samples.

Initially, the metabolomics field focused its attention on the development of analytical methods that enabled scientists to detect hundreds of compounds in a single analysis (Gavaghan et al. 2000; Roessner et al. 2000, 2001; Soga et al. 2000, 2002a, 2002b; Allen et al. 2003; Castrillo et al. 2003; Dunn et al. 2005a; Villas-Bôas et al. 2005a; and others). These methods generate large data sets, and

Topics in Current Genetics, Vol. 18
J. Nielsen, M. C. Jewett (Eds.): Metabolomics
DOI 10.1007/4735_2007_0217 / Published online: 14 April 2007

the interpretation and integration of these data with other 'omics' related data continues to represent a great challenge for the area. As a consequence, the major focus of metabolomics studies during the last few years has been on data mining and data analysis, resulting in considerable advances in bioinformatics tools applicable to metabolome analysis (Goodacre et al. 2004; Kell 2004; Smedsgaard and Nielsen 2004; Borodina and Nielsen 2005; Fell 2004; Wang et al. 2006; and others). However, the quality of the information extracted by these bioinformatics tools and techniques depends on the quality of the experimental data. Interpreting experimental data requires information about the samples used in the experiment, the conditions under which measurements were taken, the equipment used to take the measurements, etc. Important models for reporting 'omics' data that capture experimental descriptions alongside experimental results, facilitating the development of systems for storage and dissemination of experimental data have been proposed for transcriptomics i.e., MIAME (Spellman et al. 2002), for proteomics i.e., PEDRO (Taylor et al. 2003), and for plant metabolomics i.e., ArMet (Jenkins et al. 2004).

These tools offer a framework for comparison of metabolomics experimental data sets to independently verify experimental findings, and for comparative metabolomics, emulating the success of genomics in exploiting the commonalities of biological systems to extrapolate and interpolate findings. However comparisons of metabolomics data are complicated by experimental differences. Currently, a plethora of different methods are being used for sampling and extraction of metabolites from biological matrices, as well as an enormous diversity of analytical techniques and instrumentations employed for separation and detection of different classes of metabolites. Standards make an enormous contribution to most aspects of science, and the lack of a common or ubiquitous standard procedure across different research groups and laboratories limits the efficient interchangeability and comparability of metabolomics data.

The adoption of standard experimental protocols would undoubtedly facilitate the direct comparability of data, but this is only relevant for experiments of similar type with the same or similar organism. In practice, the broad range of applicability of metabolomics precludes a generalised uniform approach. The goal of comprehensive metabolomic analysis with the diversity of metabolite chemistry means that any experimental protocol represents a compromise, and different organisms, sample types and experimental goals demand different compromises. Thus progress towards method standardisation across the broad scope of metabolomics will not be achieved through the pursuit of universal standard protocols.

Therefore, method standardisation for metabolomics has to be considered in a different way. While universal protocols may not be possible, specification of performance standards of experimental procedures for metabolite analysis is a worthwhile goal. Towards this goal we examine the sources of experimental variability with the aim of achieving a greater understanding of the validity of the metabolomics data and providing for better biological interpretation.

To assess the validity of metabolomics data, it is important to have a clear understanding of the particular information each different type of measurement can provide, and the quality of that information. In particular, it is necessary to know which types of metabolites are measurable by the technique(s) used, what samples were collected, how they were harvested, processed, and stored, which molecular mass cut-off was adopted, and how the data was extracted and processed before analysis, and the implications at each stage of the process for data quality. From the laboratory perspective, sources of variation can be categorized into pre-analytical (sampling and sample preparation), intra-analytical (instrumentation), and post-analytical (data mining and handling).

Therefore, in this chapter we analyze the various factors that influence metabolite measurements (pre-, intra-, and post-analytical), drawing attention to approaches that minimize analytical variation in order to improve laboratory performance and to procedures for acquiring measures of data quality together with the data to enable scientists to compare data across studies and laboratories, and to make metabolomics data storage more meaningful.

2 Pre-analytical variability

Variability in biological sciences can be inherent to the biological material or can be introduced by human (scientist) manipulation. Thus, pre-analytical variability is of two main sources: biological variability or variability introduced by sample handling. The latter can be sub-divided further into variability introduced during sampling and variability introduced during sample processing. In the following, we review the main aspects of each class.

2.1 Biological variability

The major pre-analytical variability is the inherent biological variability of cells and organisms. For multi-cellular organisms the metabolome data typically shows high variability between individuals. Biological variability between individuals of the same genotype is typically the major source of variations in metabolite levels analyzed by GC-MS and recent research suggests patterns of correlations of individual variation may be more informative about the metabolic phenotype than is the mean metabolic profile (Morgenthal et al. 2006). Although measurements of populations of unicellular and asexual organisms tend to present less biological variability, their metabolism responds very promptly to minimal alterations in their environment, such as oxygen and substrate availability, temperature, pH, etc. Small variations in these environmental conditions very often result in considerable variability in metabolite levels. Therefore for metabolome analysis, the numbers of samples and replicates per treatment have to be as large as possible.

Large numbers of samples and replicates are particularly important in studies involving higher organisms such as plants and animals, and this can be limited by costs and feasibility. Biological variability of around 40% is usually reported for large-scale profile of plant metabolites (von Roepenack-Lahaye et al. 2004), and this is expected to be similar or even greater for animal samples. For example, to obtain a significant ($p < 0.05$) two-fold difference in metabolite level of samples presenting 40% biological variability, it is necessary to have at least five individuals per trial/treatment (based on student test principles). Even more replicates are desirable for comparative correlation analysis. While multiple samples from each individual are in principle desirable to minimize the technical variability that will be discussed further in this chapter, in practice greater biological replication is probably preferable.

On the other hand, the biological variability between replicates of microbial or plant/animal cell cultures is expected to be minimal, as all measurements are of population means. Villas-Bôas et al. (2005a, 2006a) showed that in contrast to plant and animal metabolomics, sample-to-sample variability exceeds replicate flask-to-flask variability in microbial metabolomics. The variability of metabolomics data from microbial or cell cultures is mostly due to technical variability (e.g. cultivation conditions, growth phase, sampling and sample processing, etc.). Therefore, for microbial and cell cultures the numbers of cultures/flasks can be reduced, but large longitudinal sample sizes (many samples from one culture) is required.

2.2 Variability introduced during sampling

Sampling is a critical step in metabolome analysis and must be carefully considered. There are several studies demonstrating how fast the turnover of metabolites inside the cells is (de Koning and van Dam 1992; Villas-Bôas et al. 2005b, 2007a; and many others). Cellular metabolism, particularly primary or central metabolism, rapidly adjusts to minimal changes in the environment, and since metabolite levels are coupled through metabolic networks (Nielsen 2003); changes propagate rapidly, generating marked changes in metabolite profiles in matter of seconds.

Table 1 lists the main sources of variability during sampling. Light-dependent bio-reactions are widely spread throughout living organisms. The metabolism of plants and photosynthesizing microbes are the most influenced by light, but the metabolism of animals also changes significantly between night and day periods. In addition, plant metabolism is not only dependent on light intensity but also on the wavelength of the light. For instance, upper leaves cast shadows over lower leaves resulting in significant differences in metabolite profiles for each leaf of the same plant. Thus, sampling must be performed in a short time-window and, ideally, under same light intensity (e.g. same period of the day/night), selecting leaves or other plant organs with a similar light-exposure.

Oxygen and CO_2 ratio is another important source of variability in metabolomics data, mainly because the sampling processing is likely to change the ratio of these gases in the cell environment. Microbial or cell cultures are exposed to a distinct atmosphere during sampling either by decreasing aeration of aerobic cultures or by introducing oxygen to anaerobic cultures. The change in O_2/CO_2 ratio may result in considerable changes in cell metabolism. Therefore, a quick sampling procedure, followed by a fast quenching of biochemical and chemical reactions is imperative to obtain a truly representative sample.

Table 1. Main sources of variability during sampling biological materials

	Plants	**Animals**	**Microbial and cell culture**
Light	Important source of variability during sampling	Metabolism changes according to day/night regimes	Important for photosynthesizing microorganisms/cells
O_2/CO_2 ratio	Minor importance during sampling	Important source of variability during sampling	Important source of variability during sampling
Nutrients/ substrates	Minor importance during sampling	Fasting period before sampling is recommended	Important source of variability during sampling

Another source of variability is the diet and/or nutrient/substrate availability prior to sampling. These source of variability can be minimized during sampling by feeding animals with identical diets, by monitoring the feed intake prior to sampling, and by sampling after a fasting period. Similarly, the metabolism of plants and microorganisms are also a result of nutrient and substrate availability, and the growth phase of the organism. Microbial batch cultures must be harvested at similar growth phases to minimize differences in the extent of substrate utilization. Plant samples must be properly defined by organ and growth stage as well as by growth conditions for comparability.

2.2.1 Sampling: the act of transferring a biological material to a laboratorial vessel

Based on the diverse sources of variability during sampling biological material for metabolome analysis, it is clear that this step has to be accomplished within a very short time window, in a reproducible manner, and assuring that all biochemical and chemical reactions will be quenched simultaneously or immediately after sampling. There are several limitations on achieving this goal that are specific to different biological material. Animals need to be sacrificed or submitted to surgery before their organs are removed, and both procedures induce instantaneous biochemical alteration of cellular metabolism, resulting in a distinct metabolite profile compared to the *in vivo* and non-stressed metabolic state. In addition, the organs of interest must be removed from the body, which may take seconds or even minutes to be achieved. A better alternative is to work with body fluids such as blood, cerebrospinal fluid, urine, milk, etc.

The choice of body fluids over tissues is done with the assumption that the metabolites found in most body fluids are largely reflective of the physiological state of the organ that produces or is bathed in that fluid (Wishart 2007). Hence, urine reflects processes going on in the kidney, bile in the liver, cerebrospinal fluid in the brain, and so on. The blood is a special body fluid as it potentially reflects all processes going on in all organs. According to Wishart (2007), this can be both a blessing and a curse, as metabolite perturbations in the blood, while easily detectable, cannot be easily traced to a specific organ or a specific cause. In terms of data variability, the choice of body fluid over tissues is also advantageous in that fluids are far easier to process and usually do not require extraction of metabolites from within cells, thus reducing sample handling.

Similarly, sampling plant material is also limited by the time required to remove the plants from the soil/substrate batch or to sample specific plant organs; as well as by the light dependency of plant metabolism (Roessner 2007). Although, contrary to animals, we can potentially quickly freeze a whole plant body in liquid nitrogen without any ethical issues, special care has to be taken about the time point when plant samples are harvested. In general, as a rule, all samples should be harvested at the same time point or in a very small timeframe. This may become difficult when a large set of plants is to be investigated. Otherwise, it is recommended that plant material is sampled in a randomized way in order to capture daytime differences in metabolite profiles within the variability throughout the data set (Roessner 2007).

Microbial or cell cultures have to be sampled at the same or very similar growth phase. Therefore, sample replicates have to be harvested consecutively in a very short time. The metabolism of cells in culture is much more vulnerable to the environment than cells within a tissue. Consequently, the metabolism of microorganisms and cells in cultures respond very rapidly to environmental changes. Therefore, the time required for transferring microbial and cell cultures from their culture environment (flasks, bioreactors, etc) to a sampling vessel is a critical factor that deserves special attention.

Villas-Bôas (2007a) review several techniques for fast transfer of culture samples from the cultivation flasks or reactor to the quenching solution and the different techniques vary with respect to speed and practicability. Research on sampling systems to measure microbial metabolite dynamics on a subsecond time scale has been reported during recent years (Theobald et al. 1993, 1997; Weuster-Botz 1997; Schaefer et al. 1999; Lange et al. 2001; Buziol et al. 2002; Visser et al. 2002; and others). Ingenious devices have been developed, which present *pros* and *cons* and vary from manual sampling to fully automated (computer-aided) devices. A global overview of the main sampling techniques developed to date can be found in Villas-Bôas (2007b).

2.2.2 Quenching: stopping the cell metabolism

Besides quick harvesting, the sampling process can only be accomplished if the cellular metabolism is stopped and no further biochemical and/or chemical reactions take place in the sample. Therefore, the samples have to be quenched at the time of harvesting. As discussed in Villas-Bôas et al. (2005b) and Villas-Bôas (2007a), a rapid inactivation of metabolism is usually achieved through rapid changes in temperature or pH. This is usually done by placing the biological sample in contact with a cold (< -40°C) or hot (> 80°C) solution or with an acidic (pH < 2.0) or alkaline (pH > 10) solution. This process must be sufficiently fast to avoid changes in metabolite levels caused by alteration in the environment of the cells, ideally in a time window of a second.

Different biological samples require different techniques to achieve a proper quenching. When quenching plant or animal tissues, an important factor to be considered is that the size of the sample should be compatible with the quenching technique used, as well as the quenching agent. Cell tissues are often distributed in several layers, where the peripheral cells tend to be quenched before the central ones, increasing the sample variability. Therefore, tissue thickness as well as a reproducible sample size should be seriously considered when planning experiments. The most reasonable way to achieve efficient quenching of plant or animal tissue is by rapid freezing in liquid nitrogen. As liquid nitrogen is an inert and highly volatile substance (boiling point -196°C) it can be rapidly eliminated from the sample by evaporation. Alternatively, cold methanol solution or acidic treatments using perchloric or nitric acid can be used as quenching agents; however, their efficiency is controversial and no validation of these methods to quench plant or animal tissues has been reported so far.

On the other hand, microbial or cell cultures are generally characterized by a high dilution ratio between biomass and extracellular medium, and this has large affects on the quenching process. The most common quenching methods for this type of sample are based on aqueous solutions containing an organic solvent, usually methanol or ethanol, buffered or non-buffered, set to an extreme temperature (very cold or very hot), or acidic solutions, typically perchloric acid. Sometimes liquid nitrogen is also used as a quenching agent (Villas-Bôas 2007a).

However, the greatest challenge in quenching microbial or cell cultures is the separation between intra- and extracellular metabolites. Culture media are usually very rich in nutrients and intracellular metabolites are usually in an osmotic equilibrium with the extracellular medium (Villas-Bôas 2007a). Due to the low biomass: medium ratio in typical microbial and cell cultures, the concentration of extracellular metabolites greatly exceeds the concentration of intracellular ones in a given sample; therefore, the fractionation of cell biomass from spent culture medium is highly desirable.

However, microbial cells are sensitive to most quenching agents developed to date, and this factor is very rarely taken into consideration during microbial metabolomics studies. Bacterial cells, for instance, are known to loose intracellular metabolites by leakage due to cell wall damage when in contact with any quenching solutions currently in use (Villas-Bôas 2007a). Wittmann et al. (2004) proposed a protocol for fast separation of bacterial cells from extracellular media using fast filtration under vacuum and washing the biomass with four volumes of cold saline solution (0.9% NaCl) at -0.5°C (the whole filtration step including the washing can be finished in less than 45 s). This method seems to permit authentic quantification of intracellular amino acid pools. However, this procedure may be less suitable for precise analysis of metabolites with a faster turnover, for example, phosphorylated intermediates.

The most widely spread method for quenching yeast cell cultures makes use of cold methanol solution as the quenching agent and was originally proposed by de Koning and van Dam (1992). This method gained great popularity due to its ability to separate cells from extracellular metabolites, without apparent damage of the

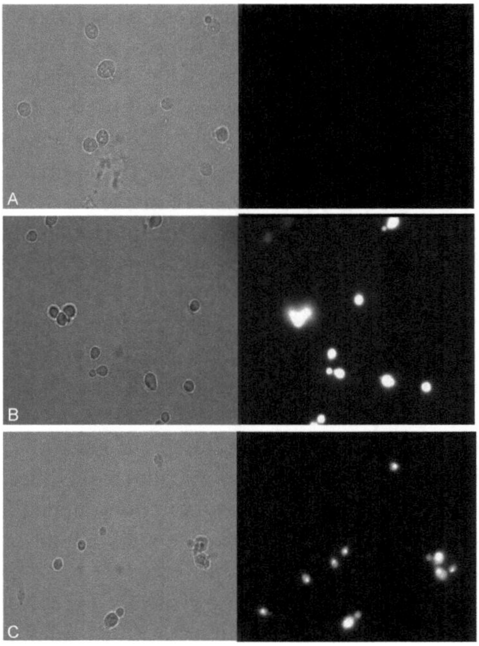

Fig. 1. Yeast cells assayed with propidium iodide to test the membrane integrity during quenching with cold methanol solution at -40°C. (A) Control cells centrifuged and resuspended in saline solution. (B) Cells quenched with 60% (v/v) methanol solution at -40°C. (C) Cells quenched with 60% (v/v) buffered methanol solution at -40°C. "Reproduced from Yeast, vol. 22, Global metabolite analysis of yeast: evaluation of sample preparation methods, page 1162, Copyright (2005), with permission from John Wiley & Sons, Inc.

yeast cell envelope. However, it was recently demonstrated that yeast cells, similarly to bacterial cells, are sensitive to cold methanol solution either buffered or non-buffered (Fig. 1), and leakage of several intracellular metabolites has been observed after quenching *S. cerevisiae* cultures with cold methanol solution, following the original protocol (Villas-Bôas et al. 2005c). Several organic and amino acids are practically washed out of the yeast cells after being in contact with the cold methanol solution. However, by decreasing the time the yeast cells stay in contact with the methanol solution (e.g. applying quicker centrifugation) the leakage of intracellular metabolites can be minimized significantly, but a few metabolites may undergo a higher leakage under faster centrifugation, for example, lactate, citramalate, myristate (Villas-Bôas et al. 2005c).

Other quenching agents such as strong inorganic acids (Cook et al. 1976; Larsson and Törnkvist 1996) or alkalis (Cook et al. 1976) or even boiling ethanol (Gonzales et al. 1997), naturally damage the microbial cell envelopes provoking the leakage of intracellular metabolites to the extracellular medium. In addition, extreme pHs and high temperatures can potentially degrade several metabolites as demonstrated by Hajjaj et al. (1998), Maharjan and Ferenci (2003) and Villas-Bôas et al. (2005c).

Alternatively, the analysis of intracellular and extracellular metabolites can be combined. Usually the extracellular metabolites are determined separately in the samples of spent culture media and their levels are subtracted from the pool (intra + extra) in order to get an estimation of the intracellular levels, but this approach limits considerably the detection of intracellular compounds and gives rise to high variability of estimates of intracellular metabolites that typically make up a small fraction of the total metabolite pool. Therefore, microbial and cell culture metabolomics desperately call for the development of an efficient quenching procedure that allows a reliable intracellular metabolite analysis.

2.3 Variability introduced during sample processing

There are very few systematic studies on variability introduced during sample processing for metabolome analysis. However, sample processing represents significant source of variability in metabolomics data. We can assume that the more steps in sample processing, the more between-sample and between-batch variability will be observed. Therefore, from the metabolomics point of view, the sample processing stage should ideally contain as few steps as possible and should prevent any loss by chemical or physical chemical degradation, remaining enzymatic activity in the samples, or mechanical losses. However, this is a virtually impossible task. Except for the few occasions where the samples do not need to be processed before analysis (e.g. during direct injection mass spectrometry of extracellular samples or NMR analysis), as soon as the sample starts to be processed the original metabolite profile starts to get "adulterated". Several metabolites are thermo-labile or light-sensitive, reduced metabolites can also be easily oxidized in the presence of air or any other oxidative compounds, and even exchange of key functional groups can be observed (e.g. phosphate groups of phosphorylated compounds) (Villas-Bôas 2007a). Therefore, any sample processing procedure in metabolome analysis should employ mild conditions (e.g. low temperatures, inert solvents, etc) and quick procedures to minimize variability.

Since we cannot avoid alteration in metabolite levels during sample processing, losses could be estimated by introducing internal standardization during the early stages of sample processing. However, based on the internal standardization concept, an efficient internal standard should mimic the analyzed compounds as much as possible (Freisleben et al. 2003). This is another challenge in metabolomics because we usually aim to analyze as many metabolites as possible in a single sample, which poses severe limitations for choosing an efficient internal standard.

Mashego et al. (2004) published a new and innovative method for accurate quantification of changes in concentrations of intracellular metabolites, named MIRACLE, where each metabolite concentration is quantified relative to the concentration of its own internal standard (an U-^{13}C-labeled equivalent), thereby eliminating most pre- and intra-analytical variability (with no spiking or standard addition needed). Despite the elegancy and efficacy of the method, the cost required for each experiment is high due to the need for *de novo* synthesizes all U-^{13}C-labeled metabolites using U-^{13}C-labeled substrates. These can be very expensive if a large number of treatments/mutants have to be investigated, and the method is essentially restricted to microbial and cell culture studies. Nonetheless, the MIRACLE method is undeniable the most powerful and robust innovation to date for reliable estimation of changes in intracellular metabolite levels and the development of micro-scale cultivations may minimize the costs of the method.

Besides internal standardization, additional precautions can be taken at different steps of sample processing in order to minimize sample variability. In the following, we review the main sources of variability introduced during sample processing, mainly focussing on the analysis of intracellular metabolites, where sample processing is greater.

2.3.1 Sample storage, homogenization, and concentration

Sample storage is usually an unavoidable step in sample processing. Commonly, samples are stored to be processed later after sampling and quenching, and once extraction of intracellular metabolites has been completed, the samples can be stored before analysis and even after analysis. According to Villas-Bôas (2007a), there are two main alternatives for storage of quenched plant/animal tissue samples before extraction of intracellular metabolites: (i) shock freezing at -80°C or (ii) freeze-dry and storage under vacuum at low temperature. Their advantages and disadvantages are presented in Table 2. Microbial and cell cultures as well as body fluids can also be stored after quenching by just freezing at -80°C. Storage at -20°C, however, is not recommended because there are indications of some biochemical reactions and enzymatic activities being able to take place at very low temperature, even down to -20°C (Junge et al. 2006; Roessner et al. 2006), specially if the sample is not completely frozen due to high salt contents, or presence of organic solvents. Although these reactions are supposed to be rare, slow and mostly reported on psychrophilic organisms, the biochemistry of below zero Celsius is poorly understood, and is therefore worth preventing.

Metabolite extracts, being in solution (aqueous/organic) or freeze-dried, are often stored before or even after analysis-in case the samples need to be re-analyzed. Therefore, the integrity of the compounds in the samples must be assured. Chemical degradation is an important source of variability in stored metabolite samples. Particularly thermo- and photo-labile metabolites can be degraded quickly if kept for long time at room temperature or exposed to light. Phosphorylated compounds, some sulphur derivatives, and reduced metabolites can be degraded or oxidised rapidly at room temperature. Similarly photo-degradation is a process that may result in high variability in the level of photosensitive metabolites. For example, *S*-adenosyl-*L*-methionine, which is a methyl donor metabolite; a cofactor for enzyme-catalyzed methylations, including catechol *O*-methyltransferase (COMT) and DNA methyltransferases (DNMT), is a very unstable compound that can be degraded very rapidly at temperatures above 0°C or when exposed to light. Therefore, the storage of metabolite extracts at low temperature (< -20°C) and

Table 2. Recommended procedures for three key steps in sample processing

Key step	Recommended procedures	Advantages	Disadvantages	Special care	Key references
Sample storage	A – Shock freezing at -80°C	A – Maintain sample integrity	A – Non-inactivation of enzymes in the samples	A – Care must be taken to avoid partially thawing samples	Fiehn 2002 Villas-Bôas et al. 2007a
	B – Freeze-dry and storage under vacuum	B – Ensure inactivation of enzymes in the sample	B – May potentially lead to irreversible adsorption of metabolites on cell walls and membranes and loss of volatile compounds	B – Dried samples must be stored in dry environment (e.g. desiccators) and at low temperatures	
Sample homogenization	A - Mortar & pestle	A – Cheap and efficient	A - Laborious	A,B & C - The samples must be ground under very low temperatures (e.g. under liquid nitrogen) and special care must be taken to ensure that all tissue is ground homogenously	Roessner et al. 2006; Villas-Bôas et al. 2007a
	B – Ball mills	B – Semi-automatic and temperature can be controlled	B & C – Bad performance with soft tissues		
	C – Ultraturax®	C - Semi-automatic and efficient for hard tissues (e.g. plant roots)			
Sample concentration	A – Lyophilization (= freeze-dry)	A – Gentle technique under low temperature	A - Time-consuming, low recovery of some metabolites and loss of volatile compounds	A – Large surface area is preferable to thick ice layers to obtain fast drying; process preferably carried out in a refrigerated chamber; and vacuum must be broken using an inert gas (e.g. nitrogen or argon)	Villas-Bôas et al. 2005a, 2007a
	B – Solvent evaporation under vacuum	B – Fast technique with good recovery of primary metabolites	B – Not recommended for large volume aqueous samples or for volatile compounds	B – Heating must be avoid and vacuum must be broken using an inert gas (e.g. nitrogen or argon)	

preferably in the dark is highly recommended. This will also avoid further chemical interactions between active compounds in extracellular samples (Villas-Bôas 2007a).

Homogenization before metabolite extraction is another important step in sample processing of plant and animal tissues in regards to variability in metabolite levels. Plant and animal tissues are heterogeneous and usually contain rigid cell walls (plants) or adipose and connective tissues and cartilage (animals), which call for sample homogenization to minimize difference between samples. It is extremely important that the homogenization process takes place at very low temperature, far below the freezing point of the tissues (e.g. under liquid nitrogen), to prevent cell defrosting and consequent re-activation of cellular enzymes (Table 2). According to Roessner et al. (2006), many enzymes are resistant to freezing and can be quickly activated after defrosting. For instance, the enzyme invertase, which efficiently cleaves sucrose to glucose and fructose, is resistant not only to freezing but also to the presence of chloroform, and can be active even at -20°C, provoking significant alterations on sugar profiles if samples are not properly frozen.

On the other hand, many metabolites are present at fairly low levels in the samples and additional sample dilution is often observed during sample preparation procedures, which imposes a requirement for sample concentration prior to analysis in order to improve detection. However, losses by degradation, evaporation and metabolite-class discrimination are also observed at this stage and again choices need to be made guided by the objectives of the study that is being carried out.

Freeze-drying, or lyophilization, and solvent evaporation under vacuum are commonly used methods to remove water or organic solvents from samples (Table 2). However, loss of analytes during lyophilization is often observed and the losses are certainly related to discrimination during resuspension as well as evaporation of low boiling point compounds. Different metabolites have different solubilities in the solvent used for resuspension, and therefore, discrimination during re-dissolving freeze-dried compounds in a very small volume of solvent is likely to happen. In addition, semi-volatile compounds such as 2-oxovaleric acid (bp = 88°C), acetic acid (bp = 117°C), lactic acid (bp = 122°C), pyruvic acid (bp = 156°C), etc can be partially lost by evaporation. Organic solvent evaporation under a vacuum seems to be a more reliable method for concentration of samples containing primary metabolites (Villas-Bôas et al. 2005b), but this technique is dependent on the type of extraction procedure used, since this procedure is not well suited for aqueous samples and loss of volatile metabolites is also observed.

Table 3 compares the recovery of metabolite standards after lyophilization and solvent evaporation, based on the results reported in Villas-Bôas et al. (2005c). Accordingly, recoveries of organic acids, nucleotides, sugars, and a peptide were higher than 80% after lyophilization. The fatty acid tested was poorly recovered, and the basic amino acid lysine was practically not recovered at all. The recovery of sugar-alcohols and sugar-phosphates was lower than 75%. On the other hand, concentration of samples by solvent evaporation under vacuum presented an excellent recovery for all amino acids, organic acids, nucleotides, sugars and sugar alcohols (Table 3). Only sugar-phosphates and glutathione (peptide) presented a

Table 3. General classification of different extraction methods based on the recovery of spiked metabolite standards (n=3) from yeast according to Villas-Bôas et al. (2005c)

	Classes of metabolites							
Method	Amino acids	Organic acids	Fatty acids	Nucleo tides	Pep-tides	Sugars	Sugar al-cohols	Sugar phos-phates
CMB	*****	****	Nr	***	***	***	****	**
BE	****	****	***	*	**	*	***	nr
PCA	**	*	*	*	***	***	****	nr
KOH	****	**	****	**	***	**	****	nr
MW	****	****	***	****	**	**	****	nr
PM	*****	*****	****	*****	***	**	****	nr

Extraction method: CMB, chloroform:methanol:buffer; BE, boiling buffered ethanol; PCA; perchloric acid; KOH, potassium hydroxide; MW, cold methanol solution 50% (v/v); PM, pure cold methanol.
***** > 80%; **** > 60%; *** > 40%; ** > 20%; * > 0%; nr, not recovered.
"Reproduced from Yeast, vol. 22, Global metabolite analysis of yeast: evaluation of sample preparation methods, page 1163, Copyright (2005), with permission from John Wiley & Sons, Inc.

comparatively lower recovery. The losses during lyophilization are quite substantial, yet this sample concentration technique is considered by most as a gentle methodology to concentrate samples. The excellent recovery obtained by solvent evaporation with reliable reproducibility (Table 3) indicates that this methodology is the best alternative for sample concentration in metabolome analysis. However, this technique is dependent on the type of extraction procedure used, which gives preference for extraction using only organic solvents.

2.3.2 Metabolite extractions

The extraction of intracellular metabolites is inevitably a time consuming step. The extraction solvent and conditions should be designed to limit any further physical and chemical alterations of the molecules and the entire extraction process should ensure minimal loss of the metabolites to be extracted. The extraction procedure aims to disrupt the cell structures liberating all or the maximum number of metabolites in their original state and in a quantitative manner to a defined extraction medium. The choice and development of efficient methods for extraction of intracellular metabolites requires an understanding of: (i) the cell wall structures, which are the first and main barrier to be broken; (ii) the chemical nature of the metabolites (i.e. physical and chemical forms, solubility, stability); and (iii) the sources of losses (especially their impact on subsequent recovery of metabolites). A complete study of these three factors influencing metabolite extraction, as well as the various forms for extraction of intracellular metabolites is found in Villas-Bôas (2007a).

It is virtually impossible to avoid losses during metabolite extractions mainly because of the high chemical diversity and the wide dynamic range of metabolite concentrations (Jiye et al. 2005; Villas-Bôas et al. 2005b, 2007a). Choices have to be made concerning which metabolites should be measured, and often analysis of

Table 4. Global recovery of different metabolites after two different sample concentration methods (n=3) according to Villas-Bôas et al. (2005c)

Metabolites	Lyophilization (= freeze-dry)*	Solvent evaporation under vacuum*
Valine	95	100
Lysine	0	100
Glutamate	100	100
Phenylalanine	50	100
Tryptophane	50	100
Pyruvate	100	100
Lactate	100	100
Fumarate	100	100
Succinate	100	100
Citrate	100	100
Isocitrate	100	100
2-Oxoglutarate	90	100
2-Oxoadipate	92	100
Pimelate	85	100
Myristate	42	83
NADP+	48	100
NAD+	50	100
Glutathione	100	56
Xylose	66	85
Trehalose	75	100
Glycerol	70	92
Xylitol	75	85
Arabitol	68	85
Mannitol	80	62
Ribose 5-phosphate	68	42
Glucose 6-phosphate	100	100
Fructose 6-phosphate	96	84
Mannose 6-phosphate	73	41

*The overall variability was bellow 12% for most analytes.

some classes of compounds has to be sacrificed in favour of a good reproducibility of other metabolites. Alternatively, multiple extraction procedures could be applied to enable analysis of as many metabolites as possible, but still keeping the variability sufficiently low to allow reliable comparisons between samples and batches of samples.

In metabolome analysis, the intracellular metabolites are usually extracted using chemical agents to lyse the cells and extract the intracellular compounds. There are a variety of chemical agents and extraction conditions that can be applied to different class of cells. Some chemical extraction methods will dissolve selectively a targeted group of metabolites (e.g. lipids or polar compounds), while others will be able to dissolve a broader range of metabolite classes. However, discrimination of certain groups of metabolites will always be observed, which will call for the use of multiple extraction agents in combination or not with some physical or mechanical process to enhance cell permeability and extraction efficiency.

Villas-Bôas et al. (2005c) evaluated the sample preparation methods for global metabolite analysis of yeasts. This work showed a huge difference in metabolite recovery comparing different extraction methods using the same biological material spiked with different metabolite standards. The authors classified five popular extraction methods in addition to a new proposed protocol according to their ability in recover different classes of compounds as showed in Table 4. None of the methods evaluated was able to extract all class of compounds with similar efficiency. By examining the different extraction procedures the authors concluded that there is a strong influence of the extraction method on the metabolite profile of yeast cells. Similar conclusions were also achieved for filamentous fungi (Hajjaj et al. 1998), bacterial cells (Maharjan and Ferenci 2003), plant tissues (Gullberg et al. 2004) and body fluids (Jiye et al. 2005). For instance, it is evident that acidic and alkaline extractions do not suit the requirements for a global metabolome analysis. According to Maharjan and Ferenci (2003), destruction of compounds such as pyruvate, nucleotides, and phosphorylated sugars under extreme acidic and alkaline pH is well documented. Loss of several metabolites was also observed using boiling solvents, such as boiling ethanol, most likely because many metabolites are heat-labile. Sugars and nucleotides in particular, presented the poorest recovery during boiling ethanol extraction, but even some amino and organic acids were badly recovered (Villas-Bôas et al. 2005c). Maharjan and Ferenci (2003), obtained similar results using boiling ethanol to extract intracellular metabolites of bacteria, but Gullberg et al. (2004) found the use of an homogeneous (1:8:1) chloroform - methanol-water extraction solvent with brief heating to 60°C gave the best compromise for extraction of a range of metabolites of *Arabidopsis thaliana* (Col) leaf tissue.

Therefore, it is obvious the current state-of-art of intracellular metabolite extraction methods is more suitable for targeted analysis than for a global metabolite analysis, as desired for metabolomics studies. Thus, much larger effort is required to optimize and to adapt these classical extraction protocols to a wider and non-discriminative metabolomics approach. On the other hand, the work of Maharjan and Ferenci (2003) and Villas-Bôas et al. (2005c) suggested two new protocols for an efficient extraction of intracellular metabolites of bacteria and yeasts using cold methanol (pure or in aqueous solution), which are serious candidates for a high-throughput extraction procedure for global metabolite analysis. However, these methods have yet to be tested for different biological matrices.

2.3.3 Derivatization

Chemical derivatization of metabolites is a sample-processing step especially important for analysis of semi-volatile and non-volatile compounds by gas chromatography (GC), coupled or not with mass spectrometry (MS). Semi-volatile and non-volatile chemical compounds are often unstable at the high temperature required for evaporation in the GC injector (Villas-Bôas et al. 2005b).

A very large number of derivatization methods for analysis of metabolites have been reported, but only a few are currently used in metabolomics. Silylation of organic compounds is the classical and most widely used derivatization procedure for metabolome analysis by GC-MS. Silyl derivatives are generally more volatile, less polar, and thermally more stable than their parent compounds. Sugars and their derivatives (sugar-alcohols, phosphorylated sugars, amino sugars, and others) are the most important class of metabolites derivatized by silylation (Villas-Bôas et al. 2005b, 2006). Silylation is characterized as being efficient with a broad range of applications, and results in stable derivatives with good reproducibility. However, silylation reactions require anhydrous reaction conditions, and therefore, the samples have to be completely free of water to minimize between-samples variability. In addition, some classes of metabolites (e.g. amino acids) produced relatively unstable silylated-derivatives (Koek et al. 2006). Following methoxymation and per-trimethylsilyation, Koek et al. (2006) estimated derivatization efficiencies for 32 metabolites, which varied widely, ranging from 25-110%. Very low efficiencies (< 40%) were found for some amino acids (asparagine, glutamine, and tryptophan) and for phosphate esters (fructose 6-phosphate, glucose 6-phosphate, glyceraldehydes 3-phosphate, glycerol 3-phosphate) (Koek et al. 2006).

Alkylation or esterification is another derivatization technique that is often used in metabolite analysis by GC and GC-MS (Villas-Bôas et al. 2005a, 2005b). This method is primarily used for derivatization of polyfunctional amines and organic acids. The use of esterification reactions based on chloroformate derivatives (CFs) became popular recently (Villas-Bôas et al. 2003, 2005a, 2005b; Hušek and Šimek 2006). CF derivatives have several advantages compared to silylation such as fast reactions at room temperature and in aqueous medium, and the derivatives are stable for several days, which decreases between-sample variability. However, the scope of CF derivatization is smaller than silylation, limited to amino and non-amino organic acids.

Independently of the derivatization method in use, variability at derivatization step is usually related to derivative stability and sample heterogeneity or so-called 'matrix-effects'. Samples are usually derivatized and then placed in a queue to be injected into the analytical instrument. Thus, some samples wait hours before analysis. Therefore, the metabolite derivatives have to be stable enough to resist the time required for analysis. Alternatively, the samples can be derivatized one-by-one and immediately analyzed. This can be achieved by employing robotic systems coupled to analytical instruments that are able to perform automatic in-vial derivatizations (e.g. CTC devices).

On the other hand, derivatization is a chemical organic reaction and, therefore, it is under the chemical reaction principles, which are: substrate + reagent = product(s). Therefore, different substrates have different reaction rates, resulting in different product yields. If metabolite "A" has a higher reaction rate than metabolite "B", and "A" is present at much higher concentration in the sample than metabolite "B", then chances are metabolite "B" will be poorly derivatized, and might be undetected. Most of the time, the competition for the derivatizing reagent is not between two metabolites, but between the metabolites and a high-concentrated

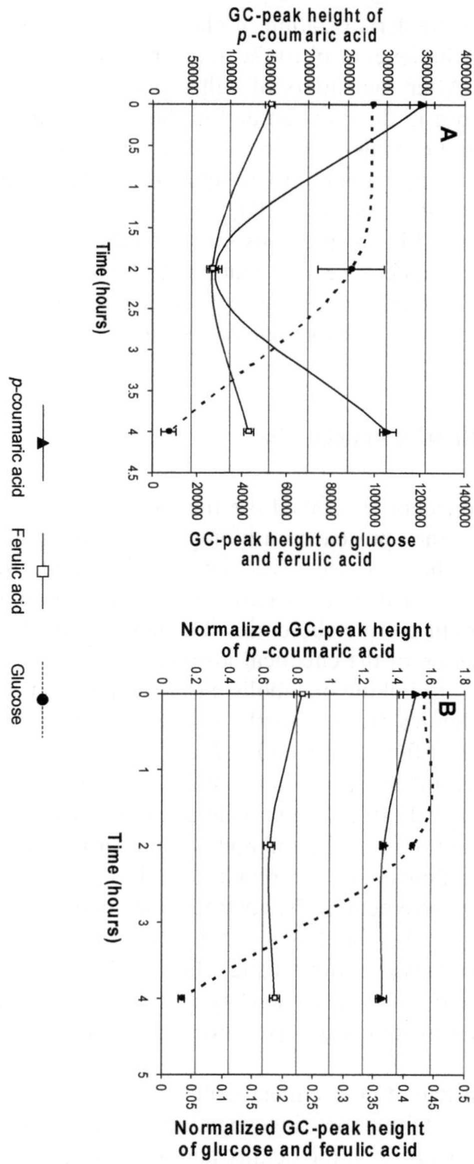

Fig. 2. Influence of glucose on the analysis of two phenolic acids by methyl chloroformate derivatization. Graphic (A) presents the analyte levels at different sampling time (n=3) based on direct GC-peak height. (Graphic (B) presents the analyte levels at different sampling time (n=3) based on the GC-peak height normalized by the peak-height of the internal standard (*trans*-cinnamic acid). The compounds were analyzed by GC-MS according to Villas-Bôas et al. 2005a). Time zero corresponds to the initial concentration as prepared.

matrix component (e.g. sugars, urea, peptides, water, etc.). Figure 2 illustrates the effect of glucose in the derivatization of phenolic acids by methyl chloroformate. The graphic shows the level of phenolic acids in a microbial culture medium analyzed by GC-MS. After four hours of cultivation, the microbial cells have consumed all the glucose in the medium and in the third sample collected the measured level of phenolic acids was greater than in the previous ones (Fig. 2A). However, by introducing an appropriate internal standard (another similar phenolic acid) during derivatization (Fig. 2B), it becomes clear that the lower levels of phenolic acids detected in the previous samples (after two hours) was due to the interference of glucose, because after that point glucose is absent. This kind of interference or matrix-effect is likely to be very common in metabolome analysis and difficult to detect on a "comprehensive" scale.

3 Intra-analytical variability

The structures of metabolites embed the functions of biosynthetic genes, and the quantities of metabolites present in a biological sample integrate the outcome of synthetic and catabolic processes over time (Nielsen and Oliver 2005). Thus an understanding and control of both qualitative and quantitative variation during analysis are important for metabolomics. Of these, qualitative accuracy, i.e., the correct characterisation of the chemical identity of metabolites is of prime importance as it is the key to linking metabolism to gene function. In a recent evaluation of several metabolomics studies Kopka (2006) found that ca. 33% of detected components are identified compounds. While the variations in anonymous unknowns may be helpful in classifying samples, and on occasion provide a useful indicator of a biological effect, the prevalence of unknown compounds is a major challenge for metabolomics. The unknown count provides a direct experimental measure of the limitations of our current knowledge of metabolism.

The unbiased measurement of the complete metabolome remains a goal for metabolomics, but in practice functional metabolomics investigations can be broadly categorised into two classes (Nielsen and Oliver 2005):

i) "Comprehensive" analyses to identify metabolites that have a key role in distinguishing between treatment groups (and which may be characterised after discovery), and

ii) Analyses restricted to those metabolites that can be measured quantitatively.

"Comprehensive" metabolomics investigations have similarities to micro-array analysis of gene expression in the quality and utility of the data generated, providing semi-quantitative relative measures of quantity, and conditional identification of the components. As with micro-arrays, such studies are the prelude to more exact and targeted studies to fully characterise the components and test particular hypotheses. With a restricted list of candidates, higher standards of quantitation and identification can be achieved in the initial analysis, but at the cost of ignoring potentially important treatment effects on the metabolome. The standards of measurement and control of variability are inevitably very different for these two approaches.

Quality standards for chemical analysis have become tightly defined as part of the legal and regulatory framework of modern societies, and validation procedures to ensure the reliability of chemical data have been defined by international scientific bodies (e.g. Thompson et al. 2002; Milman 2005). Specific analytical methods developed according to these procedures are typically defined for a single metabolite or narrow class of related compounds. The criteria include issues of applicability and fitness for purpose, selectivity, calibration and linearity, "trueness" (defined relative to reference materials), spiking/recovery, precision, range, limits of detection and quantification, sensitivity, ruggedness, matrix effects, and resulting measurement uncertainties. The scale of the ambitions of metabolomics, comprehensive unbiased analysis of the metabolome, is fundamentally incompatible with the demands of rigorous analytical method validation. Thus metabolomics analysis is inevitably a compromise as it requires extension of the scope of analytical method to cover a wide range of analytes, and cannot be optimised for the measurement of any individual component (Halket et al. 2005).

Further, questions of analytical variability in metabolomics cannot be resolved by the adoption of universal standard methodologies. Metabolomics is being applied to a diverse array of biological samples differing widely in metabolite composition, and the appropriate analytical compromise will depend on the sample and the analytical technology. Thus, the priority for standardisation in metabolomics is in the realm of agreed practical standards of quality control and measures of data quality, rather than of analytical methodology *per se*. Given the importance of correct chemical identification for metabolomics, establishing clear criteria and measures of the quality of identifications (Milman 2005) is particularly important.

We discuss these issues here for the widely used techniques of GC-MS and ESI-MS (in both direct infusion and liquid chromatography forms) on the basis of recent literature and experience in our laboratory. As we do not have direct experience with applying the experimental techniques of NMR and CE-MS to metabolomics, we have not attempted to address the associated issues of qualitative and quantitative variation and refer readers to recent reviews (Dunn et al. 2005a; Krishnan et al. 2005; Ramautar et al. 2006).

3.1 GC-MS

GC-MS methodology has been one of the primary tools in the development of metabolomics. Capillary GC provides the highest resolution of any standard chromatographic separation method, and with modern instrumentation, retention times are very consistent between runs. Electron impact ionisation and fragmentation in the source at standard voltage settings is generally reproducible between instruments and extensive libraries of spectra are available (NIST05, http://www.nist.gov/srd/nist1a.htm). The major limitation is the restriction to volatile analytes, and hence the requirement to convert many metabolites to volatile derivatives, with associated limitations and sources of chemical error (above). GC-

MS is being widely used in plant, microbial, and animal studies to measure naturally volatile compounds or volatile derivatives of non-volatiles such as TMS ethers or methylchloroformate esters, with hundreds of peaks detected in some studies (for recent reviews see Villas-Bôas et al. 2005a, 2005b; Halket et al. 2005; Kopka 2006).

3.1.1 Uncertainties in identification

The classic standard of qualitative identification by GC-MS is the combination of co-elution with an authentic standard in the same chromatographic run on a capillary GC column, or better two columns of differing type of stationary phase together with a satisfactory match to the fragmentation pattern (EI MS) and / or accurate mass (TOF MS) (Milman 2005). This can be extended with decreasing reliability using standards and samples within the same sequence of runs; run on the same instrument under similar conditions at a later date; or with appreciably less certainty, by reference to databases of retention time and mass spectrum data.

The classic analytical approach has been applied in a number of studies where a considerable list of well-defined metabolites with available authentic standards has been measured to provide insight into biological effects (e.g. Villas-Bôas et al. 2006). In cases where a more comprehensive coverage is sought, often at least a sub-set of the metabolites are characterised relative to authentic standards (e.g. Tikunov et al. 2005). Modern GC-MS instrument manufacturer software typically provides facilities for peak detection and comparison against the retention time and mass spectrum of an authentic standard previously run on the instrument, with limit criteria for particular fragment masses. Official regulatory methods specify matching criteria for both retention times and mass spectra as summarised by Milman (2005). Changes in retention time can be corrected for by reference to standards. The risk of false positive identifications with these procedures is low but non-zero. While separation of a peak from a co-injected authentic standard is unambiguous, the criteria of co-elution and spectrum-matching retain a residual ambiguity, particularly within some classes of naturally occurring isomeric compounds, for example, monoterpenes, per-silylated sugars (Wagner et al. 2003). The risk of false negatives is considerably higher due to misleading spectrum mismatches arising from co-eluting interferences. Currently this can only be overcome by manual checking (e.g. example in Table 2 of Schauer et al. 2005). These limitations can potentially be overcome by improved peak resolution, and recent development in two-dimensional GC-TOF MS (Shellie 2005; Mohler et al. 2006) offer the prospect of significant improvements in separating power.

Derivatization processes can both degrade and improve the quality of identifications. A standard approach to derivatize complex mixtures in metabolomics prior to GC-MS analysis is to convert cyclic monosaccharides to open chain methoximes prior to per-trimethylsilylation (Roessner et al. 2000; Fiehn et al. 2000). The methoximes occur as stereoisomeric pairs, but a greater number of TMS derivatives are formed from the free sugar. The process of chemical conversion can degrade identification quality as for example through the conversion of arginine to the TMS derivative of ornithine (Halket et al. 2005). The implications for unknowns are of course unknown! Conversely, where several derivatives are uniquely formed from one compound, for example, sugar methoxime isomers, the occurrence of the expected isomer peaks in the correct ratio provides an additional level of identity validation.

When more extensive profiling beyond a target list based on authentic standards is attempted, recent developments in databases for metabolomics provide the basis for a wider extension of authentication, albeit at a lower standard of proof (relative retention times or indices; and GC-MS library matches). Automated peak deconvolution programmes such as AMDIS (Stein 1999) are essential and widely used to generate target lists of components characterised by retention times and patterns of mass fragmentation (mass spectral tags or MSTs in the nomenclature of Kopka 2006). Recent developments in GC-MS data processing using hierarchical multivariate curve resolution (Jonsson 2005, 2006) may offer a more robust automated approach to spectrum extraction, with successful deconvolution of mass spectra of almost exactly co-eluting components reported.

A number of spectrum matching algorithms are available for comparing extracted mass spectra to library data, often incorporated in instrument manufacturer software. The dot product approach as implemented in AMDIS (Stein 1999) is widely used. The quality of a spectrum library search result is characterised not only by the magnitude of the match factor, but also by its distinctiveness, i.e. separation from the values for matches with other compounds (Milman 2005). However high similarities between mass spectral fragmentation patterns within classes of naturally occurring metabolites (or their derivatives) are common and Wagner et al. (2003) demonstrated unequivocally the necessity of incorporating chromatographic retention data for effective selective searching. Retention indices (Kovats 1958), which relate the retention time of a component to those of a series of alkanes run on the same column under identical conditions, provide a measure of portability between laboratories. An extensive MS/RI collection of TMS and TBDMS derivatives has been made available to the science community by the Max Planck Institute of Molecular Plant Physiology (Schauer 2005; Kopka 2005), and the wider use of combined mass-spectrum and RI matching has been advanced with the release of the NIST05 database (http://www.nist.gov/srd/nist1a.htm) incorporating RI data. For library screening, match factors of >650 for MS dot product score (AMDIS) and RI differences less that 3.0 have been suggested to identify candidates for manual evaluation (Schauer 2005). While retention indices may not be precisely portable, indices determined on columns of the same type show close co-linearity, particularly within compound classes (Kopka 2006), and library retention indices can thus be adjusted relative to authentic standards run within the laboratory. The co-occurrence of multiple derivatives of a compound can provide an added level of security of identity. However for conclusive identification by matching of retention times and mass spectra, there is no substitute for laboratory validation by standard addition experiments, especially for closely eluting isomers (Kopka 2006).

Assessing the quality of reported chemical identifications in a metabolomics investigation remains an issue. Match parameters for RI and MS might provide a guide, but these are not a direct measure of quality. Milman (2005) points out that any chemical identification involves the generation and testing of a hypothesis and the generally agreed standard of reliability is that tests based on two different techniques should reach a consistent conclusion. Thus, in principle metabolomics identifications remain provisional and subject to revision. However, prior knowledge is an important consideration in chemical identification. Analysts of metabolomes are well aware that most of the ca. 25 million known low molecular weight compounds are outside their domain of identification, and they are justifiably influenced in their assignments of chemical identity by prior knowledge of known compounds found in the species under investigation, expectations of the occurrence of ubiquitous metabolites, and knowledge of the classes of secondary metabolites likely to be present in a family. Milman (2005) suggests a Bayesian probabilistic formalisation of identification based on the joint probability of an experimental measurement result given the presence of a compound, and the prior probability a compound is present in the sample. This conceptual framework may provide a basis for a systematic approach to assigning a numerical value to the quality of compound identifications in metabolomics.

To maintain quality in chemical structure identifications the relationships between chemical structures at different stages of the analytical process must be properly recognised. The analytes determined in a GC-MS analysis are some steps removed from the metabolites in the biological system, and the form in which they are represented in public databases (Kopka 2006). Apart from inconsistencies in nomenclature, which may be resolved by reference to unambiguous computer-generated InChI codes (http://www.iupac.org/inchi), problems arise with isomeric structures such as hexoses and pentoses present as cyclic anomers in the cell and analysed as acyclic per-TMS methoxime stereoisomers. Metabolomics results need to be reported in the framework of a properly constructed chemical ontology, which clearly defines the relationship between reported estimates, actual analytes, and the underlying metabolites. The chemical ontology under construction as part of ChEBI (Brooksbank et al. 2005; http://www.ebi.ac.uk/chebi) may provide a foundation for this.

A well-structured chemical ontology would also provide for a more systematic approach towards handling the identification of the ca. 67% of unknown compounds appearing in comprehensive metabolomics analyses (Kopka 2006). The similarities of EI mass spectra within classes of metabolites or their derivatives is a constraint on unique identification, but also provides a means of partial identification (Wagner et al. 2003). Partial classification of unknown compounds to chemical classes is of value in interpreting metabolomics studies (e.g. Broeckling et al. 2005 discuss a jasmonic acid metabolite), and a properly structured chemical ontology would allow data with different levels of structural definition to be reported and analysed in a consistent way.

However the major challenge to "comprehensive" metabolomics is undoubtedly the large number of unknown or partially characterised metabolites. When the genome sequence is known the existence of many metabolites can be inferred, and for many of these, authentic compounds may be commercially available. Koek et al. (2006) derivatized and analysed ca 300 standards for postulated metabolites of *B. subtilis* and detected ca 200 metabolites as derivatives by GC-MS, although not all could be quantitatively measured by this technique. Extension beyond the range of commercially available standards has been suggested (Kopka 2006) as a "bottom up" approach to structure identification where sets of candidate molecules are synthesised. In the face of the biosynthetic diversity of nature, this is unlikely to make much impact in the short term except in specific cases, for example, completing series of isomers. The practical question is how to prioritise which metabolites should be further characterised, given the major demands on time and resources this implies. For individual laboratories, the obvious approach is to focus on those compounds showing distinctive characteristic changes between system states as discovered by data mining, but partial structural information and biological clues may also aid priority setting. Priorities may also emerge from comparative metabolomics between different experimental systems and laboratories. The development of public databases of mass spectral tags (Schauer 2005) and consistent labelling schemes for unknown compounds (Bino et al. 2004) will facilitate this process. Further efforts will be required to reconcile partial identifications from different experimental methods and laboratories to avoid reliving the confusions of the classic era of natural products research which resulted from the persistent use of different trivial names for the same compound discovered in different contexts.

3.1.2 Quantitative variability

Quantitative metabolomics analysis by GC-MS is subject to variation at each level of the process, from biological variation between replicate samples, to variation in sample preparation (quenching and extraction), derivatization as discussed previously, as well as during chromatography and detection (Villas-Bôas et al. 2005b). As an analytical tool, GC-MS ranks highly for linearity over a wide dynamic range, and for selectivity. GC with electron-impact MS provides multi-channel detection, and is a powerful tool for selective analysis within complex mixtures. Matrix effects under electron impact in the standard EI source are much less of a problem than in ESI-MS (below), and extracted ion chromatograms based on selected distinctive fragment ions can often provide baseline resolution of overlapping peaks. However different analytes in a complex mixture are likely to differ widely in recoveries, derivatization efficiencies, the stability of derivatives, and in mass fragmentation. The mass selectivity of GC-MS makes possible the comparative measurement of an analyte and a stable isotope variant, and stable isotope dilution analysis (SIDA) is well established as a good strategy for control of these sources of variation in GC-MS (e.g. Chace 2001), although analysts should be aware the chemical and physical properties of the different isotope species are not identical. For proper validation, variation should be assessed across the range of matrices the analysis is to be applied for each analyte (Thomson et al. 2002). As with unambiguous identification of all metabolites, this remains an elusive goal for metabolomics, but recent research has made substantial progress towards defining the quality of quantitative metabolome analysis and clarifying the sources of variation.

Table 5. Quality of metabolome analysis of methoximated TMS derivatives of polar compounds by GC-MS

Sample	Test of Variation	Variation (%RSD)	Difficult metabolites	Reference
Potato tubers	Reproducibility	<1% - 14.1%	glycerol	Roessner et al. 2000
	Run order effects		Gln, Trp, glycerol, (citric acid)	
M. truncatula tissue cultures	Reproducibility (249 peaks)	< 5% for 87% of peaks		Broeckling et al. 2005
A. thaliana leaf	Reproducibility (59 analytes)	5.5 - 33.4%		Gullberg et al. 2004
	Repeatability	7.8%		
	Run order effects		18 analytes	
	Reproducibility (7 analytes by SIDA)	6.9 - 9.7%		
	Repeatability (7 analytes by SIDA	3.5%		
Human plasma	Reproducibility (32 analytes)	2.4 - 29%	>15% at highest conc. Gly, Asp, laurate, glucose, Cys,cholesterol	Jiye et al. 2005
Microbial cultures	Reproducibility (32 standards)	1 - 12%	> 10% Gln, Asp, Met, Trp Glycerol-6-phosphate	Koek et al. 2006
	Linearity (standard addition)	$r2 > 0.99$	except Gln, Asp, ribose-3-phosphate	
	Reproducibility (23 analytes)	2-8%,	except > 20% phosphoenolpyruvate, 2-phosphoglyceric acid	
	Intrabatch precision (4 standards)	3 - 5%		
	(4 metabolites)	6 - 7%		
	Interbatch precision (4 standards)	8 - 11%		
	(4 metabolites)	8 - 14%		

Initial implementations of GC-MS metabolomic analysis were carried out on quadruple GC-MS instruments, but rapid-scanning time-of-flight (TOF) detection is now more commonly used despite limitations in dynamic range, as TOF detection is more sensitive and can provide more and better-aligned data points across a GC peak (Kopka 2006; Veriotti and Sacks 2003).

The pioneers of plant metabolomics took some effort to validate the reproducibility of their GC-MS analyses of MeOx TMS derivatives, and reported relative standard deviations (RSDs) of analytical replicates of the order of 5% or less for most of the metabolites they were monitoring, usually well below the biological variations within the experiment (Roessner et al. 2000; Fiehn 2000). Some limitations were identified, including a decline in measured levels of some compounds during runs, and a major discrepancy with previous literature in one case (citric acid) (Roessner et al. 2000), and Fiehn (2000) reported up to two-fold differences between external and internal calibration. Quality estimates from Roessner's study and more recent implementations and investigations of the quality of GC-MS analysis of TMS derivatives are summarised in Table 5.

The prime requirement for chemical analysis methods in biology is that analytical variability is significantly less than biological variability. Given the biological variation typically observed in metabolic profiles (Morgenthal et al. 2006), for GC-MS analyses of a wide range of TMS derivatives of metabolites this is the case as relative standard deviations (RSD) in most cases are less than a 10% (Table 5). However for some metabolites particularly some amino acids and organic phosphates, results can be quite unsatisfactory. Apart from the inability to measure some amino acids such as arginine (see above), others such as glutamine, aspartate, methionine, and tryptophan show poor reproducibility and instability during chromatographic runs.

The most detailed recent examination of sources of variation in GC-MS analysis and of measures to control this variation is that of Koek et al. (2006). For sugars and organic acids Koek at al. (2006) report generally satisfactory derivatization (60-115%), good repeatability (<5%) and reproducibility (8-14%), a large linear range (2.0 x 10^2), and low detection limits (<500 pg on column). However this is not achieved for some classes of metabolites and this is inherent in the derivatization chemistry. The formation of derivatives from MSTFA involves the displacement of an N-methyltrifluoroacetamide leaving group by the analyte, and some metabolites provide equally good leaving groups. In this case, the derivatization reaction is only driven to product by the large excess of reagent, and the products are readily degradable. This is particularly the case for amides such as asparagine, and glutamine, and for thiols, and sulfonic derivatives, with the overall trend for ease of TMS derivatization and stability of products alcohol > phenol > carboxylic acid > amine > amide. The analysis of TMS derivatives of amines, and phosphoric functional groups shows intermediate variability of derivatization efficiencies (30-110%), repeatability (1-7%) and reproducibility (10%), and higher detection limits than for sugars and organic acids (Koek et al. 2006).

Koek et al. (2006) applied a range of quality control measures, including a set of added deuterated standards to monitor extraction (phenylalaline-d_3), lyophilization (glutamic acid-d_3), derivatization (glucose-d_7 and phenylalaline-d_5) and GC-

MS analysis (alanine- d_4, dicyclohexylphthalate), and checking GC-MS performance by monitoring responses for standards. They reported that in general liners required changing after 20 samples had been injected, with occasional removal of a small section of the front end of analytical column to restore performance. With these quality control measures they claim their GC-MS method to be a comprehensive method with a very large application range, with an analytical performance with respect to stability, precision, recoveries and linear ranges that meets requirements for target analysis in biological matrices. When the procedures were applied to the analysis of the *B. subtilis* metabolome, with standard additions of ca 200 authentic standards, recoveries better than 50% were achieved for 160 metabolites. However 40 compounds could not be determined due to multiple peaks, degradation (e.g. adenosine 5'-phosphosulfate), poor recoveries (e.g. uridine 5'-monophosphate), or high volatility (acetic acid, glyoxilic acid).

Data quality can be much improved where stable isotope standards are available (stable isotope dilution analysis: SIDA) (Gullberg et al. 2004). For microbial samples, isotope standardisation can potentially be extended to the complete metabolome by the use of control samples prepared from incubation of universally ^{13}C-labelled substrates as reported by Mashego et al. (2004) for LC-MS, but this technology is very expensive and has yet to be applied to GC-MS analysis. However reproducibility statistics can be misleadingly reassuring. The gold standard for analytical validation is confirmation by an independent method Thompson et al. (2002) and Jiye et al. (2005) reported differences between GC-MS estimates for 8 compounds by SIDA and estimates of an accredited laboratory ranging from 0.7%-24.5% (median 10%).

The above studies do not address peak resolution, and the effects of multiple derivative peaks on quantitative analysis. The standard procedure for metabolomics studies by GC-MS has been to use a non-polar DB5 column with a linear gradient, which leaves some derivative peaks of isomeric metabolites unresolved and lacking distinctive measurement ions. Villas-Bôas et al. (2006) achieved an improved separation within classes of isomeric sugars with a more polar phase (ZB1701, Phenomenex) and a complex custom gradient. Where the pattern of multiple derivative peaks is known from authentic standards, this can be used to improve quantitation and in some cases to estimate quantities of components from measurements of overlapping derivative peaks.

As the diverse applications of metabolomics preclude uniform methodology and the quality of quantitative measurements will vary within and well as between investigations, evidence of quality standards should be presented together with any reported data. This includes repeat measurements of sets of standards to monitor the performance of the GC system (including variations in derivatization and in the instrument) and of blanks to detect carry-over and contamination. Any discrepancies should be remedied by appropriate housekeeping such as replacement of inlet liners and source cleaning. While stable isotope standards are the ideal, multiple internal standards as used in recent detailed method investigations (e.g. Koek et al. 2006) above provide a more reliable base for quantitation than a single standard. Reproducibility within and between batches can best be monitored by repeat analyses of control samples representative of the range of materials to be measured, for example, pooled treatment group samples. Run order effects can be corrected for by randomising run order, or by correcting peak areas relative to replicate controls across the run. Proper evaluation of matrix effects requires standard additions at the extremes of matrix composition (Thompson et al. 2002), which is impractical with large numbers of standards and impossible for unknowns. However matrix effects could in principle be monitored by measurements of pooled treatment group samples and mixtures of these samples. Finally it must be kept in mind that in metabolome analysis by GC-MS of TMS derivatives (and indeed in all metabolome analysis) the quality of data will not be uniform across all metabolites. Measures of the quality of data for individual metabolites (e.g. %RSDs within control samples) are potentially as important as the data themselves.

3.2 ESI-MS

The application of mass spectrometry in the analysis of liquid samples is wide spread in metabolomics, either through the analysis of raw extracts by direct infusion (DIMS) or through the combination with a separation system such as liquid chromatography (LC-MS) (e.g. Allen et al. 2003; Castrillo et al. 2003; Bajad et al. 2006; Chen et al. 2006). In both cases the analyte is delivered in front of the mass spectrometer dissolved in a liquid and has to enter the mass spectrometers vacuum as a charged molecule. It does not need much understanding of physical chemistry to imagine the sheer incompatibility of these systems. Many of the issues with analytical variability in DIMS and LC-MS are caused by this interface. The most commonly applied interface is electrospray ionization (ESI), which was developed at the end of the sixties by Dole and co-workers (Dole et al. 1973). Although ESI probes have been further optimized over the years, the basic principle is still the same. The liquid is delivered through a capillary in front of the mass spectrometer orifice and sprayed with an accompanying flow of nitrogen in a strong electric field. The spray droplets in the strong electric field undergo a coulomb explosion, resulting in the formation of microdroplets carrying a high charge. Solvent molecules can be further evaporated in the nitrogen stream, depending on the ESI probe design, resulting in the delivery of charged analytes (molecules or molecule clusters) at the orifice of the mass spectrometer. Inside the mass spectrometer the charged analytes can be focused by ion lenses and delivered to the actual mass analyzer. The technical setup for the focusing of ions is also different between mass spectrometers and manufacturers. The lenses can be tuned to specific standards at specific concentration in a particular solvent composition. Contamination of the lenses through intensive use will demand retuning to maintain optimal signal. This tuning process will optimize the detection of a specific compound and is therefore in conflict with the objectives of an unbiased analysis.

The spray formed at the tip of ESI probe is the first factor that has a significant effect on the analytical performance and reproducibility of a mass spectrometer. It determines the amount of molecules delivered at the orifice of the mass spectrometer as well as the proportion of these molecules that are charged. The probe tip and capillary are subject to wear and tear requiring manual adjustment to maintain an optimal spray. Probe designs differ between manufacturers. Thus we are left with the effort of each analyst to optimize the spray for and during each experiment as an inherent limit on reproducibility.

Once the analytes are charged and in the gas phase there are several options for mass analyzers to separate ions by *m/z* ratio. The different types of mass analyzers have specific advantages and problems, and it is also possible to combine types of analyzers (for a general overview of different types see Aebersold and Mann 2003). The main analyzers in use at the moment are Time of Flight (TOF) and quadruple filters (single or triple) or ion traps. The TOF system has the advantage of higher mass accuracy, which provides more specificity (limits the number of possible analytes). The ion trap and to a lesser extend triple quad can deliver fragmentation data. Both systems can be combined as in QTOF instruments. Higher mass resolution is provided by Orbitrap (Makarov et al. 2006) or FTICR (Fourier transform ion cyclotron resonance) instruments. High accuracy mass spectrometry demands optimal calibration and in the case of TOF continuous calibration.

3.2.1 DIMS

Direct infusion mass spectrometry (DIMS) is of value for metabolomics because of its speed and because it is relative lack of bias. There is, however, a toll to pay; DIMS is not suitable for accurate quantitation and cannot be used for such purposes. DIMS can only provide a rough estimate of the relative concentration of an analyte. In addition there are two main problems that grossly impair the reproducibility in DIMS, ion suppression and contamination, which will be discussed in more detail.

It is a well-known fact that co-eluting compounds in LC-MS can cause a matrix effect that suppresses the ionization of the analyte. This is for instance reviewed by Niessen et al. (2006) for pesticides. The actual physical/chemical mechanisms that underpin the suppression of ionization are not yet well understood. Based on the findings of King et al. (2000) it is mainly attributed to processes in the droplet formation of liquid phase. Under some circumstances the suppression can be caused by physical processes. Beaudry and Vachon (2006) showed that saline solution can disturb the shape of the spray, resulting in a decrease of signal.

When a raw extract containing an unknown number of different analytes is analysed using ESI there will be a complex interaction between all those analytes that is impossible to predict. Dunn et al. (2005b) studied the effect of ionization suppression in direct infusion of tomato extracts. By spiking extracts with a mixture of compounds they found that an increase of concentration gave an increase of signal, but this was not always a linear relation. Our current lack of understanding of the mechanisms and limited experimental data in DIMS should be a clear warning to expect matrix effects to compromise concentration to signal ratios in every DIMS experiment.

In most cases biological samples, like culture medium or tissue extracts, will contain large concentrations of salts as well as non-ionizable components. The infusion of such samples directly into mass spectrometer will result in a built-up of

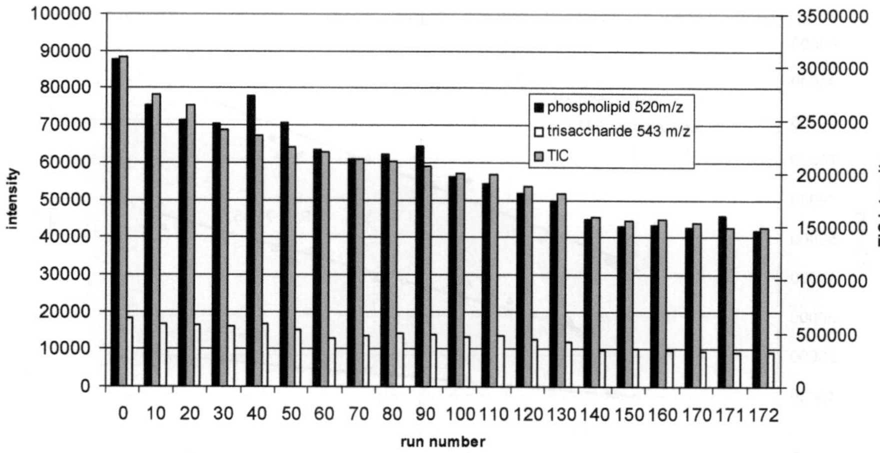

Fig. 3. Repeated analysis out of the same vial of control sample during a direct infusion experiment, on the left y-axe is the intensity of two ions 520 m/z (phospholipids, black bars), 543 *m/z* (trisaccharide, white bars), on the right the total ion current (TIC). The control sample is a mixture of IPA:H_2O extracts of ryegrass seeds infected with different strains of *Neothyphodium lolli* analysed by direct infusion using a linear ion trap mass spectrometer.

precipitate at the orifice and inside the mass spectrometer, which will decrease the sensitivity and eventually block the inlet completely. For every type of extract the infusion rate and set-up must be optimized in such a way that there is sufficient signal versus a limited decrease in sensitivity.

In Figure 3, we show how the performance of the DIMS analysis was monitored during an experiment of 170 samples. In this experiment we used a control sample (mixture of all samples in the experiments), which was analysed after every nine samples out of the same vial. This shows the lack of reproducibility in this type of experiments. There is a steady decrease in total ion current (TIC) (grey bars), this decrease in signal is also observed for most of the ions, for example, the ions with *m/z* (520) (a phospholipid) also steadily decreased but not in proportion to the TIC and a similar pattern is seen for a trisaccharide (m/z 543) as can been seen in Figure 4. The relationship between the TIC and these two ions and between the two ions is not constant but changes approximately linearly over the series of chromatograms.

These changes can be explained by complex interactions between the ions in the gas phase and contaminated regions of the interior of the mass spectrometer. Although these factors are major concerns for DIMS experiments, there are several strategies to minimize the effect on the data and data interpretation. DIMS experiments can be performed with sample randomization. However, it is important that run number is incorporated in the analysis of the data, so the effect of machine variation across the experiment can be assessed (e.g. PCA looking for run effect). As an example the results of the direct infusion experiment are analyzed by principle component analysis, and plotted as their batch number (Fig. 5). Although the data is normalized there is still a clear batch effect visible for the first three

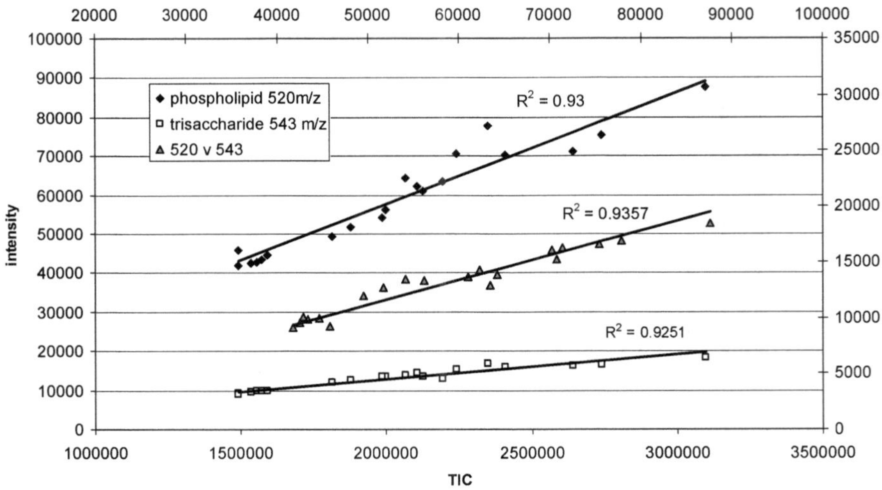

Fig. 4. Repeated analysis out of the same vial of control sample during a direct infusion experiment, showing the relation between the total ion current (TIC) and two ions 520 m/z (phospholipids, black bars), 543 *m/z* (trisaccharide, white bars) as well as the relation between the two ions. The control sample is a mixture of IPA:H2O extracts of ryegrass seeds infected with different strains of *Neothyphodium lolli* analysed by direct infusion using a linear ion trap mass spectrometer.

batches in principle component 2. After that the samples of each batch are reasonably homogenously scattered with no obvious batch effect. This type of analysis allows ions mostly contributing to the batch effect to be identified, and this can be used in further normalization or taken into account during the data analysis.

The second strategy that will help in data interpretation is normalization. The most obvious strategy is either to use one or several internal standards or the total ion current for normalization. This will have only a limited effect due to previous described problems but will reduce the major batch effects.

DIMS data comprises a *m/z* ratio and an intensity and the results are therefore ambiguous for identification. In complex mixtures there are likely to be several different metabolites contributing signals to each unit *m/z* bin. High mass accuracy as achieved with FT-MS clearly reduces the number of possible candidate metabolites considerably compared to 1 mass unit ranges, but is far from sufficient for unambiguous identification (Kind and Fiehn 2006). With ion trap and triple quad instruments, fragmentation data can also be collected in a DIMS experiment. Such an approach is rarely applied but can be highly useful in analyte identification or classification although there are costs over time and throughput, and limited libraries of spectra available, because of the variability of fragmentation between instruments and instrument settings.

It would be interesting and desirable to be able to compare direct infusion mass spectra between experiments and laboratories, to aid in the rapid classification of the metabolic state of an organism. However, the methodology is not yet enough developed to strive for inter laboratory standardization. Currently it is more useful

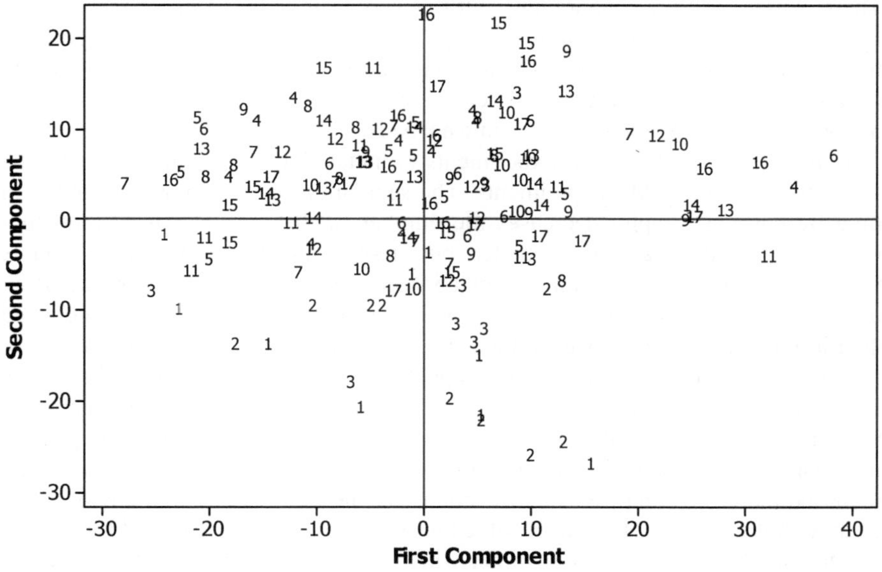

Fig. 5. Principle component analysis of normalised data from a DIMS experiment to determine batch effects. The numbers represent batch numbers, each batch consisted out of 9 samples. The samples were analysed in randomised order. Each sample is a IPA:H_2O extract of ryegrass seeds infected with different strains of *Neothyphodium lolli* analysed by direct infusion using a linear ion trap mass spectrometer.

to extend our knowledge of the main factors in matrix effects, for example, by identifying common metabolites that are only slightly influenced by matrix effects that can be used as standards or calibration points and evaluating different internal standards for quantifying analytes in direct infusion experiments

3.2.2 LC-MS

Most of the current methodologies applied in LC-MS are the result of the research efforts in pharmaceutical analysis and more recently proteomics. These applications are still the main driver in the development of this technology. There are many good reviews and books that provide a good insight into these developments (e.g. Niessen 2006).

In the section of ESIMS we discussed general problems in reproducibility using this type of ionization. The main strategy to overcome this problem in most analytical laboratories is based on the use of stable-isotope-labelled versions of the analyte as internal standard to minimize the effect of changes in spray current and matrix effects (De Leenheer and Thienpont 1992). In metabolomics studies this would call for completely labelled metabolome as an internal standard. This approach is feasible and has been successfully applied for yeast (Mashego et al. 2004). Yeast was cultured on ^{13}C labelled glucose and ethanol, and the resulting cells were used as an internal standard for the accurate quantitation of metabolites.

There results are very encouraging, but this is not the Holy Grail for metabolomics. In addition to the problem of creating fully labelled metabolomes it is possible that a stable-isotope internal standard may cause ion suppression of an analyte (Liang et al. 2003). This effect will be different for each analyte and is therefore of major concern in complete metabolome analysis.

Reproducibility of LC-MS experiments in metabolomics is confounded by the same problems of DIMS experiments. The separation of the biological sample in time by chromatography decreases the number of different analytes simultaneously entering the mass spectrometer, but there is no chromatography available that is able to resolve the complete metabolome into non-overlapping peaks. Peak overlap will cause matrix effects, although the problem will be less pronounced than in DIMS. The advantage of the decrease in the number of analytes gives rise to a new set of problems, namely the reproducibility of the chromatography. This is a serious problem due to complexity of the sample. Method development and validation of conventional LC analysis focuses on baseline separation, to optimize the quantitation of specific analytes. In contrast, LC in metabolomics aims to be unbiased therefore compromising the quality of the LC separation of any or every analyte in the sample. Baseline resolved peaks can be easily integrated for quantitation. Small shifts in retention time, which are unavoidable, will therefore not impair the analysis. In complex biological samples, where peaks overlap, this is major problem in quantitation, and as with GC-MS analysis, deconvolution of overlapping peaks is required. Recent developments in deconvolution software, and particularly the release of the publicly available software package XCMS (Smith et al. 2005) show considerable promise. Nordstrom et al. (2006) have demonstrated the ability to detect eight out of ten differences in spiked compounds in serum, some as small as 20%, utilizing the combination of high chromatographic resolution with UPLC, deconvolution with XCMS, and the use of multiple internal standards.

For the analysis of specific analytes problems with peak resolution can be circumvented by the use of MS/MS or MS^n and multiple reaction monitoring (MRM) using a triple quad or ion trap mass spectrometer. In MRM specific fragmentation (pathways) are followed for specific analytes, which can dramatically decrease signal to noise. However there is a limit in the number of MRM's that can be done in a certain time frame to acquire sufficient measurements across a peak. MRM based analysis in LC-MS/MS is, therefore, by definition a targeted analysis, applicable only on a relatively limited number of analytes per analysis. While this is of limited use for a full metabolome analysis, this approach can still be valuable for metabolomics. Bajad et al. (2006) have demonstrated the analysis of a target list of 141 microbial metabolites by LC-MS/MS and biologically important compounds such as plant hormones, which are present at trace levels far below the detection limit of any current comprehensive method, can be measured by this method (Chiwocha et al. 2003; Durgbanshi et al. 2005).

3.3 Conclusions

Given the diverse applications of metabolomics and the diverse analytical technologies applied, the main requirement for standardisation in instrumental metabolome analysis is the adoption of standards of quality control. In all, reliable "fit-for-purpose" methods with built-in quality control procedures are more important for metabolomics than standardised protocols. It should be assumed the quality of measurements will not be uniform within and between metabolomics studies, and it is important that measures of quality control form part of any metabolomics data set.

4 Post-analytical issues

The procedures for processing and quantitating analytical data embedded in modern instrument manufacturer software are largely of an adequate standard for metabolomics but are not attuned to high through-put comprehensive analysis. They handle the operations of detecting the analytical signal against baseline variations and instrumental noise; checking peak retention times are within a defined range, and checking spectra against standards. However, they are designed to meet the requirements of targeted analysis rather than those of metabolomics. In particular they typically require extensive manual checking which becomes very onerous on the metabolomics scale.

More automated procedures are required for metabolomics, and there has been considerable effort towards developing multivariate statistical methods for handling the analytical signals generated by each of the technologies of metabolomics analysis (e.g. Jonssson et al. 2005, 2006; Tikunov et al. 2005; Smith et al. 2005). This stage of the analysis appears to offer more scope for standardisation than do laboratory procedures. However, although comparisons have been reported between instrument manufacturer software and multivariate methods, and found satisfactory agreement (Nordstrom et al. 2006), there has been no report of a direct comparison of the results of applying different multivariate methods to the same set of experimental metabolomics data. The recent release of the package XCMS (Smith et al. 2005) for mass spectrometric based metabolomics data analysis (particularly LC-MS) as open-source software may encourage further development and bench-marking of software performance. The peak finding algorithm in XCMS is based on an assumed Gaussian peak shape and overlapping peaks on the same mass channel may not be resolved and detected. A major feature of the programme is the peak alignment algorithm to match peaks across sets of chromatograms by retention time and *m/z*. This is a critical factor for multivariate analysis of metabolomics data. The correct assignment of corresponding analytical signals in different samples to the same variable (or "bin") is essential as mis-assignment can severely distort multivariate analysis and may only be revealed by manual inspection of anomalies.

The proper handling of "missing data" in metabolomics also warrants more attention. As pointed out by Willse et al. (2005), zero values can bias statistical analysis, and failure to detect a component with high confidence does not imply high confidence of its absence. Omission of data points not present in every sample simplifies the mathematics. However, evidence of the absence of certain components in some samples may be of real biological significance, as with the tomato volatiles investigated by Tikunov et al. (2005) and these authors have described a procedure (multivariate mass spectra reconstruction, MMSR) by which components abundant in a single sample but absent in others can be identified.

For the analysis of treatment effects metabolomics data are often log-transformed as the focus is on relative changes in concentration. However this has the unfortunate consequence of amplifying and propagating the uncertainties in the measurements of components present at low concentrations near the detection limit. Error in chemical analysis is typically approximately proportional to the measurement, except at low levels approaching the limit of detection when it approaches a constant value due to background noise (Thompson et al. 2002). Rocke and Lorenzato (1995) demonstrated that a two-component model for analytical error gave a better approximation to observed analytical variation than standard approaches. This variance can be stabilised by a generalised log transformation and Purohit et al. (2004) have demonstrated an implementation using a maximum likelihood method for metabolomics NMR data, which resulted in improved resolution of treatment groups. Extension to GC-MS and LC-MS data sets is awaited. Nonlinearity of responses at high concentrations may also impinge on data analysis, adding a non-linear analytical component to the (often non-linear) biological variation.

The analysis and interpretation of metabolomics data is in early stages of development. Although researchers have shown that in general experimental measurement uncertainties in GC-MS and LC-MS data sets are less than the biological variation of interest, this will not be the case for all metabolites. There has been particular interest in patterns of correlation between metabolite concentrations (e.g. Steuer et al. 2003; Morgenthal et al. 2006). Correlations between variations arising within the analytical methodology are highly likely, and measures of analytical variance for individual components should be taken account of in data analysis (e.g. as weightings) to reduce the distorting effects of analytical error.

5 Final remarks

The scale of metabolomics analysis is clearly at odds with the requirements of analytical rigour, in regards to both qualitative and quantitative certainty of results. However, while "comprehensive" characterisation of metabolomes remains a goal, this is in the service of scientific understanding. The "metabolome" is metabolic phenotype and hence highly plastic: completely characterising the metabolome of a species in one experiment will not have the same power as characterising the genome of one member of a species. The limitations of current understanding of metabolism arise not only from analytical limitations, but also because due to the high degree of connectivity in metabolic networks, perturbations propagate through the network in a manner that defies intuitive interpretation (Bailey 1999). The converse of this is that detection of evidence of such perturbations (e.g. by cluster analysis) may require measurement of only a small proportion of the metabolome. Indeed classification of treatment groups by profiles of a limited number of metabolites is often successful. In terms of quantitative uncertainties, in attempting to model and interpret the behaviour of complex metabolic networks it is important that the measurement uncertainties of individual components are estimated and included in the analysis to avoid modelling analytical artefacts. In terms of qualitative uncertainties, the pragmatic answer is to extract the maximum new understanding from measurements of known compounds, and characterise those unknowns, which show the most interesting pattern of change.

Acknowledgments

The authors would like to thank the book editors for the invitation and opportunity to contribute to this work as well as AgResearch Limited and the New Zealand Foundation for Research Science and Technology for research funding.

References

Aebersold R, Mann M (2003) Mass spectrometry-based proteomics. Nature 422:198-207

Allen J, Davey HM, Broadhurst D, Heald JK, Rowland JJ, Oliver SG, Kell DB (2003) High-throughput classification of yeast mutants for functional genomics using metabolic footprinting. Nat Biotechnol 21:692-696

Bailey JE (1999) Lessons from metabolic engineering for functional genomics and drug discovery. Nat Biotechnol 17:616-618

Bajad SU, Lu W, Kimball EH, Yuan J, Peterson C, Rabinowitz JD (2006) Separation and quantitation of water soluble cellular metabolites by hydrophilic interaction chromatography-tandem mass spectrometry. J Chromatogr A 1125:76-88

Beaudry F, Vachon P (2006) Electrospray ionization suppression, a physical or a chemical phenomenon? Biomed Chromatogr 20:200-205

Bino RJ, Hall RH, Fiehn O, Kopka J, Saito K, Draper J, Nikolau B, Mendes P, Roessner-Tunali U, Beale M, Trethewey RN, Lange BM, Syrkin Wurtele E, Sumner L (2004) Opinion: Potential of metabolomics as a functional genomics tool. Trends Plant Sci 9:418-425

Borodina I, Nielsen J (2005) From genomes to *in silico* cells via metabolic networks. Curr Op Biotechnol 16:1-6

Broeckling CD, Huhman DV, Farag MA, Smith JT, May GD, Mendes P, Dixon RA, Sumner LW (2005) Metabolic profiling of *Medicago truncatula* cell cultures reveals the effects of biotic and abiotic elicitors on metabolism. J Exp Bot 56:323-336

Brooksbank C, Cameron G, Thornton J (2005) The European bioinformatics institute's data resources: towards systems biology. Nucleic Acids Res 33:D46-D53

Buziol S, Bashir I, Baumeister A, Claaβen W, Noisommit-Rizi N, Mailinger W, Reuss M (2002) New bioreactor-coupling rapid stopped-flow sampling technique for measurements of metabolite dynamics on a subsecond time scale. Biotechnol Bioeng 80:632-636

Castrillo JI, Hayes A, Mohammed S, Gaskell SJ, Oliver SG (2003) An optimized protocol for metabolome analysis in yeasts using direct infusion electrospray mass spectrometry. Phytochem 62:929-937

Chace DH (2001) Mass spectrometry in the clinical laboratory. Chem Rev 101:445-477

Chen H, Pan Z, Talaty N, Raftery D, Cooks RG. (2006) Combining desorption electrospray ionization mass spectrometry and nuclear magnetic resonance for differential metabolomics without sample preparation. Rapid Commun Mass Spectrom 20:1577-1584

Chiwocha SDS, Abrams SR, Ambrose SJ, Cutler AJ, Loewen M, Ross ARS, Kermode AR (2003) A method for profiling classes of plant hormones and their metabolites using liquid chromatography-electrospray ionization tandem mass spectrometry: an analysis of hormone regulation of thermodormancy of lettuce (*Lactuca sativa* L.) seeds. Plant J 35:405-417

Cook AM, Urban E, Schlegel HG (1976) Measuring the concentrations of metabolites in bacteria. Anal Biochem 72:191-201

De Koning W, van Dam K (1992) A method for the determination of changes of glycolytic metabolites in yeast on a subsecond time scale using extraction at neutral pH. Anal Biochem 204:118-123

De Leenheer AP, Thienpont LM (1992) Applications of isotope-dilution mass-spectrometry in clinical-chemistry, pharmacokinetics, and toxicology. Mass Spectrom Rev 11:249-307

Dole M, Cox Hl, Gieniec J (1973) Electrospray mass-spectroscopy. Adv Chem 125:73-84

Dunn WB, Bailey NJ, Johnson HE (2005a) Measuring the metabolome: current analytical technologies. Analyst 130:606-625

Dunn WB, Overy S, Quick WP (2005b) Evaluation of automated electrospray-TOF mass spectrometry for metabolic fingerprinting of the plant metabolome. Metabolomics 1:137-148

Durgbanshi A, Arbona V, Pozo O, Miersch O, Sancho JV, Gomez-Cadenas A (2005) Simultaneous determination of multiple phytohormones in plant extracts by liquid chromatography-electrospray tandem mass spectrometry. J Agric Food Chem 53:8437-8442

Fell DA (2005) Enzymes, metabolites and fluxes. J Exper Bot 56:267-272

Fiehn O, Kopka J, Dörmann P, Altmann T, Trethewey RN, Willmitzer L (2000) Metabolite profiling for plant functional genomics. Nat Biotechnol 18:1157-1161

Fiehn O (2002) Metabolomics-the link between genotypes and phenotypes. Plant Mol Biol 48:155-171

Freisleben A, Schieberle P, Rychlik M (2003) Specific and sensitive quantification of folate vitamers by stable isotope dilution assays using high-performance liquid chromatography-tandem mass spectrometry. Anal Bioanal Chem 376:149-156

Gavaghan CL, Holmes E, Lenz E, Wilson ID, Nicholson JK (2000) An NMR-based metabonomic approach to investigate the biochemical consequences of genetic strain differences: application to the C57BL10J and Alpk:ApfCD mouse. FEBS Let 484:169-174

Gonzalez B, Fronçois J, Renaud M (1997) A rapid and reliable method for metabolite extraction in yeast using boiling buffered ethanol. Yeast 13:1347-1356

Goodacre R, Vaidyanathan S, Dunn WB, Harrigan GG, Kell DB (2004) Metabolomics by numbers: acquiring and understanding global metabolite data. Trends Biotechnol 22:245-252

Gullberg J, Jonsson P, Nordstrom A, Sjostrom M, Moritz T (2004) Design of experiments: an efficient strategy to identify factors influencing extraction and derivatization of *Arabidopsis thaliana* samples in metabolomic studies with gas chromatography/mass spectrometry. Anal Biochem 331:283-295

Hajjaj H, Blanc PJ, Goma G, François J (1998) Sampling techniques and comparative extraction procedures for quantitative determination of intra- and extracellular metabolites in filamentous fungi. FEMS Microbiol Let 164:195-200

Halket JM, Waterman D, Przyborowska AM, Patel RKP, Fraser PD, Bramley PM (2005) Chemical derivatization and mass spectral libraries in metabolic profiling by GC/MS and LC/MS/MS. J Exp Bot 56:219-243

Hušek P, Šimek P (2006) Alkyl chloroformates in sample derivatization strategies for GC analysis. Review on a decade use of the reagents as esterifying agents. Curr Pharmaceutical Anal 2:23-43

Jenkins H, Hardy N, Beckmann M, Draper J, Smith AR, Taylor J, Fiehn O, Goodacre R, Bino RJ, Hall R, Kopka J, Lane GA, Lange BM, Liu JR, Mendes P, Nikolau BJ, Oliver SG, Paton NW, Rhee S, Roessner-Tunali U, Saito K, Smedsgaard J, Sumner LW, Wang T, Walsh S, Wurtele ES, Kell DB (2004) A proposed framework for the description of plant metabolomics experiments and their results. Nat Biotechnol 22:1601-1606

Jiye A, Trygg J, Gullberg J, Johansson AI, Jonsson P, Antti H, Marklund SL, Moritz T (2005) Extraction and GC/MS analysis of the human blood plasma metabolome. Anal Chem 77:8086-8094

Jonsson P, Johansson AI, Gullberg J, Trygg J, A J, Grung B, Marklund S, Sjostrom M, Antti H, Moritz T (2005) High-throughput data analysis for detecting and identifying differences between samples in GC/MS-based metabolomic analyses. Anal Chem 77:5635-5642

Jonsson P, Johansson ES, Wuolikainen A, Lindberg J, Schuppe-Koistinen I, Kusano M, Sjostrom M, Trygg J, Moritz T, Antti H (2006) Predictive metabolite profiling applying hierarchical multivariate curve resolution to GC-MS data-A potential tool for multi-parametric diagnosis. J Proteome Res 5:1407-1414

Junge K, Eicken H, Swanson BD, Deming JW (2006) Bacterial incorporation of leucine into protein down to -20°C with evidence for potential activity in sub-eutectic saline ice formation. Cryobiol 52:417-429

Kell DB (2004) Metabolomics and systems biology: making sense of the soup. Curr Op Microbiol 7:296-307

Kind T, Fiehn O (2006) Metabolomic database annotations via query of elemental compositions: Mass accuracy is insufficient even at less than 1 ppm. BMC Bioinformatics 7:234

King R, Bonfiglio R, Fernandez-Metzler C, Miller-Stein C, Olah T (2000) Mechanistic investigation of ionization suppression in electrospray ionization. J Am Soc Mass Spectrom 11:942-950

Koek MM, Muilwijk B, van der Werf M, Hankemeier T (2006) Microbial metabolomics with gas chromatography/mass spectrometry. Anal Chem 78:1272-1281

Kopka J (2006) Current challenges and developments in GC-MS based metabolite profiling technology. J Biotechnol 124:312-322
Kovats E (1958) Gas chromographische charakteriserung organischer verbindungen. I. Retentions indices aliphatischer halogenide, alkohole, aldehyde und ketone. Helvetica Chim Acta 41:1915-1932
Krishnan P, Kruger NJ, Ratcliffe RG (2005) Metabolite fingerprinting and profiling in plants using NMR. J Exp Bot 56:255-265
Lange HC, Eman M, van Zuijlen G, Visser D, van Dam JC, Frank J, Teixeira de Mattos MJ, Heijnen JJ (2001) Improved rapid sampling for *in vivo* kinetics of intracellular metabolites in *Saccharomyces cerevisiae*. Biotechnol Bioeng 75:406-415
Larsson G, Törnkvist M (1996) Rapid sampling cell inactivation and evaluation of low extracellular glucose concentrations during fed-batch cultivation. J Biotechnol 49:69-82
Liang HR, Foltz RL, Meng M, Bennett P (2003) Ionization enhancement in atmospheric pressure chemical ionization and suppression in electrospray ionization between target drugs and stable-isotope-labelled internal standards in quantitative liquid chromatography/tandem mass spectrometry. Rap Comm Mass Spectrom 17:2815-2821
Makarov A, Denisov E, Kholomeev A, Balschun W, Lange O, Strupat K, Horning S (2006) Performance evaluation of a hybrid linear ion trap/Orbitrap mass spectrometer. Anal Chem 78:2113-2120
Maharjan RP, Ferenci T (2003) Global metabolite analysis: the influence of extraction methodology on metabolome profiles of *Escherichia coli*. Anal Biochem 313:145-154
Mashego MR, Wu L, van Dam JC, Ras C, Vinke JL, van Windden WA, van Gulik WM, Heijnen JJ (2004) MIRACLE: mass isotopomer ratio analysis of U-^{13}C-labeled extracts. A new method for accurate quantification of changes in concentrations of intracellular metabolites. Biotechnol Bioeng 85:620-628b
Milman, BL (2005) Identification of chemical compounds. Trends Anal Chem 24:493-508
Mohler RE, Dombek KM, Hoggard JC, Young ET, Synovec RE (2006) Comprehensive two-dimensional gas chromatography time-of-flight mass spectrometry analysis of metabolites in fermenting and respiring yeast cells. Anal Chem 78:2700-2709
Morgenthal K, Weckwerth W, Steuer R (2006) Metabolomic networks in plants: Transitions from pattern recognition to biological interpretation. Biosystems 83:108-117
Nielsen J (2003) It is all about metabolic fluxes. J Bacteriol 185:7031-7035
Nielsen J, Oliver S (2005) The next wave in metabolome analysis. TIBTECH 23:544-546
Nordstrom A, O'Maille G, Qin C, Siuzdak G (2006) Nonlinear data alignment for UPLC-MS and HPLC-MS based metabolomics: Quantitative analysis of endogenous and exogenous metabolites in human serum. Anal Chem 78:3289-3295
Niessen WM, Manini P, Andreoli R (2006) Matrix effects in quantitative pesticide analysis using liquid chromatography-mass spectrometry. Mass Spectrom Rev (in press)
Oliver SG, Winson MK, Kell DB, Baganz F (1998) Systematic functional analysis of the yeast genome. TIBTECH 16:373-378
Purohit PV, Rocke DM, Viant MR, Woodruff DL (2004) Discrimination models using variance-stabilizing transformation of metabolomic NMR data. OMICS 8:118-130
Raamsdonk LM, Teusink B, Broadhurst D, Zhang N, Hayes A, Walsh MC, Berden JA, Brindle KM, Kell DB, Rowland JJ, Westerhoff HV, van Dam K, Oliver SG (2001) A functional genomics strategy that uses metabolome data to reveal the phenotype of silent mutations. Nat Biotechnol 19:45-50
Ramautar R, Demirci A, de Jong GJ (2006) Capillary electrophoresis in metabolomics
Trends Anal Chem 25:455-466

Rocke DM, Lorenzato S (1995) A 2-component model for measurement error in analytical-chemistry. Technometrics 37:176-184

Roessner U, Wagner C, Kopka J, Trethewey RN, Willmitzer L (2000) Simultaneous analysis of metabolites in potato tuber by gas chromatography-mass spectrometry. Plant J 23:131-142

Roessner U, Willmitzer L, Fernie A R. (2001) High-resolution metabolic phenotyping of genetically and environmentally diverse plant systems-identification of phenocopies. Plant Physiol 127:749-764

Roessner U, Patterson J, Forbes MG, Fincher G, Langridge P, Bacic A (2006) An investigation of boron toxicity in barley using metabolomics. Plant Physiol (in press)

Roessner U (2007) Plant metabolomics. In: Villas-Bôas SG, Roessner U, Hansen MAE, Smedsgaard J, Nielsen J (eds) Metabolome analysis: an introduction. John Wiley & Sons, New Jersey, USA, p. 215-237

Schaefer U, Boos W, Takors R, Weuster-Botz D (1999) Automated sampling device for monitoring intracellular metabolite dynamics. Anal Biochem 270:88-96

Schauer N, Steinhauser D, Strelkov S, Schomburg D, Allison G, Moritz T, Lundgren K, Roessner-Tunali U, Forbes MG, Willmitzer L, Fernie AR, Kopka J (2005) GC-MS libraries for the rapid identification of metabolites in complex biological samples. FEBS Let 579:1332-1337

Shellie RA (2005) Comprehensive two-dimensional gas chromatography-mass spectrometry and its use in high-resolution metabolomics. Australian J Chem 58:619

Smedsgaard J, Nielsen J (2005) Metabolite profiling of fungi and yeast: from phenotype to metabolome by MS and informatics. J Exper Bot 56:273-286

Smith CA, Want EJ, O'Maille G, Abagyan R, Siuzdak G (2005) XCMS: processing mass spectrometry data for metabolite profiling using nonlinear peak alignment, matching, and identification. Anal Chem 78:779-87

Soga T, Heiger DN (2000) Amino acid analysis by capillary electrophoresis electrospray ionization mass spectrometry. Anal Chem 72:1236-1241

Soga T, Ueno Y, Naraoka H, Ohashi Y, Tomita M, Nishioka T (2002a) Simultaneous determination of anionic intermediates for *Bacillus subtilis* metabolic pathways by capillary electrophoresis electrospray ionization mass spectrometry. Anal Chem 74:2233-2239

Soga T, Ueno Y, Naraoka H, Matsuda K, Tomita M, Nishioka T (2002b) Pressure-assisted capillary electrophoresis electrospray ionization mass spectrometry for analysis of multivalent anions. Anal Chem 74:6224-6229

Spellman PT, Miller M, Stewart J, Troup C, Sarkans U, Chervitz S, Bernhart D, Sherlock G, Ball C, Lepage M, Wiatek M, Marks WL, Gonçalves J, Mrkel S, Iordan D, Shojatalab M, Pizarro A, White J, Hubley R, Deutsch E, Senger M, Aronow BJ, Robinson A, Bassett D, Stoeckert CJ Jr, Brazma A (2002) Design and implementation of microarray gene expression markup language (MAGE-ML). Genome Biol 3: research 0046.1-0046.9

Stein SE (1999) An integrated method for spectrum extraction and compound identification from gas chromatography/mass spectrometry data. J Am Soc Mass Spectrom 10:770-781

Steuer R, Kurths J, Fiehn O, Weckwerth W (2003) Observing and interpreting correlations in metabolomic networks. Bioinformatics 19:1019-1026

Sumner LW, Mendes P, Dixon RA (2003) Plant metabolomics: large-scale phytochemistry in functional genomics era. Phytochem 62:817-836

Taylor CF, Paton NW, Garwood KL, Kirby PD, Stead DA, Yin Z, Deutsch EW, Selway L, Walker J, Riba-Garcia I, Mohammed S, Deery MJ, Howard JA, Dunkley T, Aebersold R, Kell DB, Lilley KS, Roepstorff P, Yates JR 3rd, Brass A, Brown AJ, Cash P, Gaskell Hubbard SJ, Oliver SG (2003) A systematic approach to modeling, capturing, and disseminating proteomics experimental data. Nat Biotechnol 21:247-254

Theobald U, Mailinger W, Reuss M, Rizzi M (1993) *In vivo* analysis of glucose-induced fast changes in yeast adenine nucleotide pool applying a rapid sampling technique. Anal Biochem 214:31-37

Theobald U, Mailinger W, Baltes M, Rizzi M, Reuss M (1997) *In vivo* analysis of metabolic dynamics in *Saccharomyces cerevisiae*: I. Experimental observations. Biotechnol Bioeng 55:305-316

Thompson M, Ellison SLR, Wood W (2002) Harmonized guidelines for single-laboratory validation of methods of analysis (IUPAC Technical Report). Pure Appl Chem 74:835-855

Tikunov Y, Lommen A, de Vos CHR, Verhoeven HA, Bino RJ, Hall RD, Bovy AG (2005) A novel approach for nontargeted data analysis for metabolomics. Large-scale profiling of tomato fruit volatiles. Plant Physiol 139:1125-1137

Trethewey RN (2001) Gene discovery via metabolic profiling. Curr Op Biotechnol 12:135-138

Veriotti T, Sacks R (2003) Characterization and quantitative analysis with GC/TOFMS comparing enhanced separation with tandem-column stop-flow GC and spectral deconvolution of overlapping peaks. Anal Chem 75:4211-4216

Villas-Bôas SG (2007a) Sampling and sample preparation. In: Villas-Bôas SG, Roessner U, Hansen MAE, Smedsgaard J, Nielsen J (eds) Metabolome analysis: an introduction. John Wiley & Sons, New Jersey, USA, p.39-82

Villas-Bôas SG (2007b) Microbial metabolomics: rapid sampling techniques to investigate intracellular metabolite dynamics – an overview. In: Villas-Bôas SG, Roessner U, Hansen MAE, Smedsgaard J, Nielsen J (eds) Metabolome analysis: an introduction. John Wiley & Sons, New Jersey, USA, p.203-214

Villas-Bôas SG, Delicado DG, Åkesson M, Nielsen J (2003) Simultaneous analysis of amino and nonamino organic acids as methyl chloroformate derivatives using gas chromatography-mass spectrometry. Anal Biochem 322:134-138

Villas-Bôas SG, Moxley JF, Åkesson M, Stephanopoulos G, Nielsen J (2005a) High-throughput metabolic state analysis: The missing link in integrated functional genomics. Biochem J 388:669-677

Villas-Bôas SG, Mas S, Åkesson M, Smedsgaard J, Nielsen J (2005b) Mass spectrometry in metabolome analysis. Mass Spectrom Rev 24:613-646

Villas-Bôas SG, Højer-Pedersen J, Åkesson M, Smedsgaard J, Nielsen J (2005c) Global metabolite analysis of yeast: evaluation of sample preparation methods. Yeast 22:1155-1169

Villas-Bôas SG, Noel S, Lane GA, Attwood G, Cookson A (2006) Extracellular metabolomics: a metabolic footprinting approach to assess fiber degradation in complex media. Anal Biochem 349:297-305

Visser D, van Zuylen GA, van Dam JC, Oudshoorn A, Eman MR, Ras C, van Gulik WM, Frank J, van Dedem GWK, Heijnen JJ (2002) Rapid sampling for analysis of *in vivo* kinetics using the BioScope: A system for continuous-pulse experiments. Biotechnol Bioeng 79:674-681

Von Roepenack-Lahaye E, Degenkolb T, Zerjeski M, Franz M, Roth U, Wessjohann L, Schmidt J, Scheel D, Clemens S (2004) Profiling of *Arabidopsis* secondary metabolites by capillary liquid chromatography coupled to electrospray ionization quadrupole time-of-flight mass spectrometry. Plant Physiol 134:548-559

Wagner C, Sefkow M, Kopka J (2003) Construction and application of a mass spectral and retention time index database generated from plant GC/EI-TOF-MS metabolite profiles. Phytochem 62:887-900

Wang Q, Wu C, Chen T, Chen X, Zhao X (2006) Integrating metabolomics into systems biology framework to exploit metabolic complexity: strategies and applications in microorganisms. Appl Microbiol Biotechnol 70:151-161

Weuster-Botz D (1997) Sampling tube device for monitoring intracellular metabolite dynamics. Anal Biochem 246:225-233

Willse A, Chandler DP, White A, Protic M, Daly DS, Wunschel S (2005) Comparing bacterial DNA microarray fingerprints. Statistical Appl Gent Mol Biol 4:19

Wishart D (2007) Metabolomics in human and other mammals. In: Villas-Bôas SG, Roessner U, Hansen MAE, Smedsgaard J, Nielsen J (eds) Metabolome analysis: an introduction. John Wiley & Sons, New Jersey, USA, p253-288

Wittmann C, Krömer JO, Kiefer P, Binz T, Heinzle E (2004) Impact of the cold shock phenomenon on quantification of intracellular metabolites in bacteria. Anal Biochem 327:135-139

Villas-Bôas, Silas G.

AgResearch Limited, Grasslands Research Centre, Private Bag 11008, Palmerston North 4442, New Zealand.
silas.villas-boas@agresearch.co.nz

Koulman, Albert

AgResearch Limited, Grasslands Research Centre, Private Bag 11008, Palmerston North 4442, New Zealand.

Lane, Geoffrey A.

AgResearch Limited, Grasslands Research Centre, Private Bag 11008, Palmerston North 4442, New Zealand.

Abbreviations

CE-MS: capillary electrophoresis coupled to mass spectrometry
DIMS: Direct infusion mass spectrometry
EI MS: Electron impact ionisation mass spectrometry
ESI-MS: Electrospray ionization mass spectrometry
FTICR: Fourier transform ion cyclotron resonance

GC-MS: Gas chromatography coupled to mass spectrometry
LC-MS: Liquid chromatography coupled to mass spectrometry
MRM: Multiple reaction monitoring
MS: mass spectrometry
MS/MS, MS^n: Tandem mass spectrometry
MSTFA: N-Methyl-N-(trimethylsilyl)trifluoroacetamide
NMR: Nuclear magnetic resonance
PCA: principle components analysis
QTOF: Quadrupole time-of-flight mass analyzer
RSD: relative standard deviation
RI: retention index
TBDMS: *t*-Butyldimethylsilyl
TFA: Trifluoroacetic acid
TIC: Total ion current
TMS: Trimethylsilyl
TOF: Time-of-flight mass analyzer
UPLC: Ultra performance liquid chromatography

Reporting standards

Nigel Hardy and Helen Jenkins

Abstract

Metabolomic studies generate large quantities of data. Metabolomics data sets have complex structure and will typically be subjected to a variety of processing and analysis techniques. The data sets are expensive to collect and can be expected to hold more useful information than is extracted and used by the studies, which collected them. These aspects of metabolomics have caused workers to consider, from the very early days of the field, what constitutes comprehensive and well structured metabolomics data, how it should be collected, how it should be transmitted and how, and where it should be stored. It has been generally assumed that the availability of well-curated data sets in standardised formats will pay large dividends for the science. This chapter considers the nature of reporting standards, the benefits that they can yield, existing data standardisation initiatives in metabolomics and related fields and discusses some issue surrounding their development.

1 Introduction

Metabolomic studies generate large quantities of data. Metabolomics data sets have complex structure and will typically be subjected to a variety of processing and analysis techniques. The data sets are expensive to collect, and, by their nature can be expected to hold more useful information than is extracted and used by the studies which collected them. These aspects of metabolomics have caused workers to consider, from the early days of the field, what constitutes comprehensive and well structured metabolomics data, how it should be collected, how it should be transmitted and how, and where it should be stored. It has been generally assumed that a significant effort in making available well curated data sets in standardised formats will pay dividends for the science.

In its report "Sharing Publication-Related Data and Materials: Responsibilities of Authorship in the Life Sciences" (National Research Council (US) 2003), the Committee on Responsibilities of Authorship in the Biological Sciences of the National Research Council of the US enunciates "UPSIDE" - *the uniform principle for sharing integral data and materials expeditiously.* Part of this principle reads as follows:

> "An author's obligation is not only to release data and materials to enable others to verify or replicate published findings (as journals already implic-

Topics in Current Genetics, Vol. 18
J. Nielsen, M. C. Jewett (Eds.): Metabolomics
DOI 10.1007/4735_2007_0242/ Published online: 6 July 2007

itly or explicitly require) but also to provide them in a form on which other scientists can build with further research."

Number 3 of the "five corollary principles associated with sharing publication-related data, software, and materials" is as follows:

"If publicly accessible repositories for data have been agreed on by a community of researchers and are in general use, the relevant data should be deposited in one of these repositories by the time of publication."

The report goes on to note that "these repositories help define consistent policies of data format and content". The role of the specialist community in developing technical standards that allow data and other scientific information in their field to be easily shared is emphasised. It is, therefore, expected of the metabolomics community that it should address the issues of data standards.

These issues are characterised as standards for reporting metabolomics studies and as computer implementations of those standards which integrate both with the working practices of laboratories and with other scientific fields, particularly other aspects of functional genomics. Those implementations will be i) databases for longer term storage and retrieval, ii) transmission formats for collection and dissemination, and iii) supportive data collection and manipulation tools for workers at all stages of metabolomics studies.

The community currently has no accepted reporting standards. This chapter seeks to outline the background to their ongoing development, describe the nature of that task, and highlight some of the more challenging aspects.

1.1 Data handling in metabolomics

The term metabolomics emerged in the literature in the late 1990s (Oliver et al. 1998; Tweeddale et al. 1998; Fig. 1) and shows a timeline of significant developments in the development of data handling related to it. It quickly became received wisdom than data handling would be an important aspect of the new field and that publication of the extensive data sets should become the norm (Fiehn et al. 2001). There are a number reasons for this. Across science in general, and in "post-genomic" fields in particular, publication of large and complete sets of detailed data, on web sites or generally accessible databases, for reasons of transparency and of reuse was gaining acceptance. More specifically, metabolomics began to generate large volumes of complex data (with the prospect of even larger volumes in the near future). The metabolome is very dynamic. A metabolome estimate is greatly influenced by the source and culture conditions of the biological material and by laboratory preparation and analysis procedures. The need for specific and detailed *metadata* to accompany metabolome estimates was perceived to be far greater than was then recognised for transcriptomics or proteomics data. This increases the complexity of the metabolomics data and encourages more sophisticated computer-based data handling.

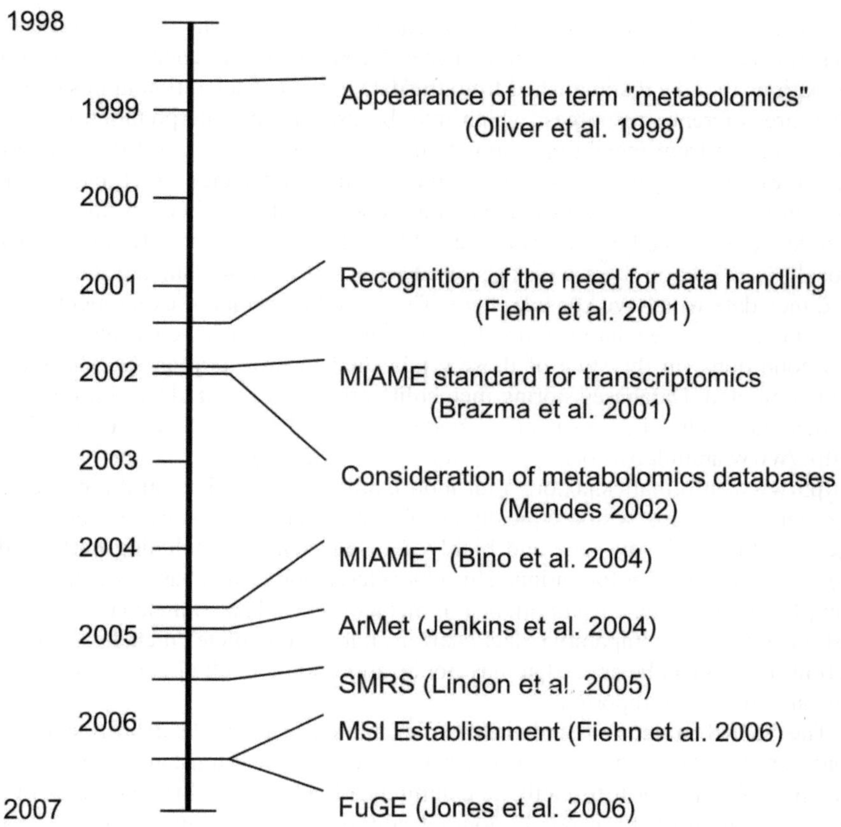

Fig. 1. A timeline of the significant developments in metabolomics data handling.

The analytical techniques used in metabolomics are fast, particularly in comparison with early transcriptomics and proteomics techniques. High throughput for screening large numbers of samples was seen as a great benefit (Spraul et al. 1997; Nicholson et al. 1999; Weckwerth 2003). The per-sample cost of metabolome estimates is low (Griffin 2006) and therefore large-scale trials involving many hundreds of metabolome estimates were quickly envisaged. Each estimate resulted in at least many tens of data points (Roessner et al. 2000), but particularly when unidentified peaks are included, this rapidly rose into hundreds. Fingerprinting approaches (see Section 3) took this into the thousands. The sheer volume of data encouraged computer-based data handling.

Metabolome estimates require significant data analysis to yield biologically relevant results. Data sets are intrinsically highly multivariate and often multifactorial, clearly making computer analysis necessary. Experimental designs may call for classification or discrimination and data collected for a specific reason may subsequently be used to answer new questions not originally addressed.

Mendes (2002) distinguishes conceptually five types of database associated with metabolomics, recognising that some implementations will belong to more that one class. Two of these types, viz. databases cataloguing all known metabolites in each biological species and databases containing reference biochemical information are to be distinguished from the data under consideration in this chapter. They are reference resources, which may be used in the interpretation of "raw" data obtained from metabolite estimation equipment, specifically for the identification of metabolites. Their development and maintenance is crucial to metabolomics, and such library databases are considered elsewhere in this volume. Mendes' other three types of database hold "metabolite profiles". Taking this term loosely (see Section 3 for a discussion of terms), it is these data and their associated metadata which we consider here. These are the results of experiments or trials. The idea of a few large comprehensive collections of data across many species and conditions (in the style of those established for transcriptomics, see Section 4.1) is mooted. Databases storing metabolite profiles for a single species, perhaps associated with species-specific portals such TAIR (Rhee et al. 2003) (http://www.arabidopsis.org/) for *Arabidopsis* or MGI (http://www.informatics.jax.org/) for mouse are suggested. The third type, the experimental database, would typically be laboratory based and store metabolite profiles together with raw data and highly detailed metadata including perhaps domestic management information. This characterisation of databases highlights two complementary issues of importance in metabolomics data: i) metadata are essential for principled comparative data analysis, but ii) "complete" metadata, even if a definition of it can be arrived at may not be appropriate in all circumstances, when data are collated or reported.

The SMRS initiative (Lindon 2005), examined in more detail in Section 3.3, considered the need to cater for a variety of reporting circumstances, specifically "the need for standards to facilitate communication between different fields of activity and to fulfil the needs of journal editors and regulatory agencies…" and noted that there are "fundamental differences between both the design and objectives of efforts focused on regulatory submission and those efforts focused on basic research". Figure 2 shows some of the stages in metabolomics data collection, manipulation and storage. It shows examples of important data transmissions and storage, which may occur. If data are managed throughout in line with agreed standards, the process will be simplified, tools can be developed, and shared. The final products (shareable data and scientific knowledge) will be of demonstrably greater value and quality.

2 Standards, models, and formats

First, an important contrast should be made between what may be termed strictly "reporting standards" which concentrate on the list of information items and their semantics which constitute a complete and adequate report of the scientific undertaking and "data models" which concentrate on the logical structure of a data set. A third concept is "data format" which is concerned with the syntax of data recording and transmission.

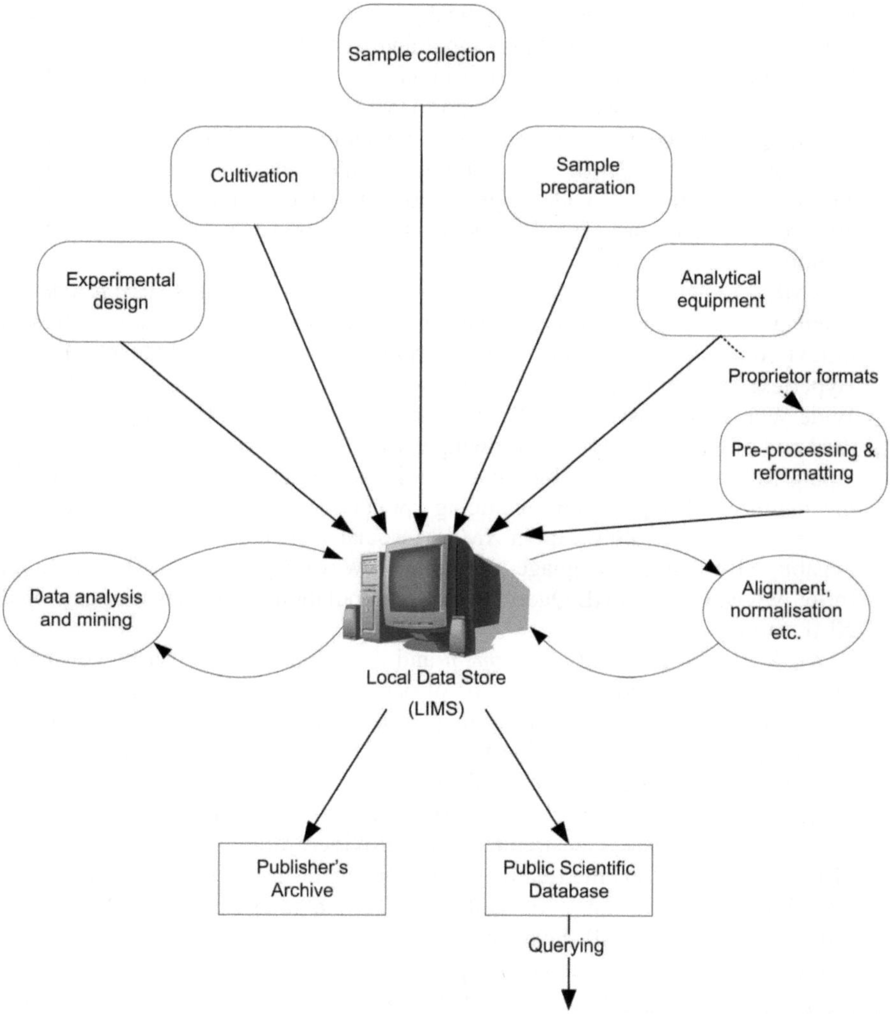

Fig. 2. Data handling in metabolomics. Curved corners: Data collection activities. Ovals: Data manipulation activities. Oblong: Data deposition activities.

A reporting standard is a document. It is written in natural language and is designed to be unambiguously understood by all parties involved: biologists, biochemists, and statisticians. It should be formal, precise, and complete. It describes what information is required but does not necessarily prescribe the representation of data items or the structuring of the data set. A data model adds these things (Hoffer et al. 2002; Hardy and Fuell 2003). Data representing some aspect of the real world is structured so as to provide as good a model as possible of that part of the real world. This provides a basis for building computer implementations, which should therefore be capable of representing all plausible states of the real world. They should also be amenable to modification to accept new developments in that world. Data models are typically not themselves computer implementations. UML – the Unified Modeling Language (Booch et al. 2005), which is largely graphical, is currently the most common formalism for building such models. Many software tools exist to support UML.

The two most common computer implementations of such data models are relational databases and XML. The relational database paradigm is long established and well supported by a wide range of tools. It delivers effective storage and powerful retrieval facilities. The data model is represented as a set of tables with constraints on their content.

XML – Extensive Markup Language - (http://www.w3.org/XML/) is a *mark-up* language and is used as a data format for transmission. It is typically represented in text containing *tags* identified by surrounding "<" and ">" brackets. HTML – Hypertext Markup Language – is today widely used to produce pages in the World Wide Web and is based on XML. XML has the advantages of being an open standard and of having a huge and growing variety of tools to support it. It can be manipulated by libraries available for all significant programming languages. Increasing numbers of applications including spreadsheets and statistical packages are accepting and generating data in XML. Associated standards, including XSL - Extensible Style sheet Language – (http://www.w3.org/Style/XSL/) and XQuery (http://www.w3.org/XML/Query/) allow manipulation and retrieval of data held in XML format.

An XML file is called a *document* and document content can be constrained. Data values can be constrained to be of particular types (integer, text, etc.); to be in specified ranges or to come from specified lists. Optional and required values can be specified and permissible numbers of repetitions of data items can be set. Constraints on co-occurrence of data, such as requiring unique values within specified groups and requiring that particular values be taken from sets defined elsewhere can be set. Constraint can be provided through a number of mechanisms. XML itself has the relatively weak mechanism of DTD – Document Type Definition – but a range of additional technologies including XMLSchema (http://www.w3.org/XML/Schema), Relax NG (http://relaxng.org/), and Schematron (http://www.schematron.com/) permit external definition of constraints using a variety of approaches. A set of constraints is known as a *schema*. An XML document can exist without validation against a schema. This is in contrast to relational databases, which impose one set of constraints on all of the data held. An XML document can be validated, on demand, against a chosen schema. It can be validated against a range of schemas for different purposes. A laboratory information system may have different constraints than a public repository or a publisher, but a document could be checked for conformance to any of these. Schemas can be built to permit data in addition to required elements. Thus, a document containing necessary data can also include additional items, which can be used by applications, which are aware of them, but can be automatically ignored by applications which are not. XML, therefore, provides for flexible data formats which can, however, be checked for conformance to fixed standards.

Table 1. Comparison of metabolomics data standards initiatives

	MIAMET	ArMet	SMRS	MSI
Principle reference	(Bino et al. 2004)	(Jenkins et al. 2004)	(Lindon et al. 2005)	(Fiehn et al. 2006)
Web site	-	http://www.armet.org/	http://www.smrsgroup.org/	http://msi-workgroups.sourceforge.net/
Status	Not under development.	In use and developing. Input to MSI.	Not under development Input to MSI.	Under development
Reporting Standard?[a]	Yes	Implicit	Yes	Yes
Data model?[b]	No	Yes	No	Yes
Data format?[c]	Suggestions for re-use	Yes	Suggestions for re-use	Yes
Data analysis?[d]	No	No	Yes	Yes
Bias?	Plant biology	Plant biology	Attempt to avoid. (Preclinical drug trials?)	Attempt to avoid

[a] Does/will the initiative provide reporting standards?
[b] Does/will the initiative provide a data model?
[c] Does/will the initiative provide data format(s)?
[d] Does/will the inititive cover data analysis and data modelling?

UML tools will often provide facilities for automatic or semi-automatic generation of relational database and/or XML implementations of models. It is also important to recognise that once a data model is agreed, other implementations are possible and data conversion between implementations based on a common model is relatively simple to implement. Code for such conversions can be automatically generated.

3 Initiatives in metabolomics data standards

A number of initial proposals for data standards have been made, from different parts of the community, with different biases and approaches. In reviewing this work, in addition to reporting standards, models, and formats discussed above, it is important to consider more closely the metabolome estimate. The nature of the final metabolome estimate can vary greatly. Terminology varies, but the early work by Fiehn (2001) distinguished between concepts important for the structure and nature of the estimate. "Metabolite profiling" provides estimates of abundance of previously determined chemical species. These species may or may not be identified but can be reliably detected. Thus the list of species is known *a priori* and abundances for those and only those components are sought. True "metabolomics" should be unbiased and detect any chemical species present. Thus the membership of the list of species is part of result, as well as their abundances. "Metabolic fingerprinting" involves collection of measurements, which are not directly related to distinguishable chemical species. There is no list of the chemistry and therefore no associated abundances.

We now consider four initiatives in the field. These are summarised in Table 1.

3.1 MIAMET

This proposal (Bino et al. 2004) is a checklist of "minimum information", explicitly inspired by the example of MIAME (see Section 4.1). It is therefore an initial reporting standard. It is presented as a "suggestion to the community". The authors and the background to the paper suggest that this community is plant biology, with significant emphasis on agronomic applications. There is significant emphasis on chromatography and mass spectrometry (MS) techniques. The list is broadly structured into 4 sections: i) Experimental design, ii) Sampling, preparation, metabolite extraction, and derivatization, iii) Metabolite profiling design, and iv) Metabolite measurement and specifications. It does not cover multivariate data analysis and data mining reporting. It is an "experiment" centred list, implying that all data will be collected under a traditional experimental design with identified experimental factors. The proposal contains some limited hints at data formats, suggesting the use of NetCDF (Network Cmmon Data Format) for raw MS data and JCAMP (Joint Committee on Atomic and Molecular Physical Data) for NMR (Nuclear Magnetic Resonance) data. Storage of fingerprint data is thus implied, but the eventual output is expected to be a list of metabolite identifiers (for both known and unknown metabolites) with relative or absolute quantification. Metabolite profiling or metabolomics are thus the main objectives. Data types and legal values are not generally specified, though some pieces of information are illustrated by lists of possible example values. Associated text documents are specifically proposed for the description of protocols, materials, and methods. This implies that the information would not be suitable for automatic manipulation.

3.2 ArMet

ArMet (Jenkins et al. 2004) is presented as a data model and has been implemented both as a relational database and as an XML schema. It is based on a systems analysis of plant metabolomics work. It does not cover data analysis or data mining. The data model presented is described as a framework since it specifies a very limited set of core data but provides a principled method of extension for laboratory, technology, or process based specialisations as a vehicle for a community-driven process of development and enhancement. There are nine components in the framework. One, the unspecialised "MetabolomeEstimate", concerns what might be termed data while eight would be considered metadata. Core ArMet relies on references to external (textual) documentation of detailed experimental protocols. This can provide a clear and complete record of all metadata, but it does not provide for automatic verification of completeness of the dataset or for automatic data retrieval on these aspects. As required, subcomponents to support core data plus greater detail in forms that can be verified and searched can be designed. Examples are published. The unspecialised "MetabolomeEstimate" must be extended to accommodate each type of analytical procedure and examples for Gas Chromatography/Mass Spectrometry (GC-MS) in both metabolite profiling and true metabolomics applications have been published (Jenkins et al. 2004, 2007). ArMet accommodates data covering all the information specified in MIAMET.

3.3 SMRS

The Standard Metabolic Reporting Structures (SMRS) working group (Lindon et al. 2005) comprised members from academia, industry, and the government. It produced a draft policy document (Lindon 2005). It is clearly a draft reporting standard. It does observe that the Standard for Exchange of Nonclinical Data (SEND) is an available and supported data format, which would handle parts of the reporting.

In contrast to MIAMET and ArMet, reporting of data analysis and data modelling (in the statistical and data mining sense) is firmly included. This stems perhaps from an emphasis on metabolomics applications in preclinical drug safety assessment where regulatory submission of complete studies is a goal. More generally, SMRS may be viewed as biased to animal work in contrast to the plant bias of MIAMET and ArMet.

The standard concerns three areas: the origin of a biological sample, the analytical technologies, and methods applied to the sample and the application of chemometrics to retrieve information from analytical data. Each is broken down in some detail into the aspects, which must be reported on.

3.4 MSI

Under the auspices of the US National Institutes of Health/National Institute of Diabetes and Digestive and Kidney Diseases a workshop on metabolomics standards was held in Bethesda, USA in August 2005 (Castle et al. 2006; Fiehn et al. 2006). Representatives of interested parties and of previous initiatives were invited. It was recognised that common agreement over as wide a community as possible was required if standards were to be accepted and applied.

The MSI consists of five working groups concentrating on experiment metadata (the biological sample context working group), datasets and their production (the chemical analysis working group), statistical analysis and data mining (the data analysis working group), nomenclature (the ontologies working group), and the development of data formats to support the transmission of datasets that comply with the MSI standards (the data exchange working group). Learning from earlier work and examples in other fields, the biological sample context group found it beneficial to establish standard reporting requirements for four contexts: plant biology, environmental, mammalian/*in vivo* experiments, and microbial/*in vitro* biology experiments. MSI work has firmly begun with the development of reporting standards. The ontologies and data exchange groups are committed to maximum collaboration with initiatives from other fields and to re-use or extension of existing models, formats, and ontologies.

4 Reporting standards in other fields

Metabolomics follows transcriptomics and proteomics onto the scene. Data standardisation initiatives are correspondingly further advanced in those fields. It is clearly important that metabolomics learns both from this work and seeks to integrate as closely as possible with it.

4.1 Transcriptomics

The longest standing standards initiative in the functional genomics field is MIAME – "Minimum information about a microarray experiment" (Brazma et al. 2001) concerned with reporting gene expression data. It is developed under The Microarray Gene Expression Data (MGED) Society whose aim is to facilitate the sharing of such data. MIAME was designed to ensure interpretability of experimental results and their potential independent verification. The "minimum" constraint is reflected in the stated requirement that the information required "should be sufficient to interpret the experiment and should be detailed enough to permit replication of experiments" (Brazma et al. 2001).

Six sections are included in the minimum description. **Experimental design** includes a description of the type of experiment, the experimental variables, the use of replicates and quality control procedures. It also associates specific samples with specific arrays. **Array design** provides a definition of all arrays used, with the genes and their physical layout. **Samples** describes the sources of biological material and any treatments applied, the extraction of the nucleic acids and the labelling. **Hybridizations** reports the laboratory conditions under which the samples and arrays meet. **Measurements** reports data from raw images through to the processed gene expression matrix. **Normalisation controls** reports parameters associated with data normalisation and array elements serving as controls. It may be noted that these six sections vary in the degree to which they are specific to transcriptomics experiments: **Array designs** is highly specific, **Samples** contains significant amounts of information, which would be relevant for other types of analysis.

A number of domain specific extensions have been developed, which adopt the MIAME requirements but extend and make more specific aspects of reporting which are crucial to particular applications. These include extensions to transcriptomics aspects to accommodate specific technologies and extensions to non-transcriptomics aspects where the application areas require additional contextual information (including toxicogenomics, Mattes et al. 2004; environmental genomics, Morrison et al. 2006b; and plant genomics Zimmermann et al. 2006). A data format to include MIAME requirements was designed under the auspices of the Life Sciences Research Task Force of the Object Management Group (OMG, http://www.omg.org/). This is MAGE-ML (Microarray Gene Expression – Markup Language), an XML document class. MAGE-OM ((Microarray Gene Expression – Object Model) is a data model, defined in UML and from which MAGE-ML can be derived. A stated aim of MIAME is to facilitate the establishment of databases and public repositories (Brazma et al. 2001). Three compliant repositories now exist: ArrayExpress (Brazma et al. 2003) (http://www.ebi.ac.uk/arrayexpress/), Gene Expression Omnibus (GEO) (Barrett et al. 2005) (http://www.ncbi.nlm.nih.gov/geo/), and the newer CIBEX (http://cibex.nig.ac.jp/).

Publication of MIAME and its adoption by both journals and public repositories has spawned the generation of a range of MAGE compliant tools and extensions to existing systems for the collection and management of microarray experiment data and for its subsequent analysis. The MGED organisation maintains a list of MIAME compliant tools:
(http://www.mged.org/Workgroups/MIAME/miame_software.html).

4.2 Proteomics

The Proteomics Standards Initiative (PSI)(Orchard et al. 2003; Taylor et al. 2006) operates under the auspices of the Human Proteome Organisation (HUPO) (Hanash and Celis 2002). One of the objectives of HUPO is to coordinate the development of standard operating procedures related to data collection, analysis, storage, and sharing. PSI output is in three areas: reporting requirement documents; controlled vocabularies and standard file formats.

The reporting requirements are known as MIAPE (Minimum Information about a Proteomics Experiment) guidelines. These are being developed in a modular form. With overall guidelines and context, technology specific documents can be developed in relative isolation to cover the aspects necessary. Requirements for both the experimental and informatics aspects of gel electrophoresis and mass spectrometry have been developed. Column chromatography, capillary electrophoresis and general sample handling, and preparation have been considered.

The "PSI MI XML" (PSI Molecular Interaction XML) format (Hermjakob et al. 2004) is a data exchange format for molecular interactions – protein-protein interactions in the first instance. It was an early output from PSI and was developed without a MIAPE requirements document. The mzDATA (Taylor et al. 2006) format for capturing mass spectrometry peak list information is supported by a number of instrument manufacturers. This handles the experimental aspects of mass spectrometry. Work is currently in progress to develop a new format, mzML (http://www.psidev.info/), to supersede both mzDATA and the mzXML format (Pedrioli et al. 2004), which was intended for essentially the same purpose. AnalysisXML is a format, which captures parameters and results of search engines for the informatics aspects of protein and peptide identification. Formats to meet other MIAPE modules are being developed.

5 Cross-domain standards

Development of standards for particular fields requires significant effort. Workers in each field have been aware throughout their development that common standards would ultimately be desirable. Pioneers of techniques naturally tend to report only in that narrow field; say in proteomics or in metabolomics. As techniques become established as widely available research tools, workers will undertake research spanning the techniques appropriate to tackle the scientific questions. Papers reporting on "multi-omics" work are now common. The consistent request from potential users is "I only want to enter the data once".

The Reporting Structure for Biological Investigation (RSBI) (Sansone et al. 2006) is a working group of MGED. It attempts to tackle "challenges associated with integrating data and representing complex biological investigations". Constituted to bring together environmental genomics, nutrigenomics, and toxicogenomics, it seeks to integrate more widely across biological domains. Starting at the highest level concepts it has defined a self-contained unit of scientific enquiry as an “Investigation”. “Studies” within an investigation are composed of “Actions” applied to “Subjects”. Subjects are the biological material under investigation. An “Assay” is the experiment (measurement procedure) carried out to produce data for computational purposes. These, and other concepts described under RSBI are fundamental to any biological investigation, necessary for its complete description and the source of significant confusion between application fields, technologies, and disciplines. (No worker in data standards has escaped the confusion of the apparently simple word “experiment” which has a range of meanings such that its use is often meaningless except within the narrowest community). Whilst still at a high level of abstraction, RSBI therefore offers an initial structure for basic concepts and potential integration.

The largest area of potential overlap in requirements (and therefore potentially of common or compatible standards) is in the description of the biological material under examination, of its classification, origins, cultivations, and preparation. Morrison et al. (2006a) argue strongly that a common concept of “Sample” across all “-omics” technologies is desirable and that sample description is technology independent until a late stage in collection and preparation. They propose an ontology based approach considering samples as entities (or “continuants” in ontological terms) to which processes happen. Samples have a temporal component (absolute or relative) such that the description at a later time point may differ from that at an earlier time due to the processes that have happened. They emphasise, however, the challenges of defining “the minimum necessary information to adequately describe a sample”. It is important to note that MIAME variants differ largely in sample description. MIAMET and ArMet have some bias towards plant work while SMRS is heavily influenced by pre-clinical trials. Diverse application area requirements were therefore to some extent obscured in that work. The breadth of requirements is more apparent across different application areas than across different technologies.

Reporting standards are being developed at a growing rate. Typically, they are being developed within a technology or application community. Synergy and re-use, rather than contradiction and repetition are hard to achieve. It became apparent that groups were working in ignorance of each other and without a framework for consistency. Minimum Information for Biological and Biomedical Investigations (MIBBI - http://mibbi.sourceforge.net/) has been established to provide a single point of information for reporting standards checklists, to foster collaborative development and to promote gradual integration. In under a year, 15 reporting checklists have registered with the site.

The Functional Genomics Experiment Object Model (FuGE) (Jones et al. 2006) is described as “a framework for creating data standards for high-throughput biological experiments”. Its aim is to provide a common data model for developing data standards across the '-omics. It is hoped that this approach will lead to data sets that represent equivalent information in a consistent way whatever 'omic approach is used to generate them. If successful, this will greatly ease the data handling issues raised by functional genomics experiments that aim to determine the function of genes using multiple ‘omic approaches.

FuGE is provided as an object model in UML and an XML Schema. The model includes support for the aspects of experiments that are common across the 'omics, for example, biological material, experiment protocols, and equipment. FuGE provides two options for supporting specific functional genomic techniques: i) the core model can be extended to support specific techniques, thereby adding specific information within the same data structures; or ii) by providing references to data held in pre-existing external data formats whilst ensuring that all the metadata that is necessary to place that data within the context of the overall experiment is included.

Both MAGE and the PSI have undertaken to use FuGE as the basis for their data standards development. FuGE is also being evaluated by the data formats working group of the MSI for use in developing metabolomics data formats.

The development of frameworks such as FuGE and the commitments made by the various standards initiatives to not only use FuGE, but to co-operate together to draw up common data standards where possible, are evidence of the desire across the 'omics for unified standards. The move towards true functional genomics and systems biology investigations emphasises the real and emerging need for such standards. The extent to which this is achievable and pragmatic remains to be shown. Certainly the development of common approaches to extending and specialising FuGE to generate data formats will promote the development of common standards. However, it is also clear that the range of experiments that are carried out and the evolving nature of many of the technologies will mean that some flexibility will be required to deal with specialist or experimental situations. A reasonable target for the near future is perhaps that a core set of common standards will be developed which may be extended following commonly understood principles to handle such situations. Publication of such extensions in repositories such as MIBBI (described above) will then encourage others to use or extend these where appropriate.

6 Issues in metabolomics standards

6.1 The detailed nature of standards

Development of data standards may be expected to comprise two aspects. First they must specify *what* is to be reported ("reporting standards" above) and secondly *how* it is to be reported ("data models" and the derived "data formats" above). A simple example might be that the standard requires the storage temperature for samples be recorded. We might name the item of data *StorageTemperature*. More specifically, it might require that the intended temperature (as set on the thermostat) be recorded or that the average storage temperature (as monitored by the equipment) be recorded. Further, the full context of that reading is required. Which samples, of what material, collected and prepared under what circumstances are concerned and how do they relate to the complete experiment and specifically to the analytical runs which generate metabolomics data. Generally, therefore, a very precise definition of the meaning of the reported value is required. Failure to carefully specify what should be reported can lead to diligent and faithful recording of values, which are not comparable. The semantic definition of standards is perhaps the most demanding aspect of establishing standards. Once a precise definition of *StorageTemperature* is established, we may turn to the question of how it is to be reported. The semantic definition may offer direction; common practice may suggest or dictate choices. For our temperature example, we will presumably need a number - "cool", "medium", or "warm" are unlikely to represent the concept appropriately. What accuracy is implied? Is an integer appropriate (i.e. "to the nearest whole degree") or should it be reported to one decimal place? Should the value be in Celsius or Kelvin or should either be permissible, thereby, implying that the scale must be stated in the data? This level of detail may seem excessive, but a crucial aspect of reporting standards is automatic comparability. Automated comparison, particularly within large data sets, will lack the flexibility normally provided by humans when, for example, comparing two papers containing largely verbal reports of "Materials and Methods" where -80C and 193K can be readily recognised as essentially the same *StorageTemperature.*

Taking forward this example on the assumption that the value will be a real number to one decimal place representing Celsius (i.e. there is no need to state the scale used) we have a remaining level of standardisation to consider: that of representation in a computer. At this level, there is in fact less need for standardisation and indeed multiple representations in a range of contexts will be an advantage. Our temperature could be written (word processed or even by hand) and still meet the standard. To obtain the benefits of automated processing we will typically want it in other forms. A textual representation (say in XML) can be automatically transformed into a "binary" representation (say in NetCDF, Rew et al. 1997), Microsoft Excel® or a relational database management system) and visa versa. The rules of the standard (one decimal place etc.) can be enforced in either format. The lack of rigid constraint at this level is a huge benefit of a standard, which is tightly defined at the other levels. Specialised or general purpose software can be developed and used for data collection, transmission, storage, manipulation, and analysis and each conforms to the standard but each can hold and manipulate the data in formats supportive of the task in hand.

Returning to the question of values rather than their representation, the *StorageTemperature* example requires a number and, together with stating that it must be given to one decimal place this defines legal values. That is, given a piece of data, we can verify that it is acceptable. "-80.0" is acceptable; "-80.01" is not; "cold" is not. It is crucial that a data value be constrained. Fundamentally, if any piece of data cannot be provided, automated comparability is lost. In practical terms automatic data validation is of huge practical value in quality control – i.e. it helps avoid typos (Jenkins et al. 2005). Many data values, particularly in metadata, will not be numbers. How will they be reported? Ontology terms are required.

6.2 Controlled vocabularies and ontologies

Well defined, unambiguous, and recognised terms for describing many aspects of metabolomics are crucial to common understanding and interpretation of data. Numerical and yes/no data items require careful description of their semantics. Where data of other types is required, the possible values that can be reported must additionally be defined.

Terms for some concepts have long or well established mechanisms for their management. Biological species naming has a long history, well described mechanisms for the creation and changing of names and computer support through online databases such as the NCBI Taxonomy Browser (Wheeler et al. 2000)(http://www.ncbi.nlm.nih.gov/Taxonomy/taxonomyhome.html/). Terms for varieties, strains, or ecotypes are harder to establish. These may be maintained for model organisms, for test subjects (e.g. rats) or for economically significant species but this tends to be on an *ad hoc* basis in each case. Terms for chemical species are discussed below and demonstrate the difficulties in a large field but one, which might have been expected to be well established. Metabolomics requires terms for many more obscure concepts that these.

A controlled vocabulary is a set of terms for a particular concept, which are well defined and probably believed to be complete. They are the "authoritative" words or phrases, which should be used. Problems associated with synonyms, homonyms, and spelling variants are handled, in addition to providing precision of description through the definition. ArMet specifies the use of controlled vocabularies for a number of data items. Specific vocabularies are not mandated; rather, each term must be associated with an "authority" which is a reference to the vocabulary from which it is taken. It expects that subcomponents would follow this model. This loose approach was seen as a pragmatic option in the absence of good and generally accepted vocabularies. It does reduce automatic comparability of data.

An ontology is a significant development on controlled vocabularies. Terms in an ontology are structured, typically at least hierarchically. Related terms are grouped and more general terms are represented as "parents" of more specific terms. The Plant Ontology (PO) (Bruskiewich et al. 2002) is a good example which covers plant growth and development stages and plant structure. An "inflorescence" is defined (as part of the "shoot") and additionally the potential parts of an inflorescence are described. These include a "flower" which is further declared to have parts. The PO further accommodates "ear" and "tassel" as specialist terms for inflorescences in *Zea mays*. Organs and tissues (perhaps as sources of biological material for metabolome estimates) can therefore be unambiguously described according to this ontology. Comparability of data is thereby enhanced. The structure also permits more flexible searching of data. Principled comparisons of data pertaining to say, "flower" and "tassel" may be possible.

Data formats require term lists. An instructive and early example of their use is in the PSI's molecular interaction format (Hermjakob et al. 2004) (see Section 4.2). External, general purpose, ontologies (including the NCBI taxonomy) are used where possible and in addition five specialist controlled vocabularies were developed to handle the specialist terms for molecular interaction.

The MSI initiative has an ontology working group. This is operating in close cooperation with the OBI (Ontology for Biomedical Investigations) community. OBI was formerly known as the Functional Genomics Investigation Ontology (FuGO) (Whetzel et al. 2006). OBI expects to develop terms, which are of general applicability and terms that are relevant only to particular applications. It thus offers the opportunity to maximise integration across technologies while supporting specialisms.

6.3 Chemical identity

Profiling and true metabolomics techniques are said to produce "peak lists". Each peak is a signal from the analytical machine and is generally assumed to reflect the presence of a particular metabolite. It can typically be quantified in relative or absolute terms. Identification of peaks is the ultimate objective in metabolomics. Goodacre et al. (2004) comment that in comparison with transcriptomics and proteomics, once quantification of specific metabolites in parallel and in various sample matrices has been leant, "a more or less universal approach that spans the species barriers can be adopted." The two or more dimensional data produced by the analytical machine is therefore transformed by numerical processing and library lookup into a simple list of name/quantity pairs. What are these names? We would like them to be chemical identities. A number of registries and catalogues exist (for example CAS (Chemical Abstracts Service), KEGG (Kyoto Enzyclopedia of Genes and Genomes), PubChem, ChEBI (Chemical Entities of Biological Interest)). The operating procedures of laboratories and (semi-)automated library lookup mechanisms will produce one or more of these for each putative metabolite. Thus, the choice of naming convention is of great importance to practitioners. Confusion between synonyms and differing levels of precision in naming compounds add to the difficulty. InChI (International Chemical Identifier; Freemantle 2002), the IUPAC (International Unit of Pure and Applied Chemistry) algorithm for generating a unique label for a given chemical structure is gaining acceptance as an objective mechanism for reporting chemical identities. It does not rely on a registry (the same label can be independently generated by different workers) and handles a range of degrees of specificity (for example, indicating isomers or not). A drawback is its lack of human interpretability, but lookup facilities are appearing and graphical structure can be extracted from it.

In practice, profiling data will necessarily include peaks whose true meaning, in the sense of a chemical identity, is not known. These are the so-called "unknown peaks". A particular experimental protocol will reliably yield estimates of these peaks and they can be recognised across runs but no chemical identity can be associated with them. Indeed, it may not be certain that such a peak represents one and only one chemical species. We require labels for these peaks. Further, two other features are necessary. Firstly, subsequent identification of at least some such peaks is to be expected. Data sets deposited for publication or regulatory purposes will be frozen (with their unknown peak labels) but augmentation of those sets, for re-use, by association of the labels with chemical identities must be facilitated. A subsidiary issue is that any naming scheme must permit multiple laboratories to label (potentially the same peaks) without confusion. Secondly, these peaks are characterised only as signals from particular analytical equipment types under specific circumstances and conditions that must be included in reporting.

Bino et al. (2004) propose a scheme for naming the unknowns. This combines the functions of creating unique identifiers and of reporting the characteristic signal. The scheme has significant technology dependence and is not rigorously defined. It requires community co-ordination to allocate laboratory identifiers. Jenkins et al. (2007) propose a specific scheme for GC-MS based on retention time and a maximum of 20 mass-to-charge ratio/ion abundance pairs. The need for a unique label (more than adequately supported by the web concept of a URI (Uniform Resource Identifier)) and the need for associated data suggest the potential of Life Science Identifiers (LSIDs) (Object management Group 2004) or other URL (Uniform Resource Location) based mechanisms for creating permanent labels in this context.

7 Conclusions

Development of reporting standards in metabolomics is necessary and beneficial. Data sets which can be consistently understood must be deposited for publication and regulatory purposes. They may also be made available for re-analysis and for re-use in new scientific contexts. The metabolomics community is undertaking this development. It will have specialised requirements. It will also have requirements in common with other disciplines and biological and biomedical research in general stands to benefit greatly from common standards for data interchange. Establishment of acceptable standards is a significant challenge.

References

Barrett T, Suzek TO, Troup DB, Wilhite SE, Ngau W-C, Ledoux P, Rudnev D, Lash AE, Fujibuchi W, Edgar R (2005) NCBI GEO: mining millions of expression profiles - database and tools. Nucleic Acids Res 33:D562-566

Bino RJ, Hall RD, Fiehn O, Kopka J, Saito K, Draper J, Nikolau BJ, Mendes P, Roessner-Tunali U, Beale MH, Trethewey RN, Lange BM, Wurtele ES, Sumner LW (2004) Potential of metabolomics as a functional genomics tool. Trends Plant Sci 9:418-425

Booch G, Rumbaugh J, Jacobson I (2005) The Unified Modeling Language User Guide, 2 edn. Boston, MA: Addison-Wesley

Brazma A, Hingamp P, Quackenbush J, Sherlock G, Spellman P, Stoeckert C, Aach J, Ansorge W, Ball CA, Causton HC, Gaasterland T, Glenisson P, Holstege FCP, Kim IF, Markowitz V, Matese JC, Parkinson H, Robinson A, Sarkans U, Schulze-Kremer S,

Stewart J, Taylor R, Vilo J, Vingron M (2001) Minimum information about a microarray experiment (MIAME) - toward standards for microarray data. Nat Genet 29:365-371

Brazma A, Parkinson H, Sarkans U, Shojatalab M, Vilo J, Abeygunawardena N, Holloway E, Kapushesky M, Kemmeren P, Lara GG, Oezcimen A, Rocca-Serra P, Sansone SA (2003) ArrayExpress - a public repository for microarray gene expression data at the EBI. Nucleic Acids Res 31:68-71

Bruskiewich R, Coe EH, Jaiswal P, McCouch S, Polacco M, Stein L, Vincent L, Ware D (2002) The plant ontology (TM) consortium and plant ontologies. Comp Funct Genomics 3:137-142

Castle AL, Fiehn O, Kaddurah-Daouk R, Lindon JC (2006) Metabolomics standards workshop and the development of international standards for reporting metabolomics experimental results. Brief Bioinform 7:159-165

Fiehn O (2001) Combining genomics, metabolome analysis, and biochemical modelling to understand metabolic networks. Comp Funct Genomics 2:155-168

Fiehn O, Kloska S, Altmann T (2001) Integrated studies on plant biology using multiparallel techniques. Curr Opin Biotechnol 12:82-86

Fiehn O, Kristal B, Ommen BV, Sumner LW, Sansone S-A, Taylor C, Hardy N, Kaddurah-Daouk R (2006) Establishing reporting standards for metabolomic and metabonomic studies: A call for participation. OMICS 10:158-163

Freemantle M (2002) Unique labels for compounds. Chem Eng News 80:33-35

Goodacre R, Vaidyanathan S, Dunn WB, Harrigan GG, Kell DB (2004) Metabolomics by numbers: acquiring and understanding global metabolite data. Trends Biotechnol 22:245-252

Griffin JL (2006) The Cinderella story of metabolic profiling: does metabolomics get to go to the functional genomics ball? Philos Trans R Soc Lond B Biol Sci 361:147-161

Hanash S, Celis JE (2002) The human proteome organization a - mission to advance proteome knowledge. Mol Cell Proteomics 1:413-414

Hardy N, Fuell H (2003) Databases, data modelling and schemas: database development in metabolomics. In: Harrigan GG, Goodacre R (eds) Metabolic profiling: its role in biomarker discovery and gene function analysis. Norwell, MA: Kluwer Academic Publishers

Hermjakob H, Montecchi-Palazzi L, Bader G, Wojcik R, Salwinski L, Ceol A, Moore S, Orchard S, Sarkans U, von Mering C, Roechert B, Poux S, Jung E, Mersch H, Kersey P, Lappe M, Li YX, Zeng R, Rana D, Nikolski M, Husi H, Brun C, Shanker K, Grant SGN, Sander C, Bork P, Zhu WM, Pandey A, Brazma A, Jacq B, Vidal M, Sherman D, Legrain P, Cesareni G, Xenarios L, Eisenberg D, Steipe B, Hogue C, Apweiler R (2004) The HUPOPSI's Molecular Interaction format - a community standard for the representation of protein interaction data. Nat Biotechnol 22:177-183

Hoffer JA, George JF, Valacich JS (2002) Modern systems analysis and design, 3rd edn. New Jersey: Prentice Hall

Jenkins H, Beckmann M, Draper J, Hardy N (2007) GC-MS peak labeling under ArMet. In: Nikolau BJ, Wurtele ES (eds) Concepts in plant metabolomics. Springer, p 297

Jenkins H, Hardy N, Beckmann M, Draper J, Smith AR, Taylor J, Fiehn O, Goodacre R, Bino RJ, Hall R, Kopka J, Lane GA, Lange BM, Liu JR, Mendes P, Nikolau BJ, Oliver SG, Paton NW, Rhee S, Roessner-Tunali U, Saito K, Smedsgaard J, Sumner LW, Wang T, Walsh S, Wurtele ES, Kell DB (2004) A proposed framework for the

description of plant metabolomics experiments and their results. Nat Biotechnol 22:1601-1606
Jenkins H, Johnson H, Kular B, Wang T, Hardy N (2005) Toward supportive data collection tools for plant metabolomics. Plant Physiol 138:67-77
Jones AR, Pizarro A, Spellman P, Miller M (2006) FuGE: functional genomics experiment object model. OMICS 10:179-184
Lindon JC (ed) (2005) Standardisation of reporting methods for metabolic analyses: a draft policy document from the standard metabolic reporting structures (SMRS) group. http://www.smrsgroup.org/documents/SMRS_policy_draft_v2.3.pdf
Lindon JC, Nicholson JK, Holmes E, Keun HC, Craig A, Pearce JTM, Bruce SJ, Hardy N, Sansone SA, Antti H, Jonsson P, Daykin C, Navarange M, Beger RD, Verheij ER, Amberg A, Baunsgaard D, Cantor GH, Lehman-McKeeman L, Earll M, Wold S, Johansson E, Haselden JN, Kramer K, Thomas C, Lindberg J, Schuppe-Koistinen I, Wilson ID, Reily MD, Robertson DG, Senn H, Krotzky A, Kochhar S, Powell J, van der Ouderaa F, Plumb R, Schaefer H, Spraul M (2005) Summary recommendations for standardization and reporting of metabolic analyses. Nat Biotechnol 23:833-838
Mattes WB, Pettit SD, Sansone SA, Bushel PR, Waters MD (2004) Database development in toxicogenomics: Issues and efforts. Environ Health Perspect 112:495-505
Mendes P (2002) Emerging bioinformatics for the metabolome. Brief Bioinform 3:134-145
Morrison N, Cochrane G, Faruque N, Tatusova T, Tateno Y, Hancock D, Field D (2006a) Concept of sample in OMICS technology. OMICS 10:127-137
Morrison N, Wood AJ, Hancock D, Shah S, Hakes L, Gray T, Tiwari B, Kille P, Cossins A, Hegarty M, Allen MJ, Wilson WH, Olive P, Last K, Kramer C, Bailhache T, Reeves J, Pallett D, Warne J, Nashar K, Parkinson H, Sansone S-A, Rocca-Serra P, Stevens R, Snape J, Brass A, Field D (2006b) Standard annotation of environmental OMICS data: application to the transcriptomics domain. OMICS 10:172-178
National Research Council (US) (2003) Sharing Publication-Related Data and Materials: Responsibilities of Authorship in the Life Sciences. Washington, DC: National Academies Press, (http://www.nap.edu/catalog/10613.html)
Nicholson JK, Lindon JC, Holmes E (1999) 'Metabonomics': understanding the metabolic responses of living systems to pathophysiological stimuli via multivariate statistical analysis of biological NMR spectroscopic data. Xenobiotica 29:1181-1189
Object management Group (2004) Life sciences identifiers, v1.0. In: Object management Group. MA: Needham, p 39
Oliver SG, Winson MK, Kell DB, Baganz F (1998) Systematic functional analysis of the yeast genome. Trends Biotechnol 16:373-378
Orchard S, Hermjakob H, Apweiler R (2003) The Proteomics Standards Initiative. Proteomics 3:1374-1376
Pedrioli PGA, Eng JK, Hubley R, Vogelzang M, Deutsch EW, Raught B, Pratt B, Nilsson E, Angeletti RH, Apweiler R, Cheung K, Costello CE, Hermjakob H, Huang S, Julian RK, Kapp E, McComb ME, Oliver SG, Omenn G, Paton NW, Simpson R, Smith R, Taylor CF, Zhu WM, Aebersold R (2004) A common open representation of mass spectrometry data and its application to proteomics research. Nat Biotechnol 22:1459-1466
Rew R, Davis G, Emmerson S, Davies H (1997) NetCDF user's guide for C. In: Unidata Program Center, University Corporation for Atmospheric Research, Boulder, Colorado, p 159

Rhee SY, Beavis W, Berardini TZ, Chen GH, Dixon D, Doyle A, Garcia-Hernandez M, Huala E, Lander G, Montoya M, Miller N, Mueller LA, Mundodi S, Reiser L, Tacklind J, Weems DC, Wu YH, Xu I, Yoo D, Yoon J, Zhang PF (2003) The *Arabidopsis* information resource (TAIR): a model organism database providing a centralized, curated gateway to *Arabidopsis* biology, research materials and community. Nucleic Acids Res 31:224-228

Roessner U, Wagner C, Kopka J, Trethewey RN, Willmitzer L (2000) Simultaneous analysis of metabolites in potato tuber by gas chromatography-mass spectrometry. Plant J 23:131-142

Sansone S-A, Rocca-Serra P, Tong W, Fostel J, Morrison N, Jones AR (2006) A strategy capitalizing on synergies: the Reporting Structure for Biological Investigation (RSBI) working group. OMICS 10:164-171

Spraul M, Hofmann M, Ackermann R, Nicholls AW, Damment JP, Haselden JN, Shockcor JP, Nicholson JK, Lindon JC (1997) Flow injection proton nuclear magnetic resonance spectroscopy combined with pattern recognition methods: Implications for rapid structural studies and high throughput biochemical screening. Anal Comm 34:339-341

Taylor CF, Hermjakob H, Julian RK, Garavelli JS, Aebersold R, Apweiler R (2006) The work of the Human Proteome Organisation's Proteomics Standards Initiative (HUPO PSI). OMICS 10:145-151

Tweeddale H, Notley-McRobb L, Ferenci T (1998) Effect of slow growth on metabolism of *Escherichia coli,* as revealed by global metabolite pool ("Metabolome") analysis. J Bacteriol 180:5109-5116

Weckwerth W (2003) Metabolomics in systems biology. Annu Rev Plant Biol 54: 669-689

Wheeler DL, Chappey C, Lash AE, Leipe DD, Madden TL, Schuler GD, Tatusova TA, Rapp BA (2000) Database resources of the National Center for Biotechnology Information. Nucleic Acids Res 28:10-14

Whetzel PL, Brinkman RR, Causton HC, Fan L, Field D, Fostel J, Fragoso G, Gray T, Heiskanen M, Hernandez-Boussard T, Morrison N, Parkinson H, Rocca-Serra P, Sansone S-A, Schober D, Smith B, Stevens R, Stoeckert CJ, Taylor C, White J, Wood A (2006) Development of FuGO: an ontology for functional genomics investigations. OMICS 10:199-204

Zimmermann P, Schildknecht B, Craigon D, Garcia-Hernandez M, Gruissem W, May S, Mukherjee G, Parkinson H, Rhee S, Wagner U, Hennig L (2006) MIAME/Plant - adding value to plant microarrray experiments. Plant Methods 2:1

Hardy, Nigel

Department of Computer Science, University of Wales, Aberystwyth, Penglais, Aberystwyth SY23 3DB, United Kingdom
nwh@aber.ac.uk

Jenkins, Helen

Department of Computer Science, University of Wales, Aberystwyth, Penglais, Aberystwyth SY23 3DB, United Kingdom

The Golm Metabolome Database: a database for GC-MS based metabolite profiling

Jan Hummel, Joachim Selbig, Dirk Walther, and Joachim Kopka

Abstract

In the post-genomic era, biological science continues a transition from a predominantly qualitative towards an increasingly quantitative science. Genomic, transcriptomic, proteomic, and now metabolomic technologies significantly contribute to the generation of huge amounts of data. These data, which typically describe changes in gene expression or changes in protein and metabolite pools, cannot effectively be analysed and interpreted by computer based programming if access is only provided through traditional publication schemes. Therefore '-omics' data sets require formalised representation and access through databases. Otherwise important information will be lost which may serve as reference data for current and future science. Transcript and protein profiling is dominated by few almost comprehensive technologies. In contrast, the metabolomic field will require multiple analytical profiling approaches to cover the chemical multitude of primary and secondary metabolism. As a consequence, technology-oriented metabolomics databases start to emerge. We will use GC-TOF-MS-based metabolite profiling as an example for the prototypical design of central database objects and structures. The focus will be on the required detailed information for the archiving of metabolite fingerprinting and profiling data sets. Special consideration is given to aspects of maintaining information sufficient and necessary for the experimental reproduction of metabolite identification and quantification results. Both aspects are essential for the sustainable use of GC-TOF-MS-based metabolite profiling and for the comparison to other metabolomics technologies.

1 Introduction

In the past decades high-throughput technologies emerged in biological science. These technologies generate huge qualitative and quantitative data sets comprising genome sequencing, protein interaction, protein structure elucidation and transcript, proteome, or more recently metabolite profiling experiments. These data sets are generally of long-term interest for the science community and require computational access which is not available through traditional journal-based science publications. Today access is facilitated by database technologies which are now comparatively easy to establish and maintain. Consequently, scientific data-

Topics in Current Genetics, Vol. 18
J. Nielsen, M. C. Jewett (Eds.): Metabolomics
DOI 10.1007/4735_2007_0229 / Published online: 13 June 2007

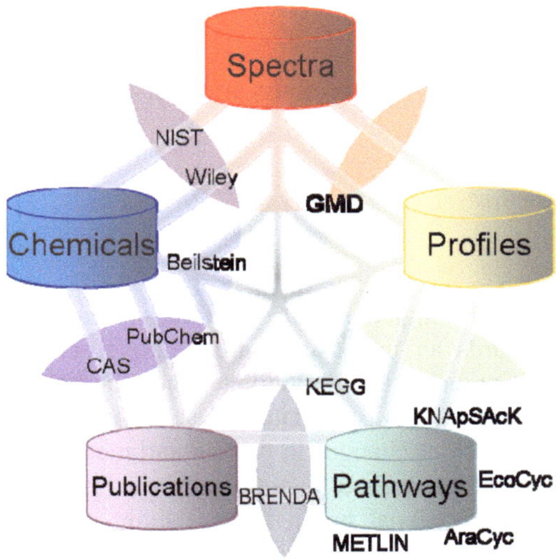

Fig. 1. Overview of leading databases supporting diverse aspects of metabolome analysis. The major types of information content relevant in metabolomics studies are indicated as nodes of the schematic pentagon. Databases are positioned according to their respective focus and are referenced in the manuscript (see Section 1); Beilstein identifies the CrossFire Beilstein database (http://www.mdl.com/products/knowledge/crossfire_beilstein/).

bases have emerged as alternate publication platforms for communicating scientific results. Laboratories involved in the generation and/or analysis of high-through-put data may, in the future, be able to effortlessly establish and develop an in-house database. Today, biological databases are ubiquitous and are growing in numbers. The molecular database collection of the year 2005 listed 719 biology related databases which are freely available to the public and reported an annual increase of 171 during 2005 (Galperin 2005). Meta-databases start to facilitate guidance and access to computer-readable data (Cary et al. 2005; Bader et al. 2006). For example, 190 web accessible resources of biological pathways and networks were available in 2006 (Bader et al. 2006). Currently, the metabolomics community does not yet support a dedicated central data repository. As the field as a whole may still be considered in its early stages, and the applied technologies are certainly relatively diverse and specialised, attempting to establish a global data management system may even seem premature. However, efforts for the standardisation of metabolomics experiments have been initiated at the first Metabolomics Standards Workshop (August 2005, Bethesda, MD, USA). Currently, the metabolome scientist has to aggregate and integrate relevant data from a large number of specialised commercial and non-commercial databases (e.g. Fig. 1).

The main information resources covering pathway, chemical substance, journal publication, NMR or mass spectrometry, and profile knowledge are highly fragmented for use in metabolomics, partially redundant and difficult to reconcile. Required resources have recently been reviewed by Arita (2004) or Mehrotra and Mendes (2006).

In the following, we will attempt an introduction to useful information resources typically needed for the present stage of metabolomics studies (Table 1) . Due to the fast-developing nature of the metabolomic and biological database field, a comprehensive analysis is not intended. The interested reader is referred to the meta-database resources mentioned above and the resources provided by the metabolomics society (http://www.metabolomicssociety.org/).

1.1 Pathway databases

In most metabolomic analyses metabolic pathway databases are the starting point of investigations. An aggregated metabolic inventory for the biological object under investigation allows estimation of the number and chemical variety of metabolites to be covered by profiling methods (e.g. Moco et al. 2006). These inventories do not only serve as a framework for the identification of yet unidentified metabolic components (see Section 4.2) from profiling experiments, inventories may ultimately be used to create metabolic models and may serve as a reference to add newly discovered metabolites, metabolic reactions, or regulatory interactions. Pathway information can be used to investigate the metabolic neighbourhood for the purpose of interpreting quantitative profiling results (e.g. PaVESy; Lüdemann et al. 2004; Schreiber and Schwobbermeyer 2005; Junker et al. 2006). Thus, the most important application may be visualisation of results within a metabolic context. This process has been used for decades without database support in journal publications to illustrate hypotheses on metabolic pathways and interactions. Today, the identification and prediction of metabolic targets of genetic modifications from a pathway context may move into the focus of computation and database utilisation.

Approximately 43 public databases exist covering metabolic pathways (Bader et al. 2006). The perhaps most frequently used metabolic resources (Bader et al. 2006) are the Kyoto Encyclopedia of Genes and Genomes (KEGG), currently holding information for approximately 40,000 pathways corresponding to 300 reference pathways (Kanehisa 1997, 2006; Kanehisa and Goto 2000); the BioCyc family of databases (Karp et al. 2005; Krummenacker et al. 2005; EcoCyc ~190 pathways, Keseler et al. 2005; AraCyc ~230 pathways, Zhang et al. 2005); and BRENDA (Schomburg et al. 2002a, 2002b, 2004). The BRENDA database provides comprehensive crosslinks of metabolic and enzyme kinetic information based on the generally accepted enzyme classification and nomenclature of the international union of pure and applied chemistry and of biochemistry and the international union of molecular biology (IUPAC-IUBMB: http://www.chem.qmul.ac.uk/iupac/jcbn/).

Table 1. Biased overview of existing databases and internet resources.

Database	Address
Meta-databases	
The Metabolomics Society	http://www.metabolomicssociety.org/
The Molecular Biology Database Collection	http://nar.oupjournals.org/
Pathguide	http://pathguide.org
Pathway databases	
AraCyc	http://www.arabidopsis.org/biocyc/index.jsp
BioCyc	http://www.biocyc.org/
BRENDA	http://www.brenda.uni-koeln.de/
ChEBI	http://www.ebi.ac.uk/chebi/
EcoCyc	http://ecocyc.org/
KEGG	http://www.genome.jp/kegg/
Cheminformatics databases	
CrossFire Beilstein	http://www.mdl.com/products/knowledge/crossfire_beilstein/
IUPAC-IUBMB	http://www.chem.qmul.ac.uk/iupac/jcbn/; http://www.iupac.org/
NIST	http://www.nist.gov/srd/nist1a.htm; http://chemdata.nist.gov/mass-spc/Srch_v1.7/index.html
PubChem	http://pubchem.ncbi.nlm.nih.gov/
SciFinder (CAS)	http://www.cas.org/SCIFINDER/
Wiley	http://eu.wiley.com/WileyCDA/WileyTitle/productCd-0471755958.html
Databases dedicated to metabolite profiling	
The Fiehn laboratory	http://fiehnlab.ucdavis.edu/compounds/
GMD	http://csbdb.mpimp-golm.mpg.de/csbdb/gmd/gmd.html
KNApSAcK	http://kanaya.naist.jp/KNApSAcK/
METLIN	http://metlin.scripps.edu/
Other resources	
BioMart	http://www.biomart.org/
BioMoby	http://www.biomoby.org/
BioPax	http://www.biopax.org/
CML	http://www.xml-cml.org/
InChI	http://www.iupac.org/inchi/
MapMan	http://gabi.rzpd.de/projects/MapMan/
Resource Description Framework (RDF)	http://www.w3.org/RDF/)

1.2 Cheminformatics databases

Metabolites are chemical substances of biological origin. Metabolomic analyses require information of physicochemical properties, such as solubility, boiling points, liquid partitioning coefficients, sum formula, exact mono-isotopic molecular mass and structures of metabolites. The best sources for these general purpose data are resources that are not restricted to metabolism. These resources are ideally linked to the original publications from which the data have been aggregated. One example is the BRENDA database of enzyme properties mentioned above. Additional and more complex properties of bio-molecules are synonymous metabolite names, information on chemical synthesis or preparation, information on bioactive substances from application studies, and occurrence in patents. The two most frequented resources are the commercial SciFinder database of the chemical abstract service (CAS) issuing the widely used CAS identifiers for chemical compounds and the publicly available PubChem access to chemical aspects contained in journal publications.

The CAS organisation provides information on approximately 29 million synthetic and natural chemical substances and has recently been extended to cover nucleic acids and proteins comprising a total of about 57 million biological sequences. Over 25 million documents from journal publications and patent literature have been curated (Schwall and Zielenbach 2000; Whitley 2002; Ben Wagner 2006). In contrast to the costly SciFinder access, PubChem (http://pubchem.ncbi.nlm.nih.gov/) represents the competing public domain project (Kremsky 2005). PubChem is hosted by the National Center for Biotechnology Information (NCBI) and is cross-referenced to the previously established services of PubMed and GenBank. PubChem allows structure and text searches on approximately 8 million substances, efficient crosslinks to other chemical databases and chemical vendors. Furthermore, valuable access is provided to bioactivity and toxicology studies (Shang and Tan 2005).

1.3 Databases dedicated to metabolite profiling

Databases focusing on the non-targeted metabolite profiling analysis of the broad spectrum of low molecular weight compounds present in biological systems have been initiated for the perhaps most widely applied GC-MS technology; for example, the metabolite profiling list of the Fiehn laboratory, http://fiehnlab.ucdavis.edu/compounds/, using the BinBase software tool (Fiehn et al. 2005) and the Golm Metabolome Database GMD, http://csbdb.mpimp-golm.mpg.de/csbdb/gmd/gmd.html (Kopka et al. 2005; Schauer et al. 2005). Other technology platforms of metabolite profiling have triggered parallel developments such as METLIN (Smith et al. 2005) and KNApSAcK (Shinbo et al. 2006) for high-resolution Fourier transform mass spectrometry (FTMS), tandem mass spectrometry (MS/MS), and LC/MS data, and the consortium for metabonomic toxicology (COMET) for NMR studies (Lindon et al. 2005).

Besides methodology-oriented databases which mediate between spectrometric means of metabolite identification and the requirement to perform multi-parallel metabolite profiles, databases started to emerge which focus on specific model organisms, e.g., yeast (Smedsgaard and Nielsen 2005), *Escherichia coli* (Sundararaj et al. 2004), or the MoToDB covering the LC-MS based metabolite profiling of tomato plants (Moco et al. 2006).

1.4 The Golm Metabolome Database (GMD)

GMD started as a collection of annotated and non-annotated, but repeatedly observed mass spectra from biological samples and was extended to contain, in addition, retention time behavior (Wagner et al. 2003). Subsequently, the concept of mass spectral tags (MSTs; see Section 5.1) was developed (Kopka 2006a, 2006). This concept became necessary because commercially available mass spectral libraries, such as the National Institute of Standards and Technology (NIST) standard reference database, NIST05 (http://www.nist.gov/srd/nist1a.htm) with the NIST MS search software version 2.0 (http://chemdata.nist.gov/mass-spc/Srch_v1.7/index.html) and Wiley mass spectral library 2005 (http://eu.wiley.com/WileyCDA/WileyTitle/productCd-0471755958.html), only contained a small fraction of those compounds that were frequently observed in profiling experiments of biological samples. The MST concept allows handling and referencing of yet unidentified metabolic components from GC-MS profiling experiments. Also, MST collections allow later identification using pure authentic reference substances. Today, multiple laboratories have added to and complemented the initial library content (Schauer et al. 2005). The current focus is the analysis of approximately 1000 commercially available reference substances representing metabolites for the purpose of enhanced MST and metabolite identification. In addition, integration of metabolite profiling data with external databases such as KEGG or visualization tools, e.g., MapMan (Thimm et al. 2004), is in preparation.

2 Database objects

A database for the purpose of archiving metabolite profile data such as GMD (Kopka et al. 2005; Schauer et al. 2005) has one main objective: to allow queries for specific metabolites and their quantitative behaviour in biological samples under different experimental conditions. Six primary objects of such a database are conceivable:

The **metabolite**: The metabolite is a chemical substance or compound imported by organisms and typically transformed by enzyme catalysed biochemical reactions. Proteins, transcripts and genes are generally not considered metabolites.

The biological **sample**: A sample is a biological object, which is subject to an analysis. The sample can be described by nomenclature of biological phylogeny, such as family, species, subspecies, cultivar or ecotype, by morphology, such as an organ, tissue or cell type, and by developmental stage.

The **experiment**: The experiment is a manipulation to which biological objects are exposed. These experimental manipulations can be divided into genotype modification and the change of environmental conditions, such as nutritive, biotic, and abiotic stresses. The latter type of experiment comprises physical data, such as duration of the experiment or temperature.

The **method**: Method information describes the procedures of sampling, extraction, fractionation, and details of the quantitative analytical technology. This information needs to be available on demand for trouble-shooting.

The **ownership** of database content: databases compile information from multiple sources. A link to traditional publications and proper acknowledgement of otherwise unpublished database information is advisable.

The **profile**: As in classical physiology, profile data may be equivalent to the exact concentration of one or multiple metabolites. In a current metabolite profiling experiment, the quantitative behaviour is typically expressed as x-fold changes relative to a set of samples kept under control conditions.

Regarding the above mentioned aspects, transcriptome and proteome profiles are highly similar to metabolome profiling experiments. Therefore, the transfer of earlier concepts such as the MIAME (**m**inimum **i**nformation **a**bout **m**icroarray **e**xperiements) standard used in gene expression experiments (Brazma et al. 2001) or the MIAPE (**m**inimum **i**nformation **a**bout a **p**roteomics **e**xperiment) protein standard (Orchard et al. 2004) to metabolite profiling experiments is obvious and feasible. Consequently, the MIAMET (**m**inimum **i**nformation **a**bout a **m**etabolomics **e**xperiment) standard has been suggested (Bino et al. 2004) and efforts to standardise metabolomic experiments are underway (Jenkins et al. 2004). The MeMo (Spasić et al. 2006) or BinBase (Fiehn et al. 2005) studies were among the first to adopt these standards. The aim of the mentioned standardisation efforts was the establishment of minimally required descriptions with the long term objective to make experiments repeatable, comparable, and the results combinable. Considering a database such as GMD, all details of measurement, chromatogram generation, and reasoning of metabolite identification has to be implemented in addition to the MIAMET requirements (Kopka et al. 2006a, 2006b). While the present concepts of metabolomic databases are highly diverse in information details, general lessons can be learned. For future systems-oriented analyses, comparisons, and between-laboratory verification, it will be essential to keep metabolite sample and experiment information tightly associated within the database and exchangeable between analytical platforms and diverse databases. One of the hardest aspects to capture is the sample information because of the broad variety of parameters required for the full description of the experimental setup. Most importantly, different experimental objectives allow either to neglect some implicit general experimental descriptions or will require additional detail.

3 Information exchange between databases

The above mentioned exchangeability of information between analytical platforms and diverse databases is related to the bioinformatics workflow concepts proposed recently in form of the Taverna system (Oinn et al. 2004). The number of publicly available computational tools and information repositories accessible as web services is growing steadily. The user has to orchestrate these web services in workflows which are part of the respective science-driven analyses. Web services architecture is a novel computing framework which uses existing internet communications and data exchange standards (Booth et al. 2003). This technology and the so-called Resource Description Framework (RDF) standard (http://www.w3.org/RDF/) facilitate the co-ordinated use of diverse computational tools and information repositories (Stein 2002). Other technologies exist that create abstraction layers. These layers hide the different physical structures and data formats. Examples of these approaches are the BioMOBY and the BioMart systems (Wilkinson and Links 2002; Durinck et al. 2005). So far, these concepts which integrate data resources with data analysis software have been applied mainly for annotating genes with ontological information. In the metabolomic field, extended statistical computations based on crosslinked data from different database repositories will be required, for example, for the structural elucidation of the immense number of unknown compounds (e.g. Kind and Fiehn 2006). System frameworks such as BioMart may prove to be highly valuable for the technical realisation of such approaches. However, in addition to effective information exchange technologies, aspects of divergent and multiple database ontologies have to be considered. As will become apparent in the following paragraphs, in two central metabolomic work flows the synonymous metabolite naming is currently the most significant barrier precluding easy data exchange in the metabolomic field. Therefore, establishment and consistent use of global, unique identifiers is urgently required for the referencing of metabolites (see Section 6).

4 The main work flows of metabolite profiling

While much can be learned from transcript profiling databases (Ball et al. 2005; Craigon et al. 2004) and the analysis of transcript profiles (e.g. Zimmermann et al. 2004, 2005), metabolite profiling projects require more than the above generalised objects. In transcript analyses, virtually the full set of genes can be determined. Based on full genome annotations, networks of biochemical pathways and constituent metabolites can be predicted. But the full finite set of metabolites present in a biological sample is not known (Sumner et al. 2003). Indeed, one of the most striking observations from metabolite profiling studies is the presence and currently even the predominance of yet unidentified metabolites (e.g. Kopka et al. 2005; Schauer et al. 2005). As a consequence, three types of data analyses prevail in current metabolomic studies (e.g. Fiehn 2002; Steinhauser and Kopka 2007); (1) the fingerprinting analysis, defined as the comprehensive "non-biased" analysis of all signals obtained by one analytical technology, (2) the profiling analysis, which utilises only the subset of identified analytical signals, namely those signals which can be delineated to represent a specific metabolite, and (3) the optional exact quantification which is possible for a subset of analytical signals for which a quantitative calibration has been performed.

All technological platforms for metabolite profiling are under constant development. The main aim is the elucidation of the metabolite identity of all fingerprinting signals and thus the enhancement of the metabolic coverage of profiling experiments. Therefore, metabolite profiling databases need to incorporate essentially two work flow schemes. One work flow, which will in the following be called *the metabolite profiling work flow*, comprises the routine generation of metabolite profiles from samples and biological experiments into quantitative numerical data matrices. One dimension of such a matrix describes the samples of an experiment and the other dimension contains the discrete analytical signals. These signals initially represent fingerprint information which can then be transformed into metabolite information through an identification process.

The second work flow creates the information necessary for this identification process and will in the following be referred to as the *metabolite mapping work-flow*. Metabolite mapping establishes the link between pure authenticated metabolic reference substances and analytical signatures.

Each of the multiple analytical platforms which are used for the profile analysis of metabolites utilises different chemical properties and signatures for metabolite identification; for example, nuclear magnetic resonance, UV-VIS spectral absorption, chromatographic retention, and mass spectral fragmentation (e.g. Sumner et al. 2003; Kopka et al. 2004). Because of the high technological and information diversity, we will focus on GC-MS fingerprints and profiles.

4.1 The metabolite profiling work flow: from sample to metabolite fingerprint and profile

The metabolite profiling work flow (Fig. 2) starts with the generation of a set of biological samples belonging to an experiment. Sample and experiment are objects which are shared with other profiling technologies (see above). In the case of GC-MS profiling, the set of biological samples is subjected to a sequence of methods for metabolite extraction, metabolite partitioning, chemical derivatisation, GC-injection, chromatographic separation, ionization, and mass detection. Technical details have been reviewed elsewhere (e.g. Kopka 2006b; Erban et al. 2007). The result of one profiling experiment is a set of at least three-dimensional GC-MS chromatograms, comprising information of mass to charge ratio, chromatographic retention time index, and ion abundance, or, in other words, quantitative response. These chromatogram files are further processed by so-called mass spectral deconvolution algorithms (Ausloos et al. 1999; Halket et al. 1999; ChromaTOF™ software; LECO, St. Joseph, MI, USA, http://www.leco.org/), which perform baseline correction and remove electronic or chemical noise and extract mass spectral tags, i.e., mass fragments with common chromatographic retention (Kopka 2006a). Depending on the algorithm, the observed absolute abundance of each mass fragment, a pre-selection of expected mass fragments, or the sum of all mass fragments can be reported. Quantitative analysis can be performed by integrating over the chromatographic peak area or peak height. These algorithms are typically applied sequentially to every single chromatogram. Therefore, processing results of all chromatograms comprising one experiment must be aligned with respect to mass and retention in order to generate a two-dimensional numerical data matrix. This data matrix represents the fingerprinting result of an experiment.

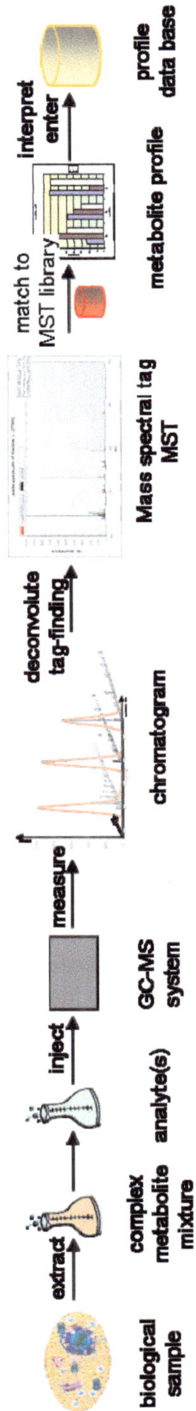
biological sample
extract
complex metabolite mixture
analyte(s)
inject
GC-MS system
measure
chromatogram
deconvolute
tag-finding
Mass spectral tag
MST
match to
MST library
metabolite profile
interpret
enter
profile
data base

Fig. 2. (overleaf) The Metabolite Profiling Work Flow: from sample to metabolite fingerprint and profile. Briefly, biological samples are assayed by metabolite extraction which results in complex metabolite mixtures. These mixtures are processed by the routine profiling procedure resulting in a standardised GC-MS chromatogram. Automated mass spectral deconvolution extracts all MSTs. These are subsequently identified by comparison to MST-library reference information (see Fig. 3). Finally, quantitative profile information is acquired which may be submitted to a profile database (see Section 4.1 for metabolite fingerprinting aspects).

In a second step, metabolites need to be identified within the data set. For this purpose, all initial mass spectral deconvolution results can be used or alternatively all co-eluting mass fragments are matched after alignment (e.g. Halket et al. 1999, 2005; Stein 1999). As a rule, only a subset of all observed mass fragments can be used to best represent the quantitative behaviour of the respective metabolites within the biological sample. Mass spectral and retention time index libraries are required for this matching and identification process (Wagner et al. 2003; Schauer et al. 2005). The work flow which generates the reference data is described in the following.

4.2 The metabolite mapping work flow: from metabolite to specific and selective GC-MS mass fragment

The metabolite mapping work flow (Fig. 3) starts with the search for a chemical substance which is known to be a metabolite. Information sources of metabolites are biochemical pathway databases and traditional phytochemical or physiological publications. The second step of this work flow is the acquisition of at least one pure authenticated reference substance which represents the metabolite(s) of interest. Reference substances may be obtained through commercial sources or through chemical synthesis and purification from biological sources. Subsequently, chemical samples are prepared of each reference substance, which are then processed by the method of GC-MS metabolite profiling analysis. GC-MS analyses typically require chemical derivatisation. This process chemically modifies reference substances into at least one so-called analyte. The analyte is the product of a reaction with chemical reagents which increase the volatility of analytes and thus make GC-MS analysis of initially non-volatile substances possible (e.g. Kopka 2006b). The reaction product is submitted to GC-MS analysis and at least one chromatogram file is generated. This file is processed manually or by automated mass spectral deconvolution algorithms generating a list of mass spectra with retention attached time indices, so-called mass spectral tags (MSTs). MSTs represent the full signature of compounds, namely mass spectral fragmentation pattern and chromatographic retention which can be employed for compound identification within the GC-MS metabolite profiling platform (Kopka 2006a). Before an MST can be submitted to a library database, a manual validation procedure is performed. Only those MSTs which are specific for the reference substance are

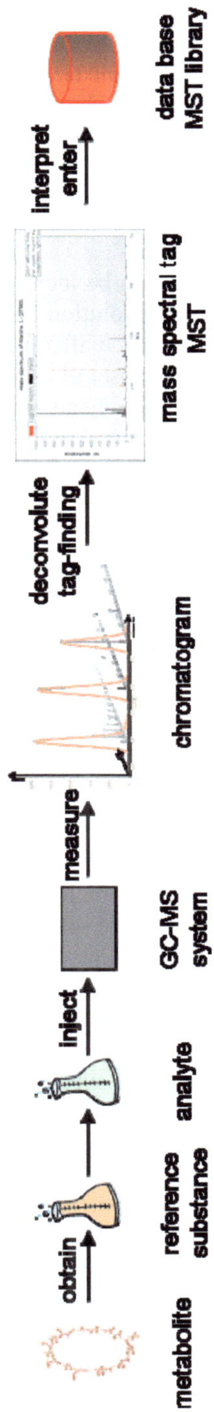
metabolite
obtain
reference substance
analyte
inject
GC-MS system
measure
chromatogram
deconvolute
tag-finding
mass spectral tag
MST
interpret
enter
data base
MST library

Fig. 3. (overleaf) The Metabolite Mapping Work Flow: from metabolite to specific and selective GC-MS mass fragments. Briefly, a pure authenticated reference substance is acquired which represents the metabolite(s) of interest. Routine chemical derivatisation and profiling analysis is applied to generate respective analyte(s), and GC-MS chromatograms. Manual validation and automated mass spectral deconvolution extracts correct mass spectral tags (MSTs), which represent analytes and respective metabolites. Finally, specific and selective mass fragments are selected from each MST and the full information submitted to a database, which comprises the complete reference information.

selected, errors of automated deconvolution are removed and laboratory contaminations excluded. The resulting subset of specific MSTs may still contain traces of chemical contaminations contained within the references substance, especially if the preparation was purified from a biological source. For this reason, the chemical structure of all possible analytes is manually predicted and the observed molecular mass and fragmentation pattern of MSTs reconciled with the structure predictions. The result is a partition of MSTs into (1) major and minor products which were verified to represent the metabolite and (2) into those MSTs which are specific to the reference substance but represent chemical contaminations.

Subsequently, the suitability of mass fragments from each MST for selective quantification is investigated. At this point in the work flow, the limitation of the employed analytical technology is determined. Each analytical technology is to some extent limited in its potential to differentiate between chemical isomers. For example, in routine GC-MS metabolite profiling D- and L- stereoisomers exhibit identical MSTs. In rare cases, chemical derivatisation may generate identical analytes from different metabolites. Also different co-eluting MSTs may share a set of common, non-selective mass fragments. The technological restrictions of the GC-MS profiling technology must, therefore, be represented in the database at the correct level of chemical detail.

A database model suited to harbour the information which is generated through the metabolite mapping work flow requires two basic objects, namely the chemical substance and the MST. Three additional supporting database objects are the chromatogram, from which MSTs originate, the chemical sample and the GC-MS method which is used to generate the chromatogram.

5 The main database objects

5.1 Modelling the "MST" database object

The MST database object is a list of mass fragments with the following attributes: the fragment mass, the absolute abundance of fragments within the mass spectral fragmentation pattern, the retention time index, which is common to all mass fragments of a MST, and the potential suitability of a mass fragment for selective quantification.

An additional characteristic aspect of the MST object is the occurrence of multiple alternate MSTs of a single analyte. These alternate MSTs can be strictly redundant. Redundant MSTs can be used to analyse the analytical reproducibility of MST properties, specifically fragment abundance and chromatographic retention. In addition, MST variants are possible, for example, due to the use of different mass spectral technologies. Mass spectral technologies affect mass spectral fragmentation pattern and fragment abundance. Also application of alternate GC separation methods may change gas chromatographic retention behaviour while the fragmentation pattern is maintained.

MST information represents a main know-how component of the GC-MS metabolite profiling technology. Libraries of MSTs comprise the analytical signatures which are used for metabolite identification and, therefore, need to be exchangeable between laboratories. In addition, information on fragment selectivity, retention time behaviour, and occurrence in standardised biological samples is essential for the laboratory routine of metabolite quantification in complex samples.

5.2 Modelling the "chemical substance" database object

The database object, chemical substance, requires thorough design, because of two aspects, which will be discussed below: (1) GC-MS has the intrinsic problem of chemical derivatisation and the requirement of reference substances for metabolite identification; (2) multiple synonyms for naming chemical compounds exist and so far no unique compound identifier has been available for the highly important purpose of exchanging metabolite definitions between databases and laboratories.

(1) Metabolites and available reference substances may not be exactly identical. For example, organic acids may be obtained either as salts or as free acids. In addition, most GC-MS methods require a chemical derivatisation step which alters the chemical identity of compounds (see above). Therefore, the abstract data object, chemical substance or compound, requires three specialised subtypes within GMD, namely the metabolite, which is represented by a reference substance und subsequently by a chemically modified analyte (Fig. 4). Multiple analytes may represent one metabolite and, in rare cases, one analyte may represent two or more metabolites. These so-called n:m relations must, therefore, be mapped into a relation-specific mapping table. The same applies to the relation of metabolites and reference substances. One metabolite may be represented by different reference substances depending on the laboratory. In addition, a reference substance, especially preparations from biological sources, might be impure and may contain a mixture of two or more different metabolites.

The three subtypes share general properties and have additional specific properties those constituting a "IS-A" relation. The supplier and purity information of reference substances or the pathway association of metabolites may serve as examples for specific properties for reference substance and metabolite, respectively. By comparison, the name, formula, and molecular weight are general properties assigned to the entity "chemical substance". The query for all properties of the specialised database items analyte, metabolite, or reference substance is possible using views, which may be joined from the general database table "chemical substance" and, for example, the specialised table "reference substance" using the

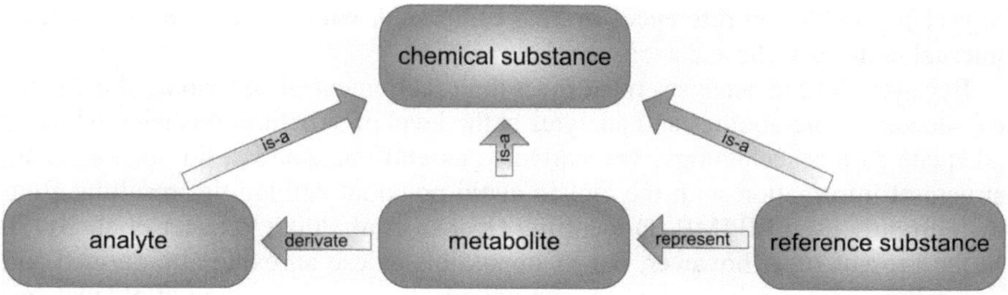

Fig. 4. The chemical substance is a key feature in metabolomic analyses. Profiling laboratories handle three types of substances: metabolites, analytes, i.e., chemical derivatives of metabolites which are required for quantitative analysis, and reference substances, which are typically commercially obtained for the validation of metabolite identity. The basic entity relationship among these substances is demonstrated.

primary key of the chemical substance used as a foreign key in the reference substance table as the joining attribute.

(2) Multiple traditional synonyms of chemical compounds exist. As a first solution to circumvent ambiguous naming, unique conventions have been provided for decades by the IUPAC nomenclature commission (http://www.iupac.org/). The issued names, however, are too complicated and, consequently, a common use has not been fully imposed in all fields of science. As a result, the access to metabolite information is hampered by synonymous use of compound names, presence of multiple database intrinsic identifier codes and the lack of complete translation tables, which map codes and names between the different information sources.

Because of the above aspects, we decided to implement a chemical substance object as integral part of our database instead of relating our database entry to a foreign repository for metabolite information. The problem of communication with external information sources is solved by utilising an Entity - Attribute - Value (EAV) table which combines foreign identifier codes and names into a source tagged synonym list for each metabolite. This list can be used for metabolite queries and for links to pathway databases, for example, by using KEGG (Kanehisa 1997, 2006; Kanehisa and Goto 2000) and MapMan (Thimm et al. 2004), CAS or other identifiers. The set of available attributes in EAV tables is not limited and, thus, information can be extended dynamically without the need of altering the database schema as demonstrated by the ArMet proposal (Jenkins et al. 2004). For example, the compound, 2-(phosphonomethylamino) acetic acid, is represented within GMD by structural information as encrypted within an InChI code, e.g., 1/C3H8NO5P/c5-3(6)1-4-2-10(7,8)9/h4H,1-2H2,(H,5,6)(H2,7,8,9)/f/h5,7-8H (cf. outlook section). This code is linked to the synonyms, "1071-83-6", "C01705", and "glyphosate", which are tagged to represent a CAS identifier, a KEGG identifier and a trivial name, respectively. Using this tagged information any of the synonyms can be transferred to query pages of the public web, such as KEGG or PubChem. The KEGG ligand database can be searched for the current pathway information or the PubChem web resource may be used for the retrieval of recently published references on this compound, without the need to maintain internal updates of these databases within GMD.

Because GC-MS analyses must meet the requirement of describing the chemical identity of metabolites and analytes at the level of structural precision which is adequate to this technology, we started to extend our database to accommodate structural information with the aim to avoid potential ambiguities resulting from the currently used MPIMP-ID identifier (Kopka et al. 2005; Schauer et al. 2005). In the present state, however, it was impossible to use an externally defined and generally accepted primary key for compounds, because annotation of old metabolite entries by available external keys such as the CAS identifier was incomplete. Therefore, we amplified the MPIMP-ID for referencing chemical substances unambiguous within GMD (http://csbdb.mpimp-golm.mpg.de/csbdb/gmd/gmd.html), while maintaining the option to query GMD using CAS and KEGG identifiers. To increase utility, full text catalogues were utilised to aid text searches.

6 Outlook

Until recently, no simple solution was available for the unambiguous description of chemical entities. Even different structure drawing algorithms may generate alternate pictures of the same compound. The problem of developing a unique and generic identifier code for chemical compounds appears now to be solved in a joined effort of the IUPAC and NIST organizations (Murray-Rust et al. 2004b). The resulting code is called InChI (IUPAC international chemical identifier) and tools for the inter-conversion of InChI codes and structure files are made available (http://www.iupac.org/inchi/).

We used basic parts of the multilayered InChI concept and will extend the GMD compound description towards fully InChI compatible codes until our arbitrarily established identifiers can be substituted by InChI codes. The main layers of the code are ideally suited to represent the technological restrictions. The primary layer contains the chemical formula, which allows calculation of the monoisotopic molecular mass, the central chemical property exploited by mass spectrometric methods. In addition atomic connectivity and numbering is given as a text string, which is much easier to manage compared to traditional mol-files. Additionally formalised information of the InChI code comprises a charge, a stereochemical, an isotopic layer and a so-called fixed-hydrogen layer which can be used if a specific tautomeric structure needs to be characterised. So far, the InChI code has been widely adopted, for example, by the ChEBI (http://www.ebi.ac.uk/chebi/) and PubChem (http://pubchem.ncbi.nlm.nih.gov/) databases or BioPax (http://www.biopax.org/) and CML (http://www.xml-cml.org/) projects for the exchange of biological pathways and chemical information, respectively. First InChI annotations of the KEGG database have been announced (http://www.iupac.org/inchi/adopters.html). Thus, this code has the potential to solve the problem caused by traditional use of different metabolite synonyms and thus provide the basis for non-ambiguous communication of metabolite and chemical compound information between data sources via customisable web services architecture (Stein 2002; Booth et al. 2003) or via Semantic Web technologies (Murray-Rust et al. 2004a). Such systems and repositories like GMD are inherently of value inherently by providing heterogeneous biological data sets in a structured and systematic description. But they may also provide a data basis for applied bioinformatics areas like biomarker identification with supervised machine learning methods (Kenny et al. 2005) and metabolic regulation modelling (Kümmel et al. 2006).

References

Sanger F, Nicklen S, Coulson AR (1977) DNA sequencing with chain-terminating inhibitors. Proc Natl Acad Sci USA 74:5463-5467

Arita M (2004) Computational resources for metabolomics. Briefings Funct Genomics Proteomics 3:84-93

Ausloos P, Clifton CL, Lias SG, Mikaya AI, Stein SE, Tchekhovskoi DV, Sparkman OD, Zaikin V, Zhu D (1999) The critical evaluation of a comprehensive mass spectral library. J Am Soc Mass Spectrom 10:287-299

Bader GD, Cary MP, Sander C (2006) Pathguide: a pathway resource list. Nucleic Acids Res 34:D504-D506

Ball CA, Awad IA, Demeter J, Gollub J, Hebert JM, Hernandez-Boussard T, Jin H, Matese JC, Nitzberg M, Wymore F, Zachariah ZK, Brown PO, Sherlock G (2005) The Stanford Microarray Database accommodates additional microarray platforms and data formats. Nucleic Acids Res 33:D580-D582

Ben Wagner A (2006) SciFinder Scholar 2006: An empirical analysis of research topic query processing. J Chem Inf Model 46:767-774

Bino RJ, Hall RD, Fiehn O, Kopka J, Saito K, Draper J, Nikolau BJ, Mendes P, Roessner-Tunali U, Beale MH, Trethewey RN, Lange BM, Wurtele ES, Sumner LW (2004) Potential of metabolomics as a functional genomics tool. Trends Plant Sci 9:418-425

Booth D, Haas H, McCabe F, Newcomer E, Champion M, Ferris C, Orchard D (2003) Web Services Architecture, http://www.w3.org/TR/ws-arch/

Brazma A, Hingamp P, Quackenbush J, Sherlock G, Spellman P, Stoeckert C, Aach J, Ansorge W, Ball CA, Causton HC, Gaasterland T, Glenisson P, Holstege FC, Kim IF, Markowitz V, Matese JC, Parkinson H, Robinson A, Sarkans U, Schulze-Kremer S, Stewart J, Taylor R, Vilo J, Vingron M (2001) Minimum information about a microarray experiment (MIAME)-toward standards for microarray data. Nat Genet 29:365-371

Cary MP, Bader GD, Sander C (2005) Pathway information for systems biology. FEBS Letters 579:1815-1820

Craigon DJ, James N, Okyere J, Higgins J, Jotham J, May S (2004) NASCArrays: a repository for microarray data generated by NASC's transcriptomics service. Nucleic Acids Res 32:D575–D577

Durinck S, Moreau Y, Kasprzyk A, Davis S, De Moor B, Brazma A, Huber W (2005) BioMart and BioConductor: a powerful link between biological databases and microarray data analysis. Bioinformatics 21:3439-3440

Erban A, Schauer N, Fernie AR, Kopka J (2007) Non-supervised construction and application of mass spectral and retention time index libraries from time-of-flight GC-MS metabolite profiles. In: Weckwerth W (ed) Metabolomics: methods and protocols. Humana Press, Totowa, pp 19-38

Fiehn O (2002) Metabolomics - the link between genotypes and phenotypes. Plant Mol Biol 48:155-171

Fiehn O, Wohlgemuth G, Scholz M (2005) Automatic annotation of metabolomic mass spectra by integrating experimental metadata. Proc Lect Notes Bioinformatics 3615:224-239

Galperin MY (2005) The molecular database collection: 2005 update. Nucleic Acids Res 33:D5-D24

Halket JM, Przyborowska A, Stein SE, Mallard WG, Down S, Chalmers RA (1999) Deconvolution gas chromatography mass spectrometry of urinary organic acids - potential for pattern recognition and automated identification of metabolic disorders. Rapid Commun Mass Spectrom 13:279-284

Halket JM, Waterman D, Przyborowska AM, Patel RKP, Fraser PD, Bramley PM (2005) Chemical derivatization and mass spectral libraries in metabolic profiling by GC/MS and LC/MS/MS. J Exp Bot 56:219-243

Jenkins H, Hardy N, Beckmann M, Draper J, Smith AR, Taylor J, Fiehn O, Goodacre R, Bino RJ, Hall RD, Kopka J, Lane GA, Lange BM, Liu JR, Mendes P, Nikolau BJ, Oliver SG, Paton NW, Rhee S, Roessner-Tunali U, Saito K, Smedsgaard J, Sumner LW, Wang T, Walsh S, Wurtele ES, Kell DB (2004) A proposed framework for the description of plant metabolomics experiments and their results. Nat Biotechnol 22:1601-1606

Junker BH, Klukas C, Schreiber F (2006) VANTED: A system for advanced data analysis and visualization in the context of biological networks. BMC Bioinformatics 7:109

Lindon JC, Keun HC, Ebbels TMD, Pearce JMT, Holmes E, Nicholson JK (2005) The Consortium for Metabonomic Toxicology (COMET): aims, activities and achievements. Pharmacogenomics 6:691-699

Kanehisa M (1997) A database for post-genome analysis. Trends Genet 13:375-376

Kanehisa M, Goto S (2000) KEGG: Kyoto Encyclopedia of Genes and Genomes. Nucleic Acids Res 28:27-30

Kanehisa M, Goto S, Hattori M, Aoki-Kinoshita KF, Itoh M, Kawashima S, Katayama T, Araki M, Hirakawa M (2006) From genomics to chemical genomics: new developments in KEGG. Nucleic Acids Res 34:D354-D357

Karp PD, Ouzounis CA, Moore-Kochlacs C, Goldovsky L, Kaipa P, Ahren D, Tsoka S, Darzentas N, Kunin V, Lopez-Bigas N (2005) Expansion of the BioCyc collection of pathway/genome databases to 160 genomes. Nucleic Acids Res 33:6083-6089

Kenny LC, Dunn WB, Ellis DI, Myers J, Baker PN and the GOPEC Consortium, Kell DB (2005) Novel biomarkers for pre-eclamsia detected using metabolomics and machine learning. Metabolomics 1:227-234

Keseler IM, Collado-Vides J, Gama-Castro S, Ingraham J, Paley S, Paulsen IT, Peralta-Gill M, Karp PD (2005) EcoCyc: a comprehensive database resource for *Escherichia coli.* Nucleic Acids Res 33:D334-D337

Kind T, Fiehn O (2006) Metabolomic database annotations via query of elemental compositions: Mass accuracy is insufficient even at less than 1 ppm. BMC Bioinformatics 7:234

Kopka J (2006a) Current challenges and developments in GC-MS based metabolite profiling technology. J Biotechnol 124:312-322

Kopka J (2006b) Gas chromatography mass spectrometry. In: Nagata T, Lörz H, Widholm JM (eds) Biotechnology in agriculture and forestry Vol 57: Saito K, Dixon RA, Willmitzer L (eds) Plant metabolomics. Springer-Verlag: Berlin Heidelberg New York, pp 3-20

Kopka J, Fernie AF, Weckwerth W, Gibon Y, Stitt M (2004) Metabolite profiling in plant biology: platforms and destinations. Genome Biol 5:109-117

Kopka J, Schauer N, Krueger S, Birkemeyer C, Usadel B, Bergmüller E, Dörmann P, Gibon Y, Stitt M, Willmitzer L, Fernie AR, Steinhauser D (2005) GMD@CSBDB: The Golm metabolome database. Bioinformatics 21:1635-1638

Kremsky J (2005) PubChem versus CAS. Chem Eng News 83:6

Krummenacker M, Paley S, Mueller L, Yan T, Karp PD (2005) Querying and computing with BioCyc databases. Bioinformatics 21:3454-3455

Kümmel A, Panke S, Heinemann M (2006) Putative regulatory sites unreveled by network-embedded thermodynamics analysis of metabolome data. Mol Syst Biol 2, (doi:10.1038/msb4100074 2006)

Lüdemann A, Weicht D, Selbig J, Kopka J (2004) PaVESy: Pathway visualization and editing system. Bioinformatics 20:2841-2844

Mehrotra B, Mendes P (2006) Bioinformatics: Approaches to integrate metabolomics and other systems biology data. In: Nagata T, Lörz H, Widholm JM (eds) Biotechnology in agriculture and forestry Vol 57: Saito K, Dixon RA, Willmitzer L (eds) Plant metabolomics. Springer-Verlag: Berlin Heidelberg New York, pp 3-20

Moco S, Bino RJ, Vorst O, Verhoeven HA, de Groot J, van Beek TA, Vervoort J, de Vos CH (2006) A liquid chromatography mass spectrometry based metabolome database for tomato. Plant Physiol 141:1205-1218

Murray-Rust P, Rzepa HS, Tyrell SM, Zhang Y (2004a) Representation and use of chemistry in the global electronic age. Org Biomol Chem 2:3192-3203

Murray-Rust P, Rzepa HS, Stein S (2004b), The INChI as an LSID for molecules in lifescience. W3C Workshop on Semantic Web for Life Sciences, 27-28 October 2004, Cambridge, Massachusetts USA, http://lists.w3.org/Archives/Public/public-swls-ws/2004Sep/att-0026/inchi.html

Oinn T, Addis M, Ferris J, Marvin D, Senger M, Greenwood M, Carver T, Glover K, Pocock MR, Wipat A, Li P (2004), Taverna: a tool for the composition and enactment of bioinformatics workflows. Bionformatics 20:3045-3054

Orchard S, Hermjakob H, Julian RK Jr, Runte K, Sherman D, Wojcik J, Zhu W, Apweiler R (2004) Common interchange standards for proteomics data: public availability of tools and schema. Proteomics 4:490–491

Schauer N, Steinhauser D, Strelkov S, Schomburg D, Allison G, Moritz T, Lundgren K, Roessner-Tunali U, Forbes MG, Willmitzer L, Fernie AR, Kopka J (2005) GC-MS libraries for the rapid identification of metabolites in complex biological samples. FEBS Lett 579:1332-1337

Schomburg I, Chang A, Schomburg D (2002a) BRENDA, enzyme data and metabolic information. Nucleic Acids Res 30:47-49

Schomburg I, Chang AJ, Hofmann O, Ebeling C, Ehrentreich F, Schomburg D (2002b) BRENDA: a resource for enzyme data and metabolic information. Trends Biochem Sci 27:54-56

Schomburg I, Chang A, Ebeling C, Gremse M, Heldt C, Huhn G, Schomburg (2004) BRENDA, the enzyme database: updates and major new developments. Nucleic Acids Res 32:D431-D433

Schreiber F, Schwobbermeyer H (2005) MAVisto: a tool for the exploration of network motifs. Bioinformatics 21:3572-3574

Schwall K, Zielenbach K (2000) SciFinder - A new generation of research tool Chem Innovat 30:45-50

Shang S, Tan DS (2005) Advancing chemistry and biology through diversity-oriented synthesis of natural product-like libraries. Curr Opin Chem Biol 9:248-258

Shinbo Y, Nakamura Y, Altaf-Ul-Amin M, Asahi H, Kurokawa K, Arita M, Saito K, Ohta D, Shibata D, Kanaya S (2006) KNApSAcK: A comprehensive species-metabolite relationship database. In: Nagata T, Lörz H, Widholm JM (eds) Biotechnology in agriculture and forestry Vol 57: Saito K, Dixon RA, Willmitzer L (eds) Plant metabolomics. Springer-Verlag: Berlin Heidelberg New York, pp 165-184

Smedsgaard J, Nielsen J (2005) Metabolite profiling of fungi and yeast: from phenotype to metabolome by MS and informatics. J Exp Bot 56:273-286

Smith CA, O'Maille G, Want EJ, Qin C, Trauger SA, Brandon TR, Custodio DE, Abagyan R, Siuzdak G (2005) METLIN - A metabolite mass spectral database. Ther Drug Monit 27:747-751

Spasić I, Dunn WB, Velarde G, Tseng A, Jenkins H, Hardy NW, Oliver SG, Kell DB (2006) MeMo: a hybrid SQL/XML approach to metabolomic data management for functional genomics. BMC Bioinformatics 7:281

Stein L (2002) Creating a bioinformatics nation. Nature 417:119-120

Stein SE (1999) An integrated method for spectrum extraction and compound identification from gas chromatography/mass spectrometry data. J Am Soc Mass Spectrom 10:770-781

Steinhauser D, Kopka J (2007) Methods, applications and concepts of metabolite profiling: primary metabolism. In: Fernie AR, Baginsky S (eds) Plant systems biology. Birkhäuser: in press

Sumner LW, Mendes P, Dixon RA (2003) Plant metabolomics: large-scale phytochemistry in the functional genomics era. Phytochem 62:817-836

Sundararaj S, Guo A, Habibi-Nazhad B, Rouani M, Stothard P, Ellison M, Wishart DS (2004) The CyberCell Database (CCDB): a comprehensive, self-updating, relational database to coordinate and facilitate *in silico* modeling of *Escherichia coli*. Nucleic Acids Res 32:D293-D295

Thimm O, Blaesing O, Gibon Y, Nagel A, Meyer S, Krüger P, Selbig J, Müller LA, Rhee SY, Stitt M (2004) MAPMAN: a user-driven tool to display genomics data sets onto diagrams of metabolic pathways and other biological processes. Plant J 37:914-939

Wagner C, Sefkow M, Kopka J (2003) Construction and application of a mass spectral and retention time index database generated from plant GC/EI-TOF-MS metabolite profiles. Phytochem 62:887-900

Whitley KM (2002) Analysis of SciFinder scholar and web of science citation searches. J Amer Soc Informat Sci Technol 53:1210-1215

Wilkinson MD, Links M (2002) BioMOBY: an open source biological web services proposal. Brief Bioinformatics 3:331-341

Zhang PF, Foerster H, Tissier CP, Mueller L, Paley S, Karp PD, Rhee SY (2005) MetaCyc and AraCyc. Metabolic pathway databases for plant research. Plant Physiol 138:27-37

Zimmermann P, Hirsch-Hoffmann M, Hennig L, Gruissem W (2004) GENEVESTIGATOR. *Arabidopsis* microarray database and analysis toolbox. Plant Physiol 136:2621-2632

Zimmermann P, Hennig L, Gruissem W (2005) Gene-expression analysis and network discovery using Genevestigator. Trends Plant Sci 10:407-409

Hummel, Jan
Max Planck Institute of Molecular Plant Physiology (MPI-MP), Am Muehlenberg 1, 14476 Potsdam-Golm, Germany

Kopka, Joachim
Max Planck Institute of Molecular Plant Physiology (MPI-MP), Am Muehlenberg 1, 14476 Potsdam-Golm, Germany
Kopka@mpimp-golm.mpg.de

Selbig, Joachim
University of Potsdam, Institute of Biochemistry and Biology, c/o MPI-MP, Am Muehlenberg 1, 14476 Potsdam-Golm, Germany

Walther, Dirk
Max Planck Institute of Molecular Plant Physiology (MPI-MP), Am Muehlenberg 1, 14476 Potsdam-Golm, Germany

List of abbreviations

GC: gas chromatography
GMD: The Golm metabolome database
MS: mass spectrometry
MST: mass spectral tag

Reconstruction of dynamic network models from metabolite measurements

Matthias Reuss, Luciano Aguilera-Vázquez, Klaus Mauch

Abstract

One of the most ambitious and challenging goals of systems biology is the identification of targets for reshaping biological systems based on quantitative predictions with the aid of mathematical models. Whereas the potential and promise of biological systems modelling is substantial, several obstacles are still encountered when addressing the issue of predictive design based on dynamic models. This is particularly because of the well known difficulties in assessing enzyme kinetics under *in vivo* conditions as a prerequisite for a sound quantitative analysis of the network via dynamic modelling. The article will describe developments and applications of tools aimed at achieving sustained improvements within this important field. Our experience in using metabolite data for reconstruction of dynamic models led to a dual approach. At the core of the modular concept is the decomposition of the networks into manageable subunits. Furthermore, a new top down approach is presented for estimating kinetic parameters for the individual reactions in whole cell metabolic networks from time series data.

1 Introduction

The use of high throughput and efficient - omic platforms has monopolized systems biology research in recent years. Given this high priority of the top down approach, it is not amazing that network reconstruction and clustering of network components as well as multivariate analysis for assessing similarities between omic profiles from different samples are gaining more and more attention. This data driven attempts to extract biological knowledge through integration information from genome-level data sources is in particularly high gear in the emerging field of metabolomics. It is certainly true that these static views capture some important aspects of the structural properties of a system. The heart of systems behaviour, however, which lies in the complex dynamics, cannot be covered by these approaches. The omic platforms are essential for this challenging goal as well, but focussing at elucidation of the dynamic behaviour of the system is more than just collecting and correlating observables by which we see it.

Quantitative analysis of the dynamic behaviour of metabolic networks depends on the mathematical description of the multiple interactions between the individ-

Topics in Current Genetics, Vol. 18
J. Nielsen, M. C. Jewett (Eds.): Metabolomics
DOI 10.1007/4735_2007_0219 / Published online: 14 April 2007

ual components (kinetic modelling). Establishing kinetic models of cellular metabolism is always embedded within the dispute if, according to the traditional biochemistry point of view, models can be designed by aggregation of *in vitro* enzyme kinetics (Teusink et al. 2000; Snoep and Westerhoff 2005) or, alternatively, *in vivo* measurements of metabolites should be applied for identification of kinetic properties. For *in vitro* kinetics we have to take the systems out of their closed metabolic cages and the observed kinetics usually describes the behaviour of test-tube isolates. There are two pivotal questions regarding the application of *in vitro* kinetics for studying the dynamics of metabolic network as real–life phenomena. (1) To what extend does the multitude of interacting processes inside the cell lead to a kinetic behaviour that differs from *in vitro* conditions? (2) What is the influence of the functioning of the entire ensemble in the living system? (Or, can we simply sum up every enzyme reaction to understand the system quantitatively?). By comparison of *in vitro*, *in situ* (permeabilized cells), and *in vivo* results for enzyme kinetics of the phosphofructokinase I system in *Saccharomyces cerevisiae*, it has been demonstrated that remarkable differences in the structure of the kinetic expressions, as well as in parameter values, can be observed in the case of such a complex metabolically regulated enzyme system (Mauch et al. 2000). Further evidence for pronounced differences between *in vitro* and *in vivo* kinetics have been presented by Aon and Cortassa (1997). These authors discussed several examples of glycolytic enzymes, in which modulation of activities and kinetic parameters were caused by interactions of the enzymes with structural polymers, such as cytoskeleton components (tubulin and actin). For detailed analysis of the important issue of impact of physiological enzyme concentrations on kinetic analysis the reader is referred to the contribution of Aragon and Sanchez (1985).

Focussing the attention to metabolomics, the key to generate dynamic metabolic network models is, therefore, to extract the kinetics of the biochemical reactions from these data and, as such, considering the reactions in their "systemic" context. This survey is intended to summarize tools and methods for identification of *in vivo* kinetics from these metabolite measurements. The tight link between the experimental observations on the one hand and modelling and simulation on the other hand is the critical issue in this model driven strategy. From there, computational experimentation - particularly in regard to appropriate perturbations – is of major importance. Furthermore, the authors of this contribution argue that any success in the identification of systems dynamics critically depends on the back and forth between simulations and wet lab experiments.

Provided that progress in the dynamic modelling and simulation of metabolic pathways and networks critically depends on such a tight link between the dry and wet lab, the organisation of this survey is a follows: first a summary of the most important tools for quantitative measurement of intracellular metabolites along with typical results is given followed by an introduction of approaches for identification of the kinetics based upon different methods for decomposition of the network. The survey also includes a critical assessment of a first attempt to identify systems dynamics for a whole cell network model.

2 Quantitative measurements of intracellular metabolites

A large number of sampling devices for measurements of intra – and extracellular metabolites have been described in the literature (Weibel et al. 1974; Reuss 1991; De Koning and van Dam 1992; Theobald et al. 1993, 1994, 1997; Weuster-Botz and de Graaf 1996; Gonzales et al. 1997; Schäfer et al. 1999; Mauch et al. 2000; Chassagnole et al. 2002; Buziol et al. 2002; Visser et al. 2002; Schmalzried et al. 2003; Castrillo et al. 2003; Mashego et al. 2006; Schaub et al. 2006). However, it is not within the scope of this paragraph to provide the reader with a comprehensive treatment on the various concepts and tools. Instead the main principles will be discussed briefly to help the reader gain a deeper insight into the related techniques, which will be introduced with a few examples.

For a quantitative analysis of the dynamic behaviour of metabolic networks it is an essential prerequisite to define the physiological state of the cells used for the experimental observations. Of course, this imperative requires experimental conditions and related process operations, which are defined and reproducible. This is not a trivial task. As indicated by Wu et al. (2005) and unpublished results from our laboratory, pools of metabolites can significantly change during long time operation of continuous cultures at a given dilution rate because of various adaptation phenomena. Thus, the corresponding constant specific growth rate does not provide a guarantee of a defined physiological state. We also need to devise methods for sampling, quenching and extraction that ensure that the results of the subsequent analysis accurately reflect the status of the living cell. The concrete design of the appropriate tools and operations in the sequence:

- process operation (steady state, fed batch, batch, transient conditions)
- sampling (steady or quasi - steady state, time span and sampling frequency during transient stimulus –response experiments)
- quenching
- extraction
- analysis

depends on the (a) the biotic variables to be measured and (b) the information to be derived from these observations. Attributes related to (a) include chemical, thermal and biological stability of the compound, turnover times, analytical methods applied, etc. The focus of the second point is the purpose of these measurements. The interest in such measurements may be related to dynamic responses of metabolite pools to extracellular disturbances (stimulus-response methodology) for identification of *in vivo* kinetics including modulation at the metabolite level. Another focal point of these measurements could be regulation phenomena involved in transcription, translation, or posttranslational processing. In the following, examples for this sequence of operations are provided from work conducted in the authors' laboratory. The main focuses of these measurements are investigations at transient conditions and application of the data for the *in vivo* diagnosis of intracellular reactions (Reuss 1991; Theobald et al. 1993; Rizzi et al. 1996; Theobald et al. 1997; Mailinger et al. 1998; Vaseghi et al. 1999, 2001; Mauch et al. 2000; Chassagnole et al. 2002). The approach, for which the name "stimulus-

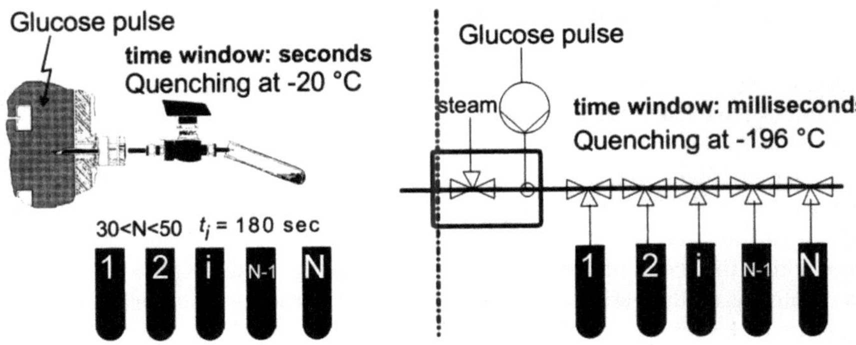

Fig. 1. Two different sampling techniques for measurement of intracellular metabolites at transient conditions: (a) manual sampling after a pulse of glucose into the bioreactor, and (b) stopped flow technique with glucose shift within the sampling valve.

response methodology" has been coined, is based on a disturbance of a steady state during continuous operation by a pulse of glucose. The response of the biological system is quantitatively characterized by measuring extra – and intracellular concentrations with high frequencies within milliseconds, seconds, or minutes after the disturbance.

There are two reasons for choosing this time scale:

- Regulation at the metabolic level occurs within seconds or even faster.
- Within this time scale, changes are caused only by metabolic regulation including posttranslational modification such as phosphorylation. Biosynthetic reactions, e.g. protein biosynthesis, can be regarded as staying in a "frozen" state.

Precise measuring of intracellular concentrations in the time window of seconds requires appropriate techniques for rapid sampling, inactivation of metabolic enzymes (quenching) and extraction of metabolites, taking into account the high metabolic turnover rates of the compound of interest. The manifold sampling systems developed for the aforementioned rapid sampling and quenching problems may be classified into two groups (Fig. 1). In the stimulus response approach depicted in Figure 1a a pulse of glucose is injected into the bioreactor with a syringe to give an initial glucose concentration of for example 1 gL^{-1} (steady state concentrations of glucose before the pulse usually are less then 20 mg L^{-1}, depending on strain, medium and specific growth rate). Samples are then rapidly taken aseptically in a sequential mode. In the original approach (Theobald et al. 1993, 1997) vacuum-sealed, precooled and membrane covered glass tubes containing an appropriate quenching fluid were used. The choice of the quenching fluid (e.g. perchloric acid solution: -20^{0}C; methanol: -70^{0}C; liquid nitrogen: -196^{0}C: hot water: +100^{0}C) depends upon the microorganism and the metabolite to be measured. Systematic investigations have indicated that the most important quenching effect is due to the temperature (Vaseghi 2000). This sampling technique can be automated to increase the frequency, precision and robustness of sampling. A semi-automated sampling device has been introduced by Schmalzried et al. (2003).

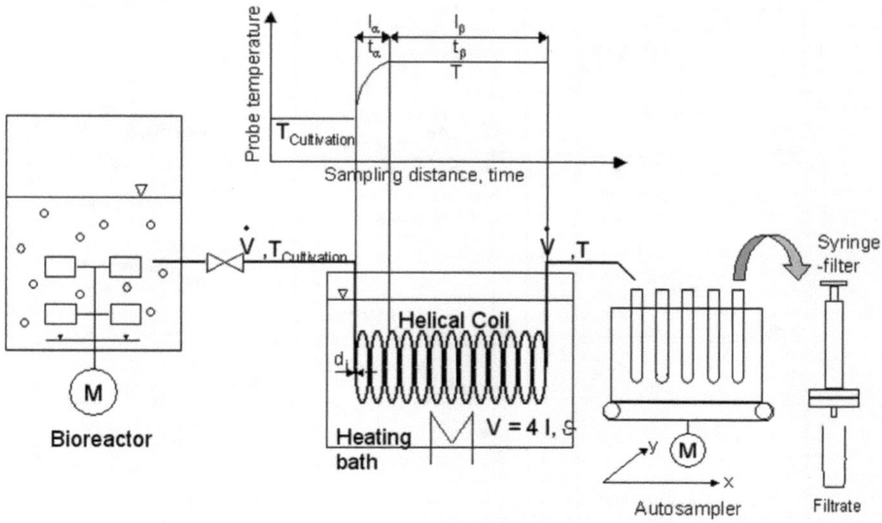

Fig. 2. Integrated sampling system with heat exchanger and a robotic system for sample collection (Schaub et al. 2006).

Also fully automated sampling with aid of robotic systems has been suggested (Schäfer et al. 1999).

A disadvantage of most of the sampling/extraction techniques is the dilution of the sample during the treatment and the contamination of the sample by solvents or other acids/bases with subsequent pH adjustment and formation of salts. This kind of treatment with chemical reagents may generate specific problems in the analytical determination of the concentration. Schaub et al. 2006 suggested a new integrated sampling technique, which overcomes most of these problems (Fig. 2). Simultaneous quenching and quantitative extraction of intracellular metabolites is realized by short-time exposure of cells to temperatures of 95^{0}C, where intracellular metabolites are released quantitatively. Based on these findings a sampling procedure has been developed which is based on a coiled single tube heat exchanger. The samples are collected with an x-y robotic driven rack of sampling tubes and filtered before analysis of metabolites.

The second approach illustrated in Figure 1b is based on the stopped-flow method used for fast measurements during enzymatic reactions. In its application to sampling from bioreactors ("BioScope": Visser et al. 2002; Mashego et al. 2006 and the "Stopped-Flow Sampling Technique": Buziol et al. 2002) a continuous stream of biosuspension leaving the bioreactor is mixed with a concentrated glucose solution in a mixing chamber. The position of the valves in the cascade illustrated in Figure 1b then determines the residence time of the biosuspension before being quenched in the corresponding sampling tube. The main features of this sampling device may be characterized as follows:

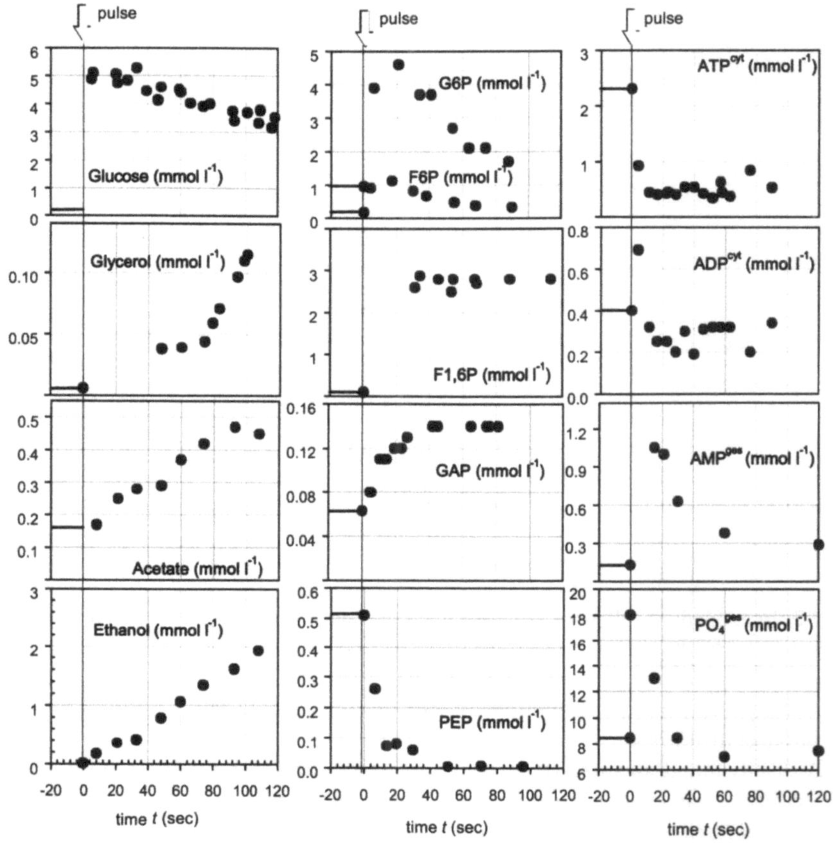

Fig. 3. Changes in the concentration of extracellular products and substrates (left hand side), intracellular metabolites (middle), and intracellular co-metabolites (right hand side) after a pulse of glucose into a biosuspension of *Saccharomyces cerevisiae*.

1. Very sharp stimuli, easy to be extended to e.g. temperature, pH and other stress.
2. The culture remains at steady state because the microorganisms are stimulated by the glucose in the mixing valve.
3. The sampling time and reaction time are decoupled. The volume of the individual samples can be chosen independently.
4. The time span between glucose stimulus and first sample can be less than 100ms (Buziol et al. 2002).

The only limitation of the technique is the problem of oxygen limitation at aerobic growth.

However this problem has been overcome by the development of the BioScope sampling system (Visser et al. 2002), which used oxygen permeable silicon tubing. The design of this system has been further improved in the so-called "second-generation" BioScope (Mashego et al. 2006). A typical example for measured intra – and extracellular metabolites from a stimulus-response experiment with the yeast *Saccharomyces cerevisiae* is summarized in Figure 3 (Theobald et al. 1997). For more details about the various techniques for sampling, quenching, extraction and analysis the authors refer to further reading.

3 Use of metabolite measurements for identification of dynamic models

To describe the dynamic systems behaviour, deterministic kinetic rate equations of the form

$$r_i = r_{\max,i} f(\mathbf{c}, \mathbf{p}) \qquad (1)$$

are formulated, where the capacity of the reaction is characterized by its maximal rate and the kinetic function f represents the kinetic properties of the reaction. Substrates, products and other metabolic effectors influencing the rate of the reaction are represented by the state vector of metabolite concentrations $\mathbf{c}$. The parameters of the reaction i are summarized in the vector $\mathbf{p}$.

In what follows three different strategies for identification of the kinetic expression and the values for the parameter by making use of metabolite measurements are outlined.

3.1 Modular decomposition of the network

The approach is similar to classical pathway modelling, in that individual rate expressions are aggregated for describing the dynamic behaviour of subsystems. A major difficulty for applying this strategy, however, is the definition of criteria for the demarcation of these modules to guarantee a certain level of autonomy (Kremling et al. 2005). Albeit a multitude of methods for decomposition of networks have been suggested, the specification and proof of existence of these modules is still a great challenge for the future. For the time being these modules are most often defined from an empirical, textbook driven decomposition of the network into subsystems performing particular physiological functions. For the difficult task of dynamic modelling the only justification of the modular concept is the fact that the approach allows for the decomposition of complex networks into manageable units. An important characteristic of the strategy of *in vivo* diagnosis, discussed in the following, is the partial integration of the subsystem into the behaviour of the network as a whole. This is an essential prerequisite for extensions of the submodules and the reassembling of modules for the design of whole cell models.

The first step to embed the behaviour of the subsystem into the metabolic network as a whole is provided by the estimation of the maximal rates of the individual reactions. Applying the rate Equation 1 to the steady state leads to

$$\tilde{r}_j = \tilde{r}_j^{\max} f_j(\tilde{\mathbf{c}}_j, \mathbf{p}_j)$$

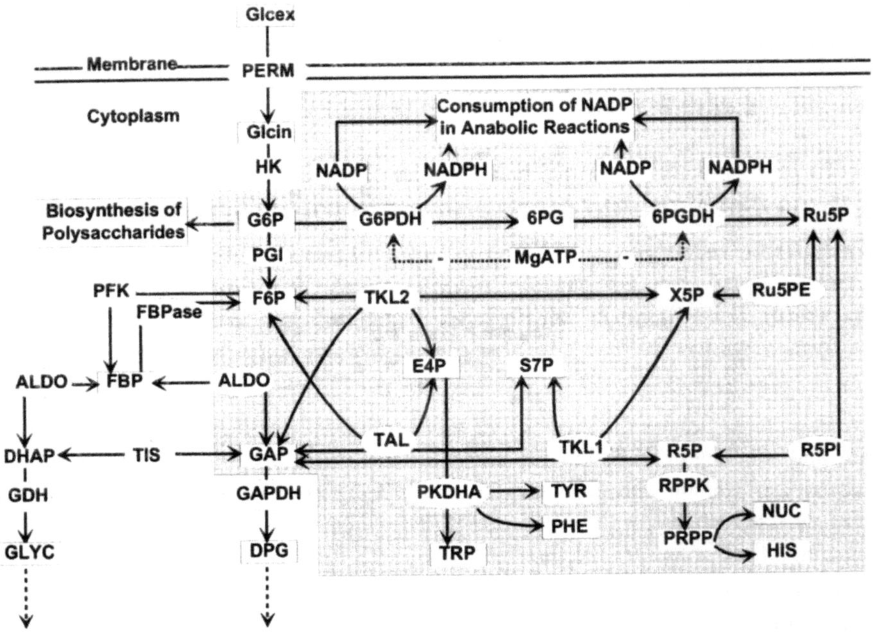

Fig. 4. The pentose-phosphate pathway.

Let us assume that reaction rate $\tilde{r}_j$ at steady state has been estimated from Metabolic flux Analysis. Let us further assume that a first estimate of the structure of the kinetics f_j and the parameter vector $\mathbf{p}_j$ is available from *in vitro* measurements. If the components of the concentration vector $\mathbf{c}_j$ influencing the rate of the reaction have been measured at steady state, the unknown maximal rates are given as

$$\tilde{r}_j^{\max} = \frac{\tilde{r}_j}{f_j\left(\tilde{\mathbf{c}}_j, \mathbf{p}_j\right)} \tag{2}$$

Further details of the strategy to identify the *in vivo* kinetics from the measured stimulus-response data are discussed in context with a concrete example presented in the following.

3.1.1 Kinetics of the irreversible reactions of the pentose phosphate shunt

Figure 4 summarizes the metabolites and enzymes of the Pentose-Phosphate (PP) shunt in *Saccharomyces cerevisiae* along with the upper part of the glycolysis. The first step towards quantitative analysis of the shunt involves the assumption that the first two reactions catalyzed by glucose-6-phosphate-dehydrogenase

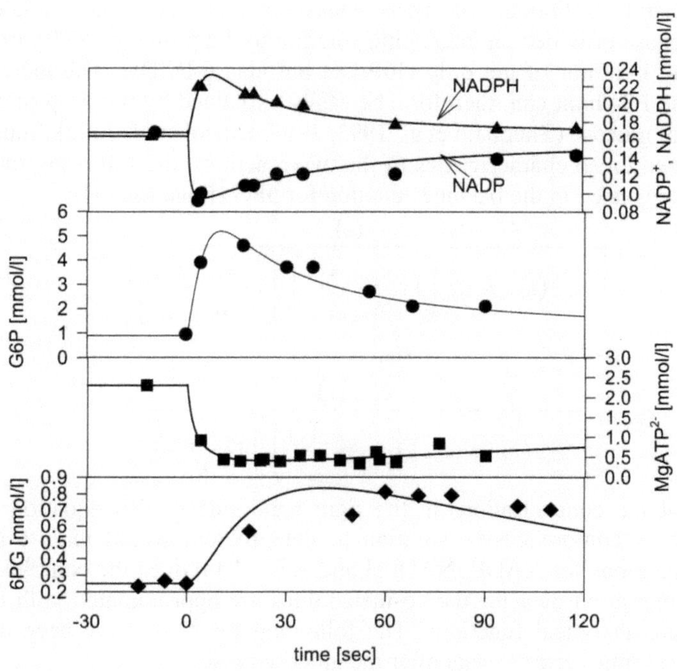

Fig. 5. Time course of NADP, NADPH, ATP, and 6PG after the pulse of glucose into a glucose limited continuous culture of *Saccharomyces cerevisiae*. The solid line in the 6PG-time course is the results from the model simulations.

(G6PDH) and 6-phospho-gluconate-dehydrogenase (6PGDH) are irreversible whereas the other reactions are assumed to be reversible and in near-equilibrium. The balance equation for 6-phospho-gluconate is given by:

$$\frac{dc_{6PG}}{dt} = r_{G6PDH} - r_{6PGDH} - \mu c_{6PG} \tag{3}$$

The last term in Equation 3 represents the dilution caused by the growth of the yeast.

The measurements required for the identification of the kinetics of the two enzymes are summarized in Figure 5. In addition to the substrates G6P and 6PG we need to know the dynamic response of the concentrations of the co-substrate $NADP^+$. Because NADPH is known as a product inhibitor for both reactions this co-metabolite must be also measured. A careful inspection of the measured data illustrates that the level of G6P at steady state already results in a substrate saturation of G6P-dehydrogenase, thus the increase in the concentration of NADPH after the pulse of glucose cannot be explained by an increase of the flux caused by the substrate G6P. This observation serves as a strong support for the strategy to treat the pentose phosphate shunt as a semi-autonomous unit as far as the link to its substrate G6P is concerned. The increase of NADPH in response to the pulse of glucose, however, still remains to be explained. A thorough consideration of the literature and additional *in vitro* measurements with the isolated G6P-dehydrogenase provided an interesting solution to the problem. ATP turned out to be a strong inhibitor of not only G6P-DH but also 6-PGDH. The increased flux through the PP-shunt can, therefore, be easily explained by the drop of ATP after the pulse of glucose (Theobald et al. 1993, 1997; Buziol et al. 2002). Summarizing the aforementioned characteristics of the two reactions, the following rate expressions are suggested in the balance equation for 6PG (Equation (3)):

$$\frac{dc_{6PG}(t)}{dt} = r^{\max}_{G6P-DH} \frac{c_{NADP}(t)}{\left[c_{NADP}(t) + K_{NADP,1}\left(1 + \frac{c_{NADPH}(t)}{K_{i,NADPH,1}}\right)\right]\left[1 + \frac{c_{MgATP}(t)}{K_{i,MgATP,1}}\right]} - r^{\max}_{6PG-DH} \frac{c_{NADP}(t)}{\left[c_{NADP}(t) + K_{NADP,2}\left(1 + \frac{c_{NADPH}(t)}{K_{i,NADPH,2}}\right)\right]\left[1 + \frac{c_{MgATP}(t)}{K_{i,MgATP,2}}\right]} - \mu c_{6PG}(t) \tag{4}$$

Notice that the concentrations at the right hand side of this equation are time-dependent. A comprehensive solution to the problem would require additional balance equations for NADP, NADPH and ATP. To reduce the complexity of the problem, measured data for the co-metabolites are approximated with the aid of approximate analytical functions. The following functions have been used to fit the observed time series of data after the glucose pulse:

$$\hat{c}_{NADP}(t) = 0.17 - \frac{1.48t}{9.7 + 16.1t + 0.48t^2} \tag{5}$$

$$\hat{c}_{NADPH}(t) = 0.16 - \frac{0.516t}{25.4 + 0.37t + 0.5t^2} \tag{6}$$

$$\hat{c}_{MgATP}(t) = 2.3 - \frac{29.8t}{29.8 + 13.4t + 0.05t^2} \tag{7}$$

These functions are only used to fit the data and do not have any mechanistic background. However, they serve to couple the dynamics of the autonomous balance equation with the dynamic response of the co-metabolites mirroring the behaviour of the cell as a whole.

The next step towards estimation of the parameters is the calculation of the maximal rates $r^{\max}_j$. Estimates from measured enzyme activities under *in vitro* conditions are questionable, because of the possible removal of effectors during cell disruption, shear sensitivity, effects of ion strength, protein-membrane and protein-protein interactions, etc. An alternative estimate is based on the rate equation and measured intracellular concentrations under steady-state conditions as described above (Equation (2)). This approach, again, guarantees that the isolated system is tightly linked to the behaviour of the whole cell, because the steady state metabolic flux distribution is a holistic attribute of the cell and does not depend on individual modules.

Eventually, the dynamic balance equation for 6-phospho-gluconate (Equation (4)) can be numerically integrated and by minimizing the error square

$$\Phi_{rel,6PG} = \sum_k \left(\frac{c_{6PG}^{calc}(t_k) - c_{6PG}^{meas}(t_k)}{c_{6PG}^{calc}(t_k)} \right)^2 \tag{8}$$

the unknown kinetic parameters can be estimated with the aid of appropriate parameter estimation algorithms.

This example not only illustrates the strategy for the *in vivo* identification of intracellular kinetics – for a more detailed description of this approach the reader is referred to the original publications of Rizzi et al. (1997) and Vaseghi et al. (1999) – it also indicates an interesting result regarding the discussion of the issue of modular structures, which is important for more complex network analysis. According to the results of the identification procedure, the flux through the PP-shunt is independent from the concentration of the substrate G6P and only depends on the ATP pool as well as the NADP/NADPH ratio. It therefore seems that the flux is adjusted to the energy state and balanced to the demand for biosynthesis. The PP-shunt thus only apparently acts as a semi-autonomous functional unit, which is carefully modulated by the energy state of the cell and the demand for biosynthesis. As such, the module is wired with the hubs of the cellular network, which participate in a very large number of links (Müller et al. 2005). These hubs eventually integrate all substrates into a single integrated web in which the existence of fully autonomous modules is prohibited. This issue will be further discussed in context with the large scale dynamic modelling of metabolic networks in Section 3.2.

3.1.2 Kinetics of the phophosfructokinase I (PFK1)

The kinetic behaviour of this enzyme has attracted a lot of attention because of its important role in the regulation of the glycolysis. The allosteric enzyme catalyses the phosphorylation of fructose-6-phosphate (F6P) to fructose-1,6-bisphosphate (F1,6BP) and is modulated by several effectors as schematically illustrated in Figure 6. The identification of the kinetic rate expression and the parameters is based on the balance equation of F6P, which reads:

$$\begin{aligned} \frac{dc_{F6P}}{dt} &= -r_{PFK1}^{\max} f_{PFK1}\left(c_{F6P}(t), c_{AMP}(t), c_{ADP}(t), c_{ATP}(t), c_{F2,6BP}(t), \bar{P}_{PFK1}\right) \\ &\quad + r_{PGI}^{\max} f_{PGI}\left(c_{G6P}(t), c_{F6P}(t), \bar{P}_{PGI}\right) + r_{TAL}(t) + r_{TKL2}(t) - \mu c_{F6P}(t) \end{aligned} \tag{9}$$

The reactions involved in this balance of F6P are shown in Figure 6. The contribution from the reactions transaldolase (TAL) and transketolase (TKL2) are computed from the model of the pentose shunt (Vaseghi et al. 1999 and paragraph 3.1.1). The rate expression for the glucoseisomerase (PGI) is described in the original paper of Rizzi et al. (1997). Again the time courses of the concentrations of the various effectors in response of a pulse of glucose have been measured (Fig. 7) and fitted with approximate functions as described above. The details of the identification of the kinetic expression for the PFK1 of the yeast *Saccharomyces*

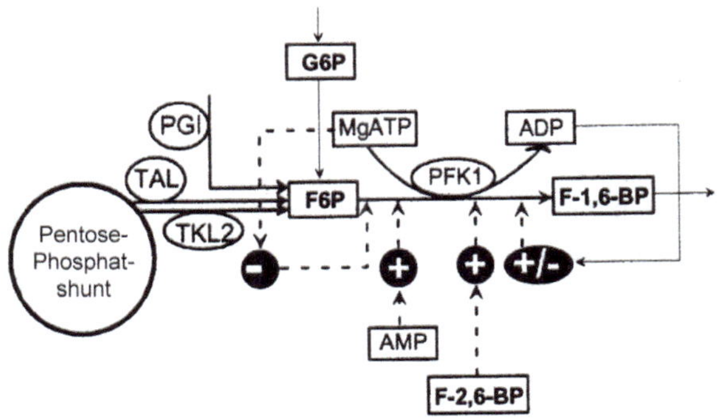

Fig. 6. The reactions around PFK1.

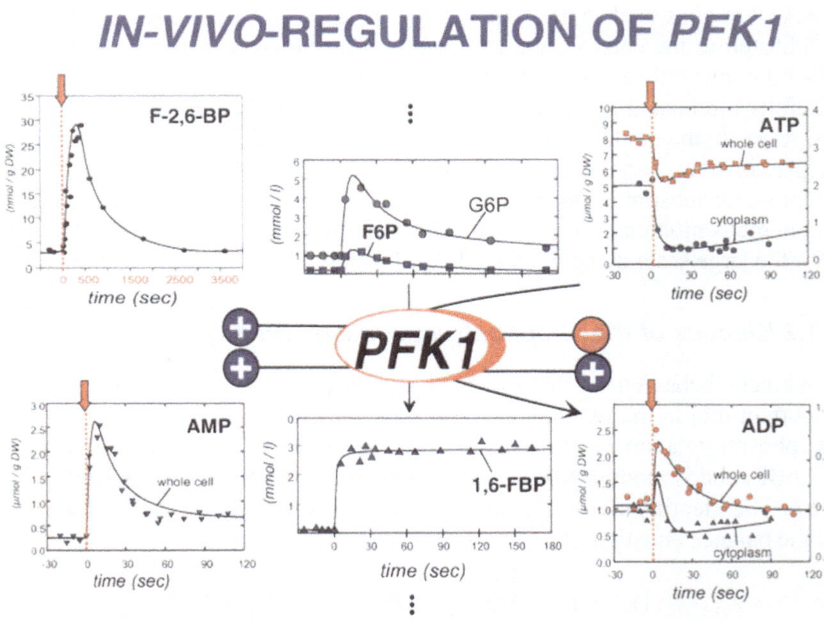

Fig. 7. Time series of data for identification of the kinetics of PFK1.

cerevisiae have been comprehensively described by Mauch et al. (2000). An important aspect of this analysis concerns the comparison between the kinetic rate expression identified from *in vitro, in situ*, and *in vivo* measurements. The *in situ* measurements were performed via permeabilisation of the yeast cells with a toluol-ethanol mixture according to a protocol suggested by Serrano et al. (1973). For discussion of the pronounced differences between *in vivo, in situ*, and *in vitro* results it is necessary to introduce some characteristics of the kinetic properties of the enzyme.

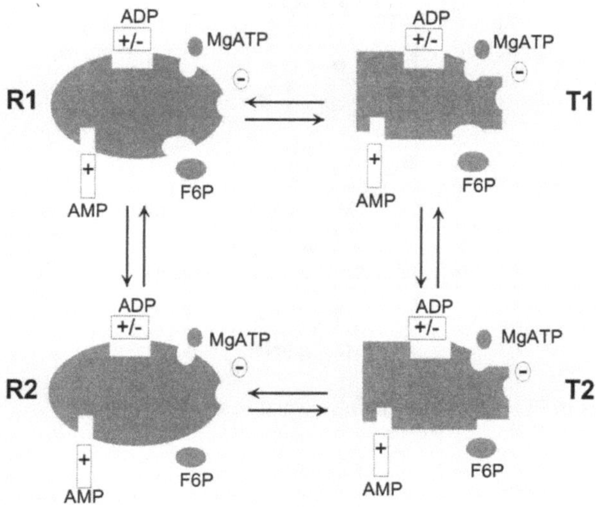

Fig. 8. The mechanism of the V- model for PFK1.

The most important attempts to model the allosteric properties of PFK1 goes back to the work of Monod et al. (1965), who considered two conformations of the enzyme: conformation T with a low and conformation R with a high affinity to the substrate F6P. Each enzyme molecule consists of eight subunits. The basic Model of Monod et al. (1965) only considers the allosteric behaviour with respect to the substrate F6P and a non-allosteric effect of ATP. The so-called Reuter-model (Reuter et al. 1979) additionally considers the modulation through AMP, ADP, and ATP and its influence upon the allosteric effects of F6P. In a similar way to the more sophisticated model by Vaseghi (Mauch et al. 2000; Vaseghi 1999) illustrated in Figure 8 this module assumes that:

- each promoter of the octameric protein appears in two basic conformations R1 and R2 as well
- as two sub-conformations R2 and T2,
  - substrate F6P is bonded to R1 and T1 with different affinities
- ATP binds to all different conformations with the same affinity
- ATP binds as inhibitor to R1 and T1
- AMP and ADP bind as activators to R1/T1 or R2/T2, respectively.

As illustrated in Figure 8 the model of Vaseghi (V-model) also considers the strong activation through F2,6BP as well as a competitive inhibition between F2,BP and F6P.

For comparison the kinetic parameters were estimated from the *in vitro* data of Hofman and Kopperschläger (1982) and compared with our own *in situ* and *in vivo* measurements. The differences in the values of the parameters are pronounced (Mauch et al. 2000). There are two striking phenomena, which attract attention:

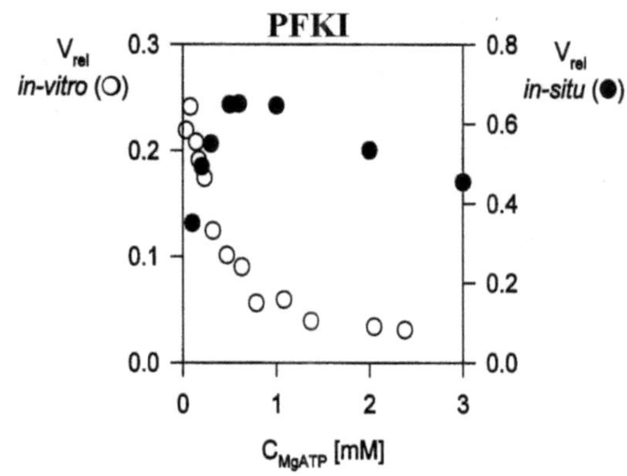

Fig. 9. Activities of PFK1 measured at *in vivo* and *in situ* conditions.

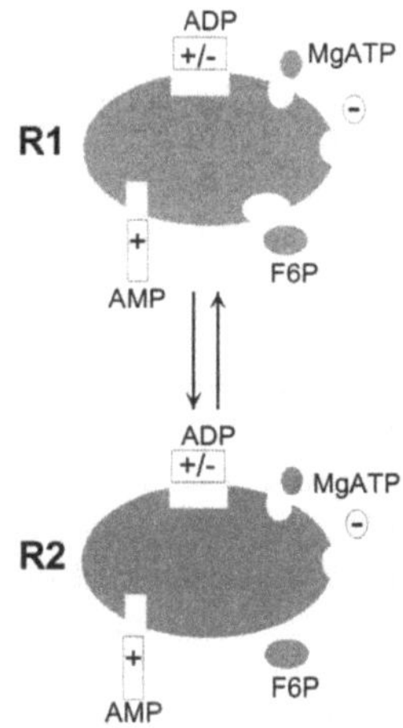

Fig. 10. The reduced V-model for PFK1 under *in vivo* conditions.

- Compared to the *in vitro* situation, the modulation strength of all the effectors are less pronounced under *in situ* and *in vivo* conditions. This is exemplarily shown in Figure 9, in which the modulation of the enzyme activity through ATP is depicted for the *in vitro* and *in situ* conditions.
- The parameter value L_0 (ratio of the tensed to the relaxed state) tends to zero at the *in situ* and *in vivo* conditions. This results leads to the interesting conclusion that the enzyme only acts in the R conformation (high affinity). On the basis of these observations, the model can be reduced to a much more simple structure illustrated in Figure 10.

To extrapolate the results obtained from the kinetic analysis of PFK1 to other enzymes, it is worthwhile to speculate about possible clues for these pronounced differences between *in vivo* and *in vitro* conditions. A reasonable explanation needs

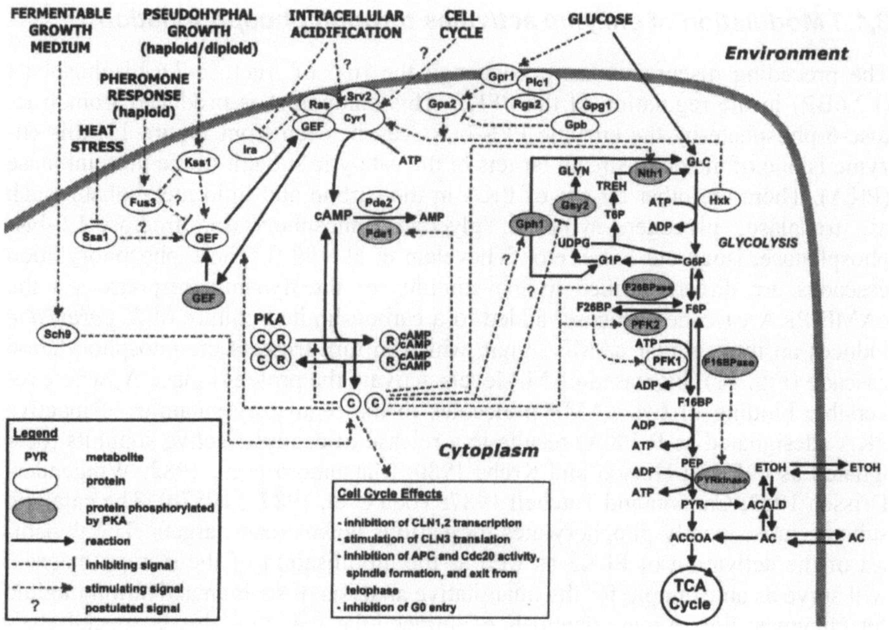

Fig. 11. The cAMP-protein kinase A signal transduction pathway and targets for phosphorylation.

to consider the well known effects of homogeneous and heterologous protein interactions inside the cell. The first effect is related to the concentration of the enzyme. Srere (1967) indicated that concentrations of glycolytic enzymes inside the cells are usually 100 times higher than those normally applied to *in vitro* assays. According to Aragón and Sanchez (1985) the concentration of PFK1 in *Saccharomyces cerevisiae* is of the order of 190-550 μg/ml whereas the concentration of *in vitro* assay is 1-10 μg/ml. The influence of high intracellular concentrations of the enzyme on the regulation of PFK1 has been shown by Aragón and Sanchez (1985). The second effect – heterologous interactions – stands for associations between enzymes and structure proteins of the cytoskeleton (Ovadi 1995). Particularly for PFK1 in *S. cerevisiae*, Kopperschäger (1999) has shown for the first time an organized association between the enzyme molecule and the microtubules under *in vivo* conditions. The impact of such associations on the kinetic behaviour of the enzymes is not known yet. However, because of the strong relationship between structure and function, these are strong candidates for the explanation of the aforementioned differences between *in vitro* and *in vivo* conditions. Because of the closer agreement between *in situ* and *in vivo* observations, the *in situ* experiments could be viewed as interesting and alternative tools for the kinetic analysis of intracellular reactions.

3.1.3 Modulation of enzyme activities through phosphorylation

The preceding discussion has accentuated the role of fructose-2,6-bisphosphate (F2,6BP) in the regulation of the PFK2. This compound is produced from fructose-6-phosphate by the enzyme PFK2. As can be seen from Figure 11, this enzyme is one of the downstream targets of the catalytic subunit of the proteinkinase (PKA). There are other targets of PKA in the carbon and lipid metabolism, such as, trehalase, glycogen synthase, glycogen phosphorylase, fructose-1,6-bisphosphatase, isocitrate lyase, etc. (Thevelein et al. 1999). These phosphorylation cascades are directly linked to the stimulus of the dynamic response via the cAMP-PKA cascade. Glucose, added to a carbon-limited culture of *S. cerevisiae* induces an intracellular cAMP signal, which in turn triggers the phosphorylation cascade (Fig. 11): increased cAMP levels activate the protein-kinase A, where reversible binding of two cAMP molecules to the regulatory subunits of inactive PKA (designated as R-PKA) results in a release of catalytic active subunits (designated as C-PKA; Hixson and Krebs 1980; Matsumoto et al. 1982; Wingender-Drissen 1983; Cannon and Tatchell 1987; Toda et al. 1987a, 1987b). The catalytic subunit subsequently phophorylates the various downstream targets. The dynamics of the activation of PFK2 as well as the mobilisation of the storage material will serve as an example for the quantitative analysis of such an activation module for enzymes. For *in vivo* diagnosis of intracellular reactions this is special challenge because the maximal rates r_{max} in the rate expression change during the transient, reflecting that enzymes do not stay any more in a "frozen" state. The situation is very similar to the more sophisticated dynamic modelling of metabolic networks linked to gene regulation.

Figure 12 summarizes additional time series of experimental data required for the analysis. The measurements of the cAMP concentrations were performed with a commercially available competitive protein-binding assay (Amersham International TRK 432). The time course of the signal indicates a rapid response to the glucose concentration, which is in qualitative agreement with the observations of Beullens et al. (1988). A further important measurement illustrates the dynamic changes of the activity of one of the target enzyme – the PFK2. The quenching and extraction methods applied to measure enzyme activities are those suggested by Francois et al. 1984. Samples are taken after the pulse of glucose and quenched in methanol at -70^0 C. After centrifugation at -9^0 C and resuspension in a buffer the cells are mechanically disrupted by intensive shaking with glass beads. To prevent temperature increase, the disruption procedure is interrupted four times to cool the sample down to 0^0 C. After centrifugation, the extract is incubated with the substrate F6P. Product concentration is determined by an assay developed by van Schaftingen et al. (1982). This assay uses the activation of pyrophosphate dependent PFK1 from potato tubes through F2,6BP (see also Vaseghi et al. 2001). From the comparison of the time courses of ATP, F6P and F2,6B P it can be concluded that the rate of formation of F2,6BP is not influenced by the two substrates of the PFK2, ATP and F6P. We may therefore assert that the formation of F2,6BP is determined by the activity of PFK2, which in turn is regulated by the cAMP cascade.

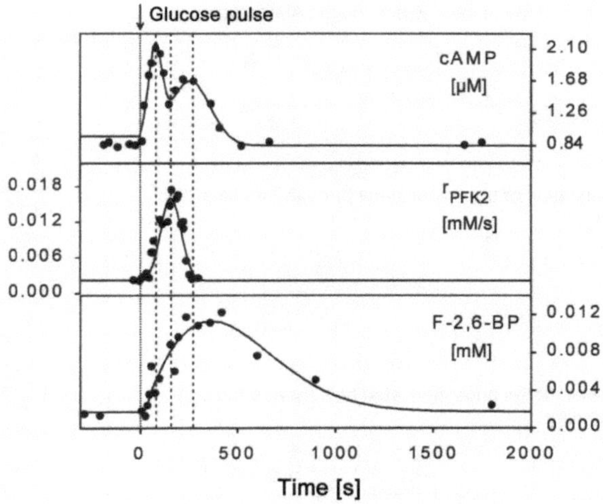

Fig. 12. Time course of cAMP, PFK2-activity and fructose-2,6-bisphosphate (F2,6BP) after a pulse of glucose into a glucose limited continuous culture at $D=0.1h^{-1}$.

The first step towards dynamic modelling of these phenomena is the definition of an appropriate module. In what follows we restrict the analysis to the downstream part, thus using the measured cAMP trace as an input signal and the measured enzyme activity as the output. A more detailed analysis of the kinetic analysis of the upstream part of the cAMP signal cascade, which deals with the dynamics of signal formation and degradation due to various feedback mechanisms, has been performed by Mueller (Mueller 2006; Mueller et al. 2005; Mueller et al. submitted). For correlation of the time course of the cAMP signal the following analytical function has been identified:

$$\hat{c}_{AMP}(t) = a_o + a_1 \exp\left(\frac{a_2 (t - a_3)^2}{a_4^2}\right) + a_5 \exp\left(\frac{a_5 (t - a_7)^2}{a_8^2}\right) \tag{10}$$

with

$$a_0 = .842e-3; a_1 = .893e-3; a_2 = -.5; a_3 = 258.45; a_4 = 92.23;$$

$$a_5 = 1.2e-3; a_6 = -.5; a_7 = 80.6; a_8 = 37.03$$

The kinetic model for the activation of the PKA rests upon the reaction scheme illustrated in Figure 13. The following rate expressions are assumed for the individual reactions:

Dissociation of the holoenzyme through binding of cAMP:

$$r_1 = k_1 c_{R_2C_2} c_{cAMP}^4 - k_{-1} c_{R_2} c_C^2 \tag{11}$$

Autocatalytic phosphorylation (irreversible)

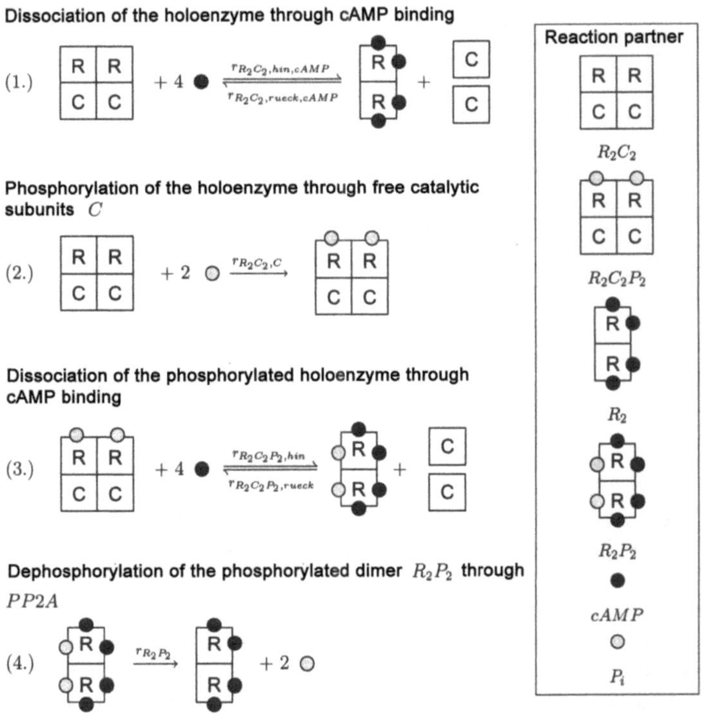

Fig. 13. Reaction scheme for the activation of PKA.

$$r_2 = k_2 c_C \frac{c_{R_2C_2}}{K_{R_2C_2M,2} + c_{R_2C_2}} \tag{12}$$

Dissociation of the phosphorylated holoenzyme through binding of cAMP (reversible)

$$r_3 = k_3 c_{cAMP}^4 c_{R_2C_2P_2} - k_{-3} c_{R_2P_2} c_C^2 \tag{13}$$

Dephosphorylation of the phosphorylated dimer (irreversible)

$$r_4 = r_{R_2P_2}^{\max} \frac{c_{R_2P_2}}{K_{R_2P_2M,5} + c_{R_2P_2}} \tag{14}$$

It is assumed that the dimer is dephosphorylated by a phosphatase, e.g. protein-phosphatase 2A (PP2A). The maximal rate may therefore be expressed as a function of the active phosphatase

$$r_{R_2P_2}^{\max} = k_{v\max,R_2P_2} c_{C_pABP} \tag{15}$$

Figure 14 depicts the assumptions for the phosphorylation mechanism of the phosphatise PP2A (trimer C_pAB). The PP2AA is composed from three subunits. The core of the enzyme is a dimer. The third subunit (B) is a monomer (Janssens

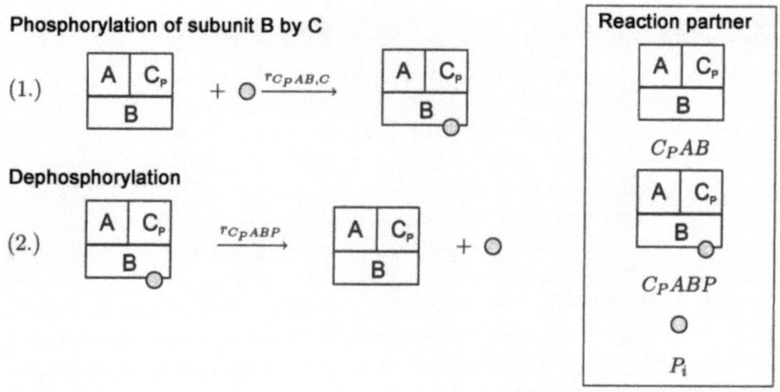

Fig. 14. Reaction scheme for activation of the PP2A.

and Goris 2001). The regulatory subunit B in *S. cerevisiae* is phosphorylated by the catalytic subunit C from the PKA (Zhao et al. 1997). This phosphorylation leads to a feed back regulation, which is important for the adaptive behaviour of the signal cascade. If the trimer is present in excess we can assume a Michaelis-Menten kinetic for the first reaction in Figure 13:

$$r_{C_pAB,C} = k_4 c_C \frac{c_{C_pAB}}{K_{C_pAB,M} + c_{C_pAB}} \tag{16}$$

For the dephosphorylation reaction of the enzyme (C_pABP) the following rate expression is assumed:

$$r_{C_pABP} = r_{C_pABP}^{\max} \frac{c_{C_pAB}}{K_{C_pABP,M} + c_{C_pAB}} \tag{17}$$

Eventually, we need rate expressions for the phosphorylation and dephosphorylation of the target enzyme, PFK2 in the chosen example.

For the activation of the PFK2 we assume a rate expression according to Michaelis Menten kinetics

$$r_{PFK2\rightarrow PFK2P} = r_{PFK2\rightarrow PFK2P}^{\max} \frac{c_{PFK2}}{K_{PFK2,M} + c_{PFK2}} \tag{18}$$

With the maximal rate depending on the concentrations of the catalytic subunit of the PKA and the phosphatase (C_pABP). In accordance to what has been reported in the literature (Yamashoji and Hess 1984) a competitive inhibition between the two proteins is assumed:

$$r_{PFK2\rightarrow PFK2P}^{\max} = v_{\max,C,1} \frac{c_C}{K_{C,3}\left(1 + \frac{c_{C_pABP}}{K_{i,1}}\right) + c_C} \tag{19}$$

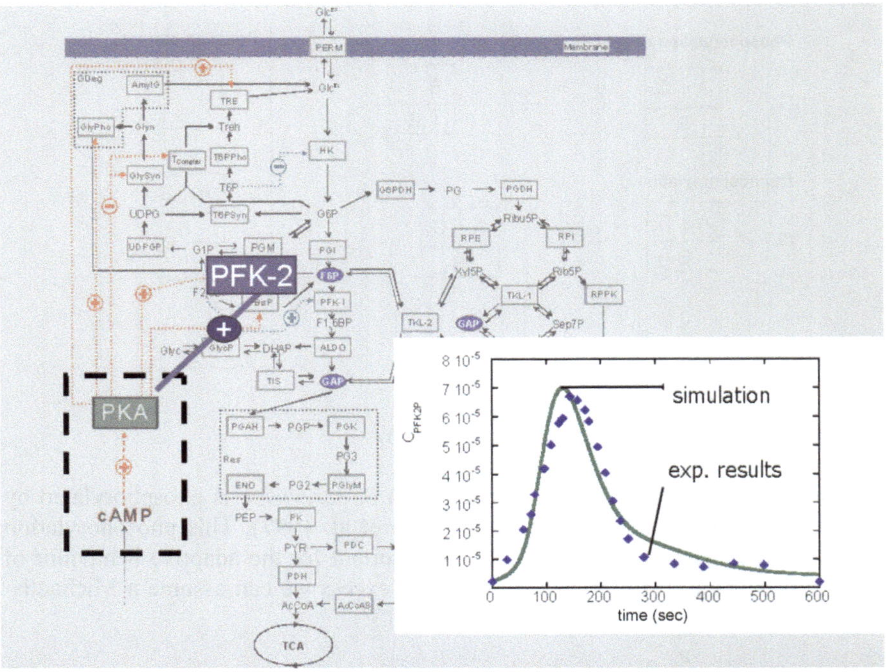

Fig. 15. Comparison between experimental observations and model simulation for the activity of the phophofructokinase 2 (PFK2).

Furthermore, a simple Michaelis Menten kinetic is assumed for the dephosphorylation of the PFK2 through the active form of the phosphorylase $\left(C_p ABP\right)$

$$r_{PFK\,2P \to PFK\,2} = r_{PFK\,2P \to PFK\,2}^{\max} \frac{c_{PFK\,2P}}{K_{PFK\,2PM,2} + c_{PFK\,2P}}$$

with

$$r_{PFK\,2P \to PFK\,2}^{\max} = k_{PFK\,2P} c_{C_p ABP} \tag{20}$$

Figure 15 shows the comparison between the measured activities of the PFK2 and the model predictions simulated from the balance equation for the enzyme:

$$\frac{dc_{PFK\,2P}}{dt} = r_{PFK\,2 \to PFK\,2P} - r_{PFK\,2P \to PFK\,2} \tag{21}$$

In a similar way as illustrated for the PFK2, it is possible to model the trehalase and, thus, to link the dynamics of the mobilisation of the storage material trehalose with the carbon flux through the glycolysis (Aguilera-Vázques 2006).

3.1.4 A critical assessment of the modular approach for dynamic modelling of metabolic networks

The modular approach discussed so far is based on a partition of metabolic networks into relatively autonomous subunits. It is widely believed that this modular structure of the networks plays a critical role in their functionality. Usually, the decomposition of metabolic networks is performed on an intuitive basis. In spite of the fact that several algorithms purely based on network topology have been proposed development of methods for exploiting the modularity, the accurate definition of these modules and questions of their autonomy is still a crucial issue and there is a clear need to develop alternative algorithms for identifying modules accurately (Murray 1999; Girvan et al. 2002; Ravasz et al. 2002; Alon 2003; Holme et al. 2003; Newman and Girvan 2004; Raddicci et al. 2004; Guimera and Amaral 2005). Because of a missing rigorous definition of these subunits the questions remains if the fundamental organisation principle of metabolic networks is modular at all or distributed or probably best described as having a little bit of both (Huss and Holme 2006).

In context with the discussion of modular approaches for dynamic modelling the issue of ubiquitous substrates or sometimes, by analogy to economy, termed currency metabolites, such as AMP, ADP, ATP, NAD, NADH, NADP, NADPH etc. needs a special consideration because they are involved in a very large number of reactions. Even if we restrict the discussion to only structural properties of modularity in most cases, the data are pre-processed by removing the high degree nodes (hubs) compounds before decomposition algorithm are applied (Schuster et al. 2002). As nicely elaborated by Huss and Hole (2006), effective modularity will increase when a currency metabolite is deleted from the network. The question, to what extent the results of, for example, "cartographic representation" of complex networks for identification of modules, can be applied to the problem of dynamic modelling remains a completely open question.

What is actually required for a systematic and rigorous decomposition of metabolic networks for the purpose of dynamic modelling are considerations of biochemical modules as dynamical entities taking into account the impact of common co-metabolites occurring in and between almost all modules. Even if we take care of a partial integration of modules in the network as a whole, as demonstrated in the aforementioned examples, there are several applications in which the predictive power of the dynamic models would be necessary (optimal re-design of metabolic systems (Schmid et al. 2006), estimation of optimal drug dosages) and rigorous balancing of the co-metabolites and metabolites in overlapping modules is required. The set up of large-scale dynamic models based on a canonical representation of reaction kinetics is therefore an important issue.

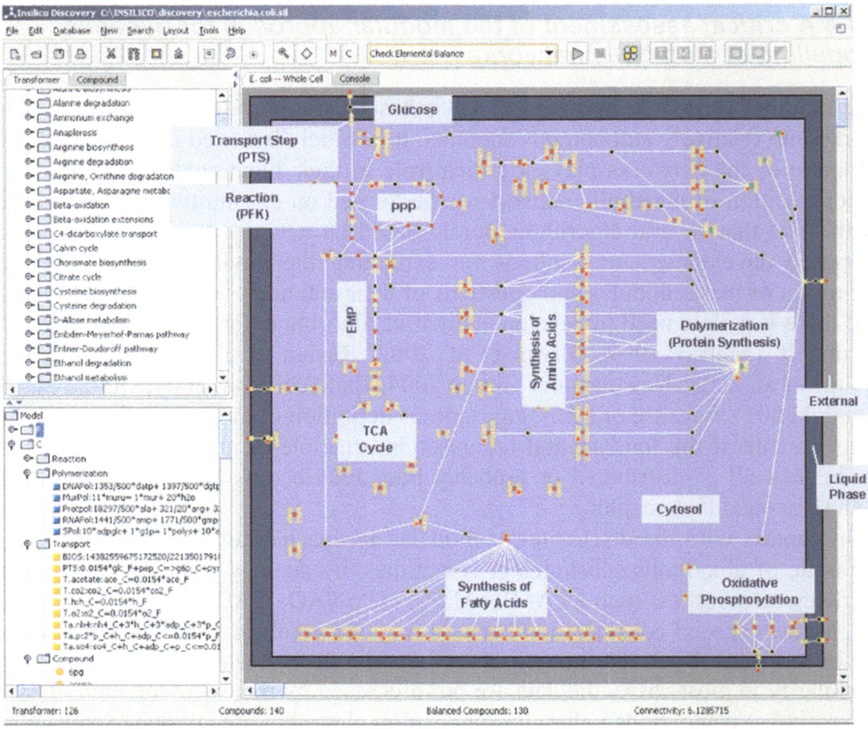

Fig. 16. Metabolic reaction network of *Escherichia coli.*

3.2 *In silico* identification of whole cell metabolite dynamics through evolutionary algorithms and parallel computing

In what follows a new top down approach for estimating kinetic parameters of individual transformation steps (reaction, transport, and polymerisation) in whole cell metabolic networks is presented. The identification again rests upon time series data of a limited number of metabolites. The computationally demanding method couples evolutionary strategies with dynamic simulation of the metabolite balance equations for the network and is implemented by high performance cluster computing. The approach is illustrated by identifying whole cell network dynamics for *E. coli* from time series data from stimulus-response experiments.

Figure 16 shows a graphical representation (Software: Insilico Discovery/Insilico Biotechnology AG) of the reaction network of *Escherichia coli* introduced by Chassagnole et al. 2002. The network comprises both catabolic and anabolic routes with protein, DNA, RNA, polysaccharides, murein, and lipids building up biomass. Sequential reaction steps and parallel routes are lumped. With 126 reactions, 130 balanced metabolites, and seven conserved moieties, the degree of freedom of the network is fixed to 126-130+7=3. The relatively small number of free fluxes in combination with about 50% of all reactions signs fixed

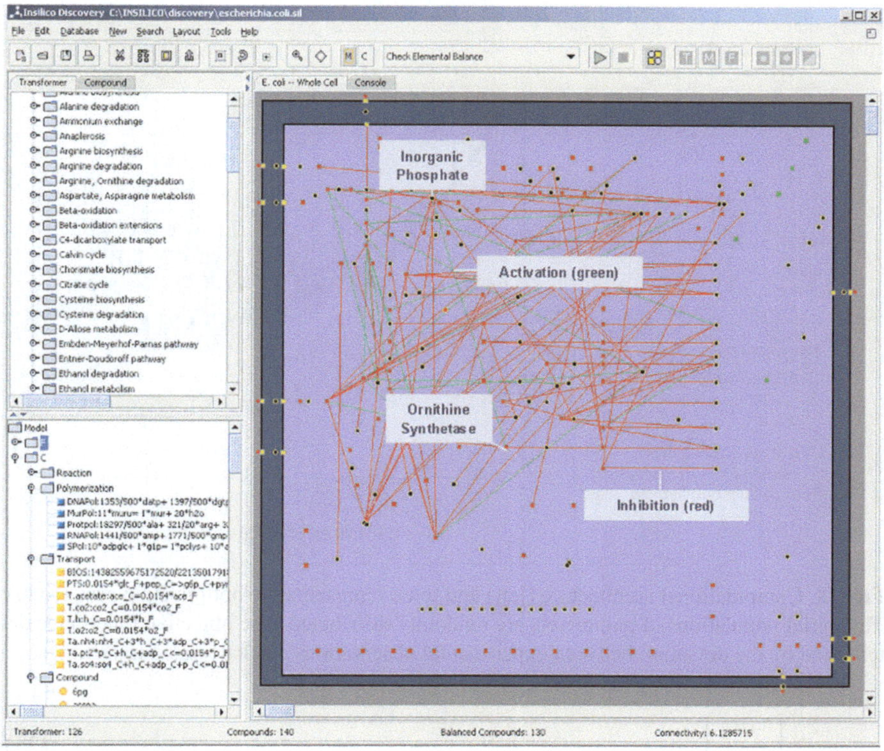

Fig. 17. Modulation network of *Escherichia coli*. Inhibition and activation in accordance to the MetaCyc database.

("irreversible reactions") leads to a total of 23 elementary flux modes. The complete list of reactions can be found in the paper of Chassagnole et al. 2002. Additional information used for the dynamic modelling regards the network of the metabolic modulation is illustrated in Figure 17. The depicted inhibitions and activations for the individual reactions have been gathered from the MetaCyc database (www.metacyc.org, Karp et al. 2004). The kinetic behaviour of the individual reactions is assigned according to the universal linlog approach (Vissser and Heijnen 2003; Visser et al. 2004):

$$r = J^0 \frac{c_E}{c_E^0}\left(1+\underbrace{\sum_i \varepsilon_{S,i} \ln\frac{c_{S,i}}{c_{S,i}^0}}_{\text{substrates}}+\underbrace{\sum_j \varepsilon_{P,j} \ln\frac{c_{P,j}}{c_{P,j}^0}}_{\text{products}}+\underbrace{\sum_k \varepsilon_{a,i} \ln\frac{c_{a,k}}{c_{a,k}^0}}_{\text{activators}}+\underbrace{\sum_l \varepsilon_{i,l} \ln\frac{c_{i,l}}{c_{i,l}^0}}_{\text{inhibitors}}\right) \quad (22)$$

The variables are defined relative to the relative reference steady state, with concentration levels c^0, fluxes J^0, and enzyme level c_E^0. The parameters are the elasticity coefficients

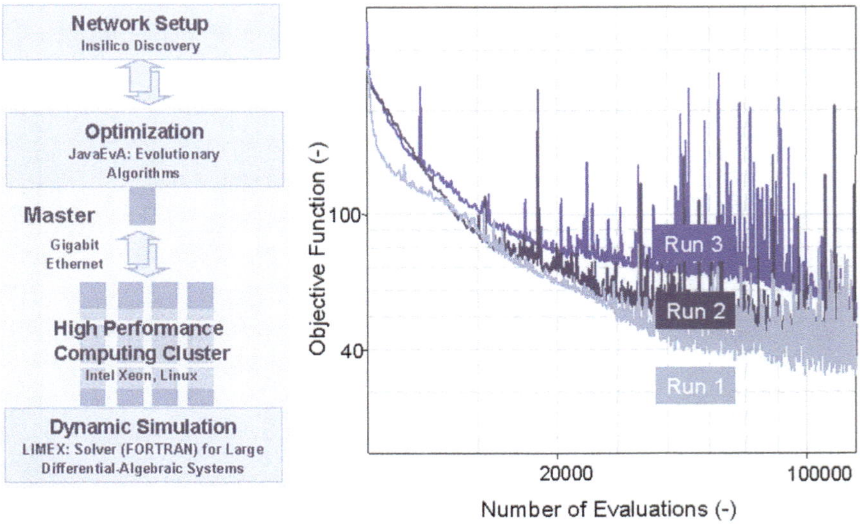

Fig. 18. Computational architecture (left) and three representative optimization runs (right). The initial populations of parameters are randomly distributed. The objective function is determined by the deviation between experimental observations and calculated trajectories.

$$\varepsilon_M = \frac{c_M}{r}\left(\frac{\partial r}{\partial c_M}\right)_0 \tag{23}$$

In total, the network holds 921 kinetic parameters (elasticities). The dynamic simulation of the non-linear and stiff system of differential equations was performed with the aid of the extrapolation solver LIMEX from the Konrad-Zuse-Centre for Information Technology in Berlin (Ehrig et al. 1999). For estimation of the parameters the evolutionary algorithm, JavaEva developed by the Computer Science Department of the University of Tuebingen (Streichert et al. 2005; www-ra.informatik.uni-tuebingen.de/software/JavaEva) has been applied. The initial populations of parameters are randomly distributed. The computational architecture and the results of three representative optimization runs are depicted in Figure 18. The objective function is determined by the deviation between experimental observations (Chassagnole et al. 2002) and calculated trajectories after the pulse of glucose. Since simulating the network dynamics is significantly more time consuming compared to a single optimization step, the simulations are run in parallel on a high performance computing cluster (Intel Xeon) whereas the optimization is carried out on a single master computer.

Typical results of comparisons between model simulations and experimentally observed time series of metabolites are shown in Figure 19. The trajectories represent simulations with kinetic parameters (elasticities) resulting from 10 independent optimization runs. With exception of PEP, NAD, and NADP, the trend of the experimental data is described well by the various simulation runs. For the first

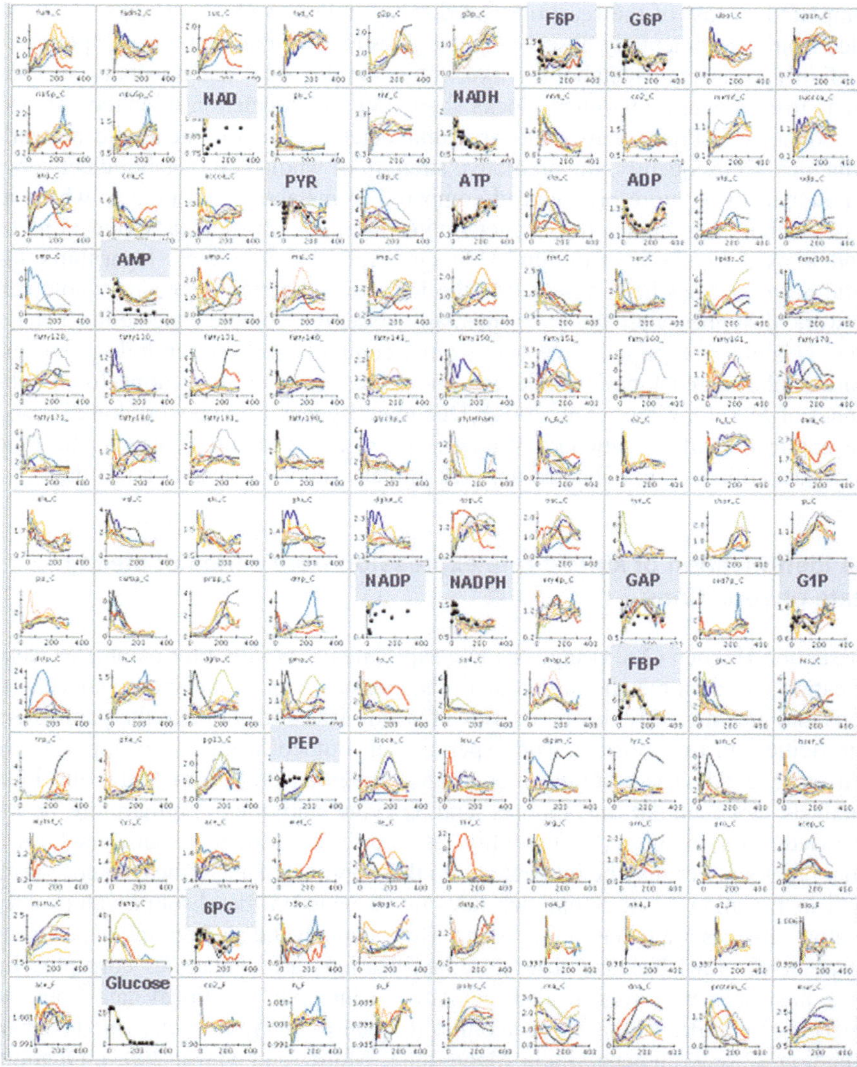

Fig. 19. Whole cell metabolic network dynamics. The trajectories represent simulations with kinetic parameters (elasticities) resulting from ten different optimization runs. Time series data represented by black circles were obtained by stimulating *Escherichia coli* at a growth rate of D= 0.1 h^{-1} with glucose (Chassagnole et al. 2002).

time, the time course of highly connected metabolites like ATP could be identified through an autonomous whole cell metabolic network model. Although only 16 metabolites are used for identifying the network dynamics, the standard parameter error of more than 500 model parameters is less than 30%. The reason for this finding is due to the fact that many reactions are directly connected to measured compounds. Most of the reactions not directly linked to measurements are separated from experimentally observed metabolites by only one or a few compounds.

Highly connected metabolites like, for example, ATP and feedback mechanisms from monomer building blocks show how the interaction of anabolism and catabolism constitutes important structural determinants for metabolic network dynamics. Consequently, the predictive strength of whole cell metabolic networks is assumed to be more pronounced compared to metabolic sub-models where in most of the cases highly linked metabolites are kept constant in applications of model based metabolic re-design. Although a rather simple linear logarithmic kinetic behaviour has been assigned to the individual reaction steps, the whole cell network describes well the experimentally observed complex metabolic behaviour. This observation is in line with similar comparisons between linlog kinetics, more sophisticated kinetic modelling and experimental observations for a sub-model of *E. coli* (Visser et al. 2004). Due to self-generating network kinetics and the application of evolutionary strategies on c, and computer clusters, the time from data collection to validated network models can be reduced to a few days.

3.3 Identification of kinetic rate expression from series of steady state observations

Visser and Heijnen (2003) suggested an experimental protocol for parameter identification based on metabolic flux analysis, measurement of metabolite concentrations and enzyme activities in a series of steady states to estimate the parameters (elasticities) in Equation 22. The experimental protocol has been exemplified by Wu et al. 2004. The resulting models thus integrate data from Metabolic Flux Analysis, intracellular metabolites and enzyme activities. Furthermore, prior knowledge on topology, reversibility, effectors, and kinetic parameters known from other sources may be integrated into this approach. Due to the reference state used in Equation 22, the approach does not depend on the absolute values only relative changes are taken into account. This is an advantage from the experimental point of view, because absolute quantifications are sometimes corrupted by large measurement errors. It also facilitates to integrate additional data from the emerging field of quantitative proteomics to quantify the ratio of enzyme levels in Equation 22 (e.g. Aebersold and Mann 2003; Moritz and Meyer 2003). Ongoing research (collaboration between the Medical Proteome Centre University Bochum, H. Meyer and the authors laboratory) is addressing this issue, by applying metabolic labelling ($^{14}N/^{15}N$) and MALDI-TOF Mass Spectrometry for relative quantification of protein levels (Franke et al. 2004) as well as ^{13}C-labelling for Metabolic Flux Analysis and measurement of intracellular metabolites (LC-MS and GC-MS) for analysis of time series of samples during fed batch fermentation with constant feeding rate (decreasing specific growth rate).

4. Summary and outlook

To obtain a true picture of the dynamics of metabolic networks and to predict targets for relevant genetic manipulations or drug discovery we need quantitative descriptions of the behaviour of the relevant reactions embedded in the network of interest. The identification of reliable and accurate structure and parameters of the kinetic expressions imposes severe challenges and there is a great need for more rigorous approaches. It should be noted that in the past a great deal of research has been based on bottom-up methods in which kinetics from *in vitro* measurements have been aggregated to obtain models of pathways or modules. There are two major problems related to this approach. First of all, there are numerous indications that *in vitro* kinetics may considerably differ in structure and parameters from *in vivo* situations. Moreover, despite impressive progresses made in more rigorous definitions of modularity in context with biochemical networks, open questions remain as whether and to what extent large-scale dynamic models generated from aggregation of these modules reflect reality. It needs to be stressed that while the definition of functional modules greatly facilitates model development for cellular processes, one must not forget that the definition of these modules usually is not unique and mainly serves to delineate systems boundaries when developing models for small subsystems. Assigning highly connected compounds, such as adenine ore pyridine nucleotides, to a specific module represents a further challenge, considering that these metabolites are produced and consumed in a vast number of intracellular reactions, in processes as diverse as respiration, biosynthetic reactions, solute transport, cytoskeletal dynamics, or cell cycle progression. When formulating a model for one of these processes as an isolated functional module it will, thus, frequently be difficult to obtain a satisfactory description of, for example, ATP dynamics solely on the basis of its role within the module when compared to *in vivo* data.

The set-up of large-scale dynamic models based on a canonical representation of reaction kinetics, as illustrated in this contribution, is a promising concept to overcome the aforementioned limitations. The approach has been illustrated by identifying whole cell network dynamics for *E. coli* from time series data from stimulus-response experiments. Such a large-scale model necessarily extends beyond the module boundaries. However, much remains to be done with respect to the problems of observability and optimal experimental design. In the long-range, it may prove useful to combine the detailed representation of a functional module with a less detailed large-scale model to describe the dynamics of highly connected module compounds. Key challenges to achieving these and other rigorous large scale dynamic network models will be further developments of appropriate computational methods and the critical linkage between modelling and experiment.

References

Aebersold R, Mann M (2003) Mass spectrometry-based proteomics. Nature 422:198-207

Aguilera-Vázquez L (2005) Modellgestützte Analyse der Dynamik des Speicherstoffwechsels in *Saccharomyces cerevisiae*. PhD Thesis, University Stuttgart, Shaker Verlag: Aachen

Alon U (2003) Biological networks: The tinkerer as an engineer. Science 26:1866-1867

Aon MA, Cortassa S (1997) Dynamic Biological Organization. Fundamentals as applied to cellular systems. London: Chapman & Hall

Aragón JJ, Sánchez V (1985) Enzyme concentration affects the allosteric behaviour of yeast phosphofructokinase. Biochem Biophys Res Commun 131:849-855

Beullens M, Mbonyi K, Geerts L, Gladines D, Detremerie K, Jans AWH, Thevelein JM (1998) Studies on the mechanism of the glucose-induced cAMP signal in glycolysis and glucose repression mutants of the yeast *Saccharomyces cerevisiae.* Eur J Biochem 172:227-231

Buziol S, Bashir I, Baumeister A, Claassen W, Noisommit-Rizzi N, Mailinger W, Reuss M (2002) New bioreactor-coupled rapid stopped-flow sampling technique for measurements of metabolite dynamics on a subsecond time scale. Biotechnol Bioeng 80:632-636

Cannon JF, Tatchell K (1987) Characterization of *Saccharomyces cerevisiae* genes encoding subunits of cyclic AMP-dependent protein kinase. Mol Cell Biol 7:2653-2663

Castrillo JI, Hayes A, Mohammed S, Gaskell SJ, Oliver SG (2003) An optimized protocol for metabolome analysis in yeast using direct infusion electrospray mass spectrometry. Phytochemistry 62:929-937

Chassagnole C, Noisommit-Rizzi N, Schmid JW, Mauch K, Reuss M (2002) Dynamic modelling of the central carbon metabolism of *Escherichia coli.* Biotechnol Bioeng 79:53-73

De Koning W, van Dam K (1992) A method for determination of changes of glycolytic metabolites in yeast on a subsecond time scale using extraction at neutral pH. Anal Biochem 204:118 -123

Ehrig R, Nowak U, Oeverdieck L, Deuflhard P (1999) Advanced extrapolation methods for large scale differential algebraic problems. In: Bunghartz HJ, Durst F, Zenger C (eds) High performance scientific and engineering computing (lecture notes in Computational Science and Engineering). Berlin: Springer-Verlag 8:233 -244

Girvan M, Newman MEJ (2002) Community structure in social and biological networks. Proc Natl Acad Sci USA 99:7821-7826

Gonzales B, Francois J, Renaud M (1997) A rapid and reliable method for metabolite extraction in yeast using boiling buffered ethanol. Yeast 13:1347-1356

Guimera R, Amaral LAN (2005) Functional cartography of complex metabolic networks. Nature 433:895-900

Hixson CS, Krebs EG (1980) Characterization of a cyclic AMP-binding protein from baker´s yeast. J Biol Chem 255:2137-2145

Hofmann E, Kopperschläger G (1982) Phosphofructokinase from yeast. Meth Enzymol 90:49-60

Holme P, Huss M, Jeong H (2003) Subnetwork hierarchies of biochemical pathways. Bioinformatics 19:532-538

Huss M, Holme P (2006) Currency and commodity metabolites: Their identification and relation to the modularity of metabolic networks. Quantitative Biology q-bio. MN/0603038, arXiv.org

Janssens V, Goris J (2001) Protein phophatase 2A: a highly regulated family of serine/threonine phophatases implicated in cell growth and signalling. Biochem J 353:417-439

Karp PD, Arnaud M, Collado-Vides J, Ingraham J, Paulsen I, Saier M (2004) The *E. coli* ecocyc data base: No longer just a metabolic pathway database. ASM News 70:25-30

Kopperschläger G (1999) personal communication

Kremling A, Stelling J, Bettenbrock S, Fischer S, Gilles ED (2005) Metabolic networks: biology meets engineering sciences. In: Alberghina L, Westerhoff HV (eds) Topics in Current Genetics: Systems Biology- Definitions and Perspectives. Berlin: Springer-Verlag 13:215-234

Mailinger W, Baumeister A, Reuss M, Rizzi M (1998) Rapid and highly automated determination of adenine and pyridine nucleotides in extracts of *Saccharomyces cerevisiae* using a micro robotic sample preparation-HPLC system. J Biotechnol 63:155-166

Masegho MR, van Gulik WM, Vinke JL, Visser D, Heijnen JJ (2006) *In vivo* kinetics with rapid perturbation experiments in *Saccharomyces cerevisiae* using a second-generation BioScope. Metabol Eng 8:370-383

Matsumoto K, Uno I, Tohe A, Ishikawa T, Oshiuma Y (1982a) Cyclic AMP may not be involved in catabolic repression in *Saccharomyces cerevisiae*: evidence from mutants capable of utilising it as an adenine source. J Bacteriol 150:277-285

Matsumoto K, Uno I, Oshima Y, Ishikawa T (1982b) Isolation and characterisation of yeast mutant deficient in adenylate cyclase and cAMP-dependent protein kinase. Proc Nat Acad Sci USA 79:2355-2359

Mauch K, Vaseghi S, Reuss M (2000) Quantitative analysis of metabolic and signalling pathways in *Saccharomyces cerevisiae*. In: Schügerl K, Bellgardt KH (eds) Bioreaction Engineering. Berlin: Springer-Verlag 435-477

Moritz B, Meyer HE (2003) Approaches for the quantification of protein concentration ratios. Proteomics 3:2208-2220

Müller D (2006) Model-assisted analysis of cyclic AMP signal transduction in *Saccharomyces cerevisiae* - cAMP as dynamic coordinator of energy metabolism and cell cycle progression. PhD Thesis, University Stuttgart, Shaker Verlag: Aachen

Müller D, Aguilera-Vázquez L, Barl T, Diaz-Cuervo H, Guerrero-Martin E, Marquetand JO, Murugan PK, Niebel A, Reuss M (2005) Integration of cyclic AMP signaling and metabolism in a single-cell model of *Saccharomyces cerevisiae*. FOSBE 2005, Santa Barabara, Proceedings, CACHE: 249-254

Murray W (1999) From molecular to modular cell biology. Nature 402:C47-C52

Newman MEJ, Girvan M (2004) Finding and evaluating community structure in networks. Phys Rev E69:026113

Radicchi F, Castellano C, Cecconi F, Loreto V, Parisi D (2004) Defining and identifying communities in networks. Proc Natl Acad Sci USA 101:2658-2663

Ravasz E, Somera AL, Mongru DA, Oltvai ZN, Barabasi AL (2002) Hierarchical organisation of modularity in metabolic networks. Science 297:1551-1555

Reuss M (1991) Structured modeling of Bioreactors. Ann NY Acad Sci 646:284-299

Reuter R, Eschrich K, Schellenberger W, Hofman E (1979) Kinetic modelling of yeast phosphofructokinase. Acta Biol Med Germ 38:1067-1079

Rizzi M, Theobald U, Querfurth E, Rohrhirsch T, Baltes M, Reuss M (1996) *In vivo* investigations of glucose transport in *Saccharomyces cerevisiae.* Biotechnol Bioeng 52:316-327

Rizzi M, Baltes M, Theobald U, Reuss M (1997) *In vivo* analysis of metabolic dynamics in *Saccharomyces cerevisiae*: II Mathematical model. Biotechnol Bioeng 55:592-608

Schaefer U, Boos W, Takors R, Weuster-Botz D (1999) Automated sampling device for monitoring intracellular metabolite dynamics. Anal Biochem 270:88-96

Schaub J, Schiesling C, Reuss M, Dauner M (2006) Integrated sampling procedure for metabolome analysis. Biotechnol Progr 22:1434-1442

Schmid JW, Mauch K, Reuss M, Gilles ED, Kremling A (2004) Metabolic design based on coupled gene expression-metabolic network model of tryptophan production in *Escherichia coli.* Metabol Eng 6:364-377

Serrano R, Gancedo JM, Gancedo C (1973) Assay of yeast enzymes *in situ.* Eur J Biochem 34:479-482

Schmalzried S, Jenne M, Mauch K, Reuss M (2003) Integration of physiology and fluid dynamics. Adv Biochem Eng 80:19-68

Snoep JL, Bruggeman F, Olivier BG, Westerhoff HV (2006) Towards building the silicon cell: a modular approach. Biosystems 83:207-216

Srere PA (1967) Enzyme concentrations in tissues. Science 158:936-937

Streichert F, Ulmer H, Zell A (2005) Java Eva: A Java based framework for evolutionary algorithms. www-ra.informatik.uni-tuebingen.de/software/javaeva

Teusink B, Passarge J, Reijenga CA, Esgalhado E, van der Weijden CC, Schepper M, Walsh MC, Bakker BM, van Dam K, Westerhoff HV, Snoep JL (2000) Can yeast glycolysis be understood in terms of *in vitro* kinetics of the constituent enzymes? Testing biochemistry. Eur J Biochem 267:5313-5329

Theobald U, Mailinger W, Reuss M, Rizzi M (1993) *In vivo* analysis of glucose-induced fast changes in yeast adenine nucleotide pool applying a rapid sampling technique. Anal Biochem 214:31-37

Theobald U, Mailinger W, Rizzi M (1994) Use of $HgCl_2$ to investigate dynamic phenomena in yeast cytoplasm. Biotechnol Techniques 8:723-728

Theobald U, Mailinger W, Baltes M, Rizzi M, Reuss M (1997) *In vivo* analysis of metabolic dynamics in *Saccharomyces cerevisiae*: I. Experimental observations. Biotechnol Bioeng 55:305-316

Thevelein JM, de Winde JH (1999) Novel sensing mechanisms and targets for the cAMP-protein kinase A pathway in the yeast *Saccharomyces cerevisiae.* Mol Microbiol 33:904-918

Toda A, Cameron S, Sass P, Zoller M, Wigler M (1987a) Three different genes in *Saccharomyces cerevisiae* encode the catalytic subunits of the cAMP-dependent protein kinase. Cell 50:277-287

Toda A, Cameron S, Sass P, Zoller M, Scott JD, McBullen B, Hurwitz M, Krebs EG, Wigler M (1987b) Cloning and characterization of *BCY1*, a locus encoding a regulatory subunit of cyclic AMP-dependent protein kinase in *Saccharomyces cerevisiae.* Mol Cell Biol 7:1371-1377

Van Schaftingen E, Lederer B, Bartons R, Hers HG (1982) A kinetic study of pyrophosphate: Fructose-6-phosphate phosphotransferase from potato tubers. Eur J Biochem 129:191-195

Vaseghi S, Baumeister A, Rizzi M, Reuss M (1999) *In vivo* dynamics of the pentose phosphate pathway in *Saccharomyces cerevisiae*. Metabol Eng 1:128-140

Vaseghi S (2000) Modellgestützte Analyse der Dynamik des Phosphofructokinas-Systems in *Saccharomyces cerevisiae*. PhD Thesis: University Stuttgart

Vaseghi S, Macherhammer F, Zibek S, Reuss M (2001) Signal transduction dynamics of the protein kinase A/Phosphofructokinase-2-system in *Saccharomyces cerevisiae*. Metabol Eng 3:163-172

Visser D, van Zuylen GA, van Dam JC, Oudshoorn A, Eman MR, Ras C, van Gulik WM, Frank J, van Dedem GWK, Heijnen JJ (2002) Rapid sampling for analysis of *in vivo* kinetics using the BioScope: a system for continuous-pulse experiments. Biotechnol Bioeng 79:674-681

Visser D, Heijnen JJ (2003) Dynamic simulation and metabolic redesign of a branched pathway using linlog kinetics. Metabol Eng 5:164-176

Visser D, Schmid J, Mauch K, Reuss M, Heijnen JJ (2004) Optimal re-design of primary metabolism in *Escherichia coli* using linlog kinetics. Metabol Eng 6:378-390

Weibel KE, Mor JR, Fiechter A (1974) Rapid sampling of yeast cells and automated assays of adenylate, citrate, pyruvate and glucose-6-phosphate pools. Anal Biochem 58:208-216

Weuster-Botz D, de Graaf AA (1996) Reaction engineering methods to study intracellular metabolite concentrations. Adv Biochem Eng 54:75-108

Wingender-Drissen R (1983) Yeast cyclic AMP-dependent protein kinase. FEBS Lett 163:28-32

Wu L, Wang W, van Winden WQA, van Gulik WM, Heijnen JJ (2004) A new framework for the estimation of control parameters in metabolic pathways using lin-log kinetics. Eur J Biochem 271:3348-3359

Wu L, Masegho MR, Proell AM, Vinke JL, Ras C, van Dam JC, van Winden WA, van Gulik WM, Heijnen JJ (2006) *In vivo* kinetics of primary metabolism in *S. cerevisiae* studied through prolonged chemostate cultivation. Metabol Eng 8:160-171

Yamashoji S, Hess B (1984) Activation of yeast 6-phosphofructo-2-kinase by protein kinase and phosphate. FEBS Lett 178:253-256

Zhao Y, Boguslawski G, Zitomer RS, DePaoli-Roach AA (1997) *Saccharomyces cerevisiae* homologs of mammalian B and B′ subunits of protein phosphatase 2A direct the enzyme to distinct cellular functions. J Biol Chem 272:8256-8262

Aguilera-Vázquez, Luciano

Depto.de Biotecnología, Universidad Politécnica de Pachuca, Ex-Hacienda de Sta. Barbara, CP 43830, Municipio de Zenpoala, Hgo, Mexico

Mauch, Klaus

Insilico Biotechnology AG, Nobelstrasse. 15, D-70569 Stuttgart

Reuss, Matthias

Institute of Biochemical Engineering and Centre of Systems Biology, University Stuttgart, Allmandring 31, D-70569 Stuttgart
reuss@ibvt.uni-stuttgart.de

Toward metabolome-based ^{13}C flux analysis: a universal tool for measuring *in vivo* metabolic activity

Nicola Zamboni

Abstract

Intracellular metabolic rates cannot be directly assessed from metabolome concentrations and vice versa. For most biological questions, stable isotope tracers must be administered and tracked to effectively determine metabolic fluxes by means of numerous computational steps. Although flux analysis targets the same analytes as metabolomics, priority is given to measuring their exact isotopic distribution rather than their concentration. In the first part of this chapter, I describe principles and issues of current ^{13}C flux analysis methods, following the entire process from experimental design, to detection of isotopic distributions, and data interpretation. Notably, current practice largely relies on the labeling patterns of protein-bound amino acids, because of their abundance and stability. In the second part, I focus on achievements, challenges, and opportunities of metabolome-based ^{13}C flux analyses, which are emerging in response to the need to tackle larger networks, higher cells, and to improve both spatial and temporal resolution.

1 Introduction

Physiological phenotypes of cells are macroscopic manifestations of their metabolic activity, that is determined by all molecular fluxes through metabolism, i.e. the fluxome (Hellerstein 2003; Sauer 2004). In nature, the fate of a cell between growth and senescence, or even life and death, is linked to its metabolic capacity to utilize heterogeneous substrates that are encountered. Whenever cellular functions have to be adjusted, for example upon shifts in external conditions, after mutations, or upon aberrant growth such as in tumors, the fluxome has to be adapted to support growth. To a large extent, adjustment of the fluxome is realized in central metabolism. These primary pathways are at the crossroad of catabolism and anabolism, and catalyze the largest metabolic fluxes in the cell. They form an intertwined reaction network capable of rearranging carbon and nitrogen from a wide range of substrates to fuel growth. Oxidation of the cofactors NADH and NADPH in respiration and biosynthesis, respectively, is flexibly balanced by modulation of fluxes through alternative routes in central metabolism.

Topics in Current Genetics, Vol. 18
J. Nielsen, M. C. Jewett (Eds.): Metabolomics
DOI 10.1007/4735_2007_0220 / Published online: 14 April 2007

Table 1. Exemplary applications of ^{13}C metabolic flux analysis

Objective of study	Type of analysis	Analytes	Platform	Organisms	Representative References
Quantitation of intracellular fluxes	Isotopomer balancing	Amino acids	GC-MS	Microorganisms	(Dauner et al. 2001)
		Amino acids	NMR	Microorganisms	(Petersen et al. 2000; Dauner et al. 2002; van Winden et al. 2003)
				Plants	(Schwender et al. 2003; Sriram et al. 2004)
		Primary metabolites	LC-MS	Microorganisms	(van Winden et al. 2005)
			NMR	Human cells	(Forbes et al. 2006)
		$^{12}CO_2/^{13}CO_2$	Online MS	Microorganisms	(Hoon Yang et al. 2006)
	Flux ratio analysis	Amino acids	GC-MS	Microorganisms	(Fischer and Sauer 2005)
		Amino acids	NMR	Microorganisms	(Maaheimo et al. 2001)
	^{13}C constrained MFA	Amino acids	GC-MS	Microorganisms	(Hua et al. 2006)
Unraveling of metabolic reaction network	Flux ratios	Amino acids	GC-MS	Microorganisms	(Cannizzaro et al. 2004; Fuhrer et al. 2005)
Unraveling of enzymatic mechanism	Isotopomer balancing	Primary metabolites	LC-MS	Microorganisms	(Kleijn et al. 2005)
Investigation of redox and energy metabolism	^{13}C constrained MFA	Amino acids	GC-MS	Microorganisms	(Sauer et al. 2004)
	Isotopomer balancing	Amino acids	GC-MS	Microorganisms	(Wittmann and Heinzle 2002)
			NMR	Microorganisms	(Dauner et al. 2001)
Profile metabolic changes	SIDMAP	Amino acids and secretes	GC-MS	Human cells	(Boren et al. 2001; Boros et al. 2003)
	Fluxome profiling	Amino acids	GC-MS	Microorganisms	(Zamboni and Sauer 2004)

How are fluxes regulated? Metabolic fluxes are the integrated result of (i) all catalytic activities of enzymes as set by kinetic properties, and concentrations of educts, products, cofactors, ions, or protons; and (ii) all non-linear regulatory interactions at the transcriptional, translational, post-translational, and allosteric level, which all influence amount and state of enzyme (Hellerstein 2003; Sauer 2004). An essential consequence is that metabolic fluxes cannot be directly quantified solely from metabolites concentrations or vice versa. To realize this with good confidence, detailed enzyme modeling together with exact data on protein amounts and modifications, and metabolite concentrations would be a precondition. For this purpose, a palette of techniques is available to reveal concentrations, interactions, and kinetic parameters. (1) The concentration of cell components is determined by transcriptomics, proteomics, and metabolomics, although several specific methods are usually necessary to obtain complete information, for example on protein levels and modification state, or on chemically diverse metabolites. (2) Some approaches exist to discover the binding between proteins (Cusick et al. 2005) or of proteins to DNA (Hoglund and Kohlbacher 2004; Bulyk 2006). Unfortunately, they can hardly be used to quantify their strength, and the comprehensive identification of interactions between DNA, transcripts, proteins, and small molecules is still far out of reach. (3) Estimation of *in vivo* kinetic parameters can be done with stimulus-response experiments (Vaseghi et al. 1999). The drawback of such procedures is that these experiments are demanding, performed locally for a reduced number of parameters, and require a priori knowledge of all possible interactions: for mid-sized and large networks, the task rapidly becomes prohibitive.

The general lack of detailed regulation and kinetic information has two main consequences. First, today's omics data can, at most, provide constraints on metabolic fluxes. For example, the absence of a protein or lack of transcription can be used to exclude that it is catalytically active. Analogously, the combination of metabolome data and thermodynamics knowledge can delineate directionality of reactions in a given state, but is insufficient to precisely assess metabolic fluxes (Kümmel et al. 2006). Second, the experimental workflow is preferably reversed: metabolic fluxes are measured together with concentrations to infer changes in enzyme activity or concentration (Wu et al. 2005), or overlap with proteome or transcript data to discover regulation circuits (Krömer et al. 2004; Shimizu 2004).

Metabolic fluxes are monitored by feeding organisms with substrates enriched in stable (i.e. non-radioactive) isotopic tracers such as ^{13}C, ^{2}H, ^{18}O, ^{34}S, or ^{15}N. Physiologists extensively employed similar labeled substrates for decades to track local metabolism of nutrients or monitor polymerization and degradation of biopolymers such as lipids, DNA, or proteins in animals and cells (Hellerstein 2003; McCabe and Previs 2004; Bequette et al. 2006), and are nowadays also employed to lead drug development (Turner and Hellerstein 2005). Only in the last decade, developments independent from physiology led to ^{13}C-based metabolic flux analysis for microbes. These methods were initially developed for purposes of strain optimization in industrial biotechnology (Stephanopoulos 1999), but have found large application and consensus in systems biology (Blank et al. 2005; Fischer and Sauer 2005; Koffas and Stephanopoulos 2005) (Table 1). In contrast to the methods utilized with animals that focus on local activities, novel ^{13}C metabolic flux analysis methods were devised to comprehensively assess carbon fluxes in large metabolic networks. Owing to the fact that microorganisms are rarely differentiated and able to grow on single carbon sources under carefully controlled conditions, an arsenal of ^{13}C flux methods was established to quantify the intracellular fluxome with different networks, substrates, and culture conditions. Modern ^{13}C flux analyses consequently enabled to investigate - from a global perspective - the link between cellular redox equilibrium, generation of energy equivalents, and metabolic phenotypes.

In this chapter, I first present the principles of metabolic flux analysis and the corollary methods that were designed to map reaction velocities in microbes with ^{13}C-labeling patterns of protein-bound amino acids. In the second part, I focus on the extension to metabolome-based ^{13}C metabolic flux analysis, that holds promise to become a universal tool to monitor the fluxome from microorganisms to animals for purposes of systems biology, understanding metabolic control in health and disease, or drug development.

2 Fundamentals of metabolic flux analysis

Metabolic flux analysis aims at measuring *in vivo* activity of metabolic reactions. In contrast to concentrations, rates are per se not directly measurable. *In vitro*, the rate of a reaction is determined via interpretation of measured concentration profiles of the substrates and products. Similarly, one can extend this approach and quantify the reaction rates in sequential and even branching reaction networks by monitoring the concentration profiles of substrates, intermediates, and products. The rate of every single reaction is then obtained by a set of material balances, one for each compound in the reaction chain. *In vivo*, however, it is experimentally impossible to measure concentration profiles for all metabolites in a cell that encompasses thousands of compounds. This problem is obviated when metabolic fluxes are measured in a metabolic steady state, meaning that fluxes and intracellular metabolite concentrations are constant over time. When this precondition is fulfilled, all intermediates pools are by definition invariant over time and in the case of linear, non-converging, non-cyclic pathways metabolic fluxes are calculable from the time profiles of all substrates and end products, while the concentrations of all balanced intermediates are neglectable.

Stoichiometric balancing has an additional inherent flaw that normally impairs complete flux estimations and that is associated to the topology of the biochemical reaction network. In most cells, especially in central carbon metabolism, alternative biosynthetic routes and reaction cycles exist and generate redundancies. Such redundancies cannot be unequivocally resolved by stoichiometric balancing, because each one introduces an additional degree of freedom where an infinite number of flux maps lead to identical overall balances. To obtain a unique solution, one approach is to select the flux distribution that satisfies all stoichiometric constraints and also maximizes an arbitrarily chosen objective function of network operation, e.g. maximize ATP overproduction or growth yield (Varma and Palsson

1994; Kauffman et al. 2003). The outcome of optimization corresponds to the most-likely flux estimate according to the arbitrary assumptions. Systematic studies have demonstrated that the chosen paradigm of network operation can differ between organisms, mutants, and environmental conditions (Küpfer et al.2007; Segre et al. 2002). Thus, objective functions have to be carefully selected to avoid biased and erroneous results.

The inherent uncertainty of *in silico* predictions and their discordance with empirical observations evidenced the importance of experimental metabolic flux determination. This was brought about by the introduction of isotopically labeled substrates. Depending upon which pathways are active in catabolism and anabolism, atoms from the substrate are scrambled and rearranged following the schemes of enzymatic reaction mechanisms. The labeling patterns of metabolites are then detected by either mass spectrometry (MS) or nuclear magnetic resonance (NMR), and quantitatively reflect partitioning of substrate through metabolic routes. They provide information independent from stoichiometric balances, and with a properly designed tracer substrate they serve to distinguish the fluxes through alternative pathways or reaction cycles. In general, ^{13}C-tracers enable to effectively resolve the redundancies occurring in central carbon metabolism, where all catabolic and anabolic pathways diverge from. In contrast to the highly interconnected central carbon metabolism, the peripheral metabolism is composed by mostly linear biosynthetic routes (e.g. amino acids or nucleotide synthesis). Since these pathways are utilized to synthesize the building blocks for growth, their *in vivo* flux is estimated with good precision by stoichiometry with detailed models of biomass composition.

Although ^{13}C metabolic flux analysis enables monitoring of pathway activity *in vivo* in most cases, it is important to stress that (i) quantitative analysis is only possible in minimal media, (ii) technical difficulty increases exponentially when multiple carbon substrates are utilized, (iii) it is not possible to discriminate between pathways or reactions that do not differ in the scrambling of labeled atoms, i.e. between isoenzymes.

3 Principles of labeling experiments

For a labeling experiment, cells are first grown on naturally labeled substrates until metabolic steady state. Once this prerequisite is fulfilled, isotopically enriched nutrients can be administered to the cells. In batch cultures, this is done either by spiking the tracer substrate to the medium, or by diluting exponentially growing cells in fresh, labeled medium. Harvesting and resuspending is preferably avoided because handling perturbs metabolic steady state. In continuous or fed-batch cultures, the feed is switched from naturally labeled medium to an equivalently concentrated tracer-enriched solution. Ideally, these operations should provoke an immediate step change of the tracer fraction in the culture medium. Although such rapid shifts can easily be attained in well-stirred systems, enrichment of label within the cellular metabolome will take considerably longer (Fig. 1). The reason

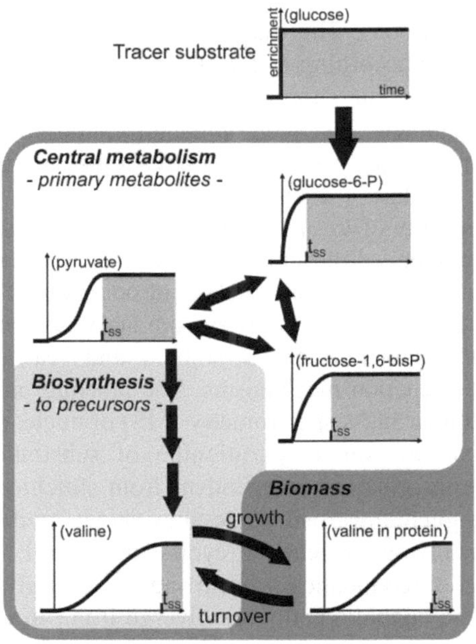

Fig. 1. Progressive propagation of labeling through intermediate pools in experiments with stable isotopic substrates. Each plot exemplarily shows time profiles of label enrichment for species of central metabolism, peripheral biosynthetic pathways leading to biomass precursors, and biomass components. Exemplary names are indicates in brackets. t_{SS} is the time necessary to attain isotopic stationarity, it is specific for each pool, and it sets the minimum labeling time that has to be respected for stationary computational methods to be applied. Delayed onset of isotopic steady state is typically observed far from tracer substrate uptake, in large pools, or when biomass turnover occur. Refer to the text for more detailed explanations.

is that starting from the entry point of the tracer, the label has to propagate through the metabolic network and progressively replace unlabeled intermediates. This is an important phenomenon, because routine application of ^{13}C flux analysis is so far solely possible from the labeling patterns of metabolites in isotopic steady state, i.e. with time-invariant labeling patterns at the time point of sampling. In theory such an isotopic steady state is never attained, but due to analytical imprecision isotopic equilibrium is experimentally observed within minutes to hours.

The time after which such an isotopic steady state is achieved depends upon the turnover rate of each pool, which is directly proportional to the flux through the pool and inversely to the concentration: larger pools slow down the process, higher fluxes accelerate it. A general and intuitive consequence is that the closer an intermediate is to the original tracer substrate, the faster it will reach isotopic steady state. Thus, for ^{13}C glucose tracers, flux analysis based on the labeling pattern of glycolytic intermediates requires shorter labeling times than with tricarboxylic acid (TCA) cycle intermediates. Biomass compartments (e.g. proteins) exhibit the longest isotopic transients, whose duration is roughly proportional to the inverse of the growth rate. Similarly slow label uptake can also be observed for the free pool of corresponding precursors (in the same example the amino acids) when biomass turnover interferes with quick onset of isotopic steady state (Fig. 1) (Grotkjaer et al. 2004).

Substantial advantages are brought about by the analysis of labeling patterns in intermediates of central metabolism. First, it decreases duration of experiments and the costs coupled to the amount of employed isotopic tracer. Second, shorter observation windows provide much more flexibility in experimental design since metabolic steady state does not have to be ensured over several hours (van Winden et al. 2005). In turn, this opens for the investigation of slow metabolic transients for which a quasi steady state can be assumed for the time span of labeling (e.g. fed batches). The analysis yields a flux map that averages pathway activities over the labeling interval. An extension is to sample the same labeling experiment at several time points during the slow flux transients to obtain time-resolved flux maps (Zamboni et al. 2005). Limitations are set by the characteristic time of monitored analytes necessary to attain isotopic steady state, which is prone to variation during metabolic transients due to changing fluxes and pool concentrations.

These underlying principles hold for every experiment involving labeling with isotopic tracers, and should carefully be considered in the design stage. In the next sections, the workflow of ^{13}C-based flux analysis from inception to evaluation and current practice is briefly reiterated.

4 Current practice of stationary ^{13}C flux analysis

4.1 Experimental design

The capability of resolving and quantifying fluxes *in vivo* is a function of (i) the tracer substrate used, (ii) the biochemical reaction network, and (iii) the analytes that are detectable. Several protocols were presented to assess a priori calculability from a dataset in the case of stoichiometric balancing (Klamt and Schuster 2002) or ^{13}C metabolic flux analysis (Möllney et al. 1999; van Winden et al. 2001; Isermann and Wiechert 2003). Notably, analytical accuracy in the detection of labeling patterns strongly influences the confidence of flux estimates. This information is frequently neglected in the aforementioned calculability tests and, thus, it is often necessary to perform more complex and detailed experiments than the simplest setup prescribed based on such tests (van Winden et al. 2001).

The selection of the tracer distribution in the substrate is paramount for effective resolution of metabolic fluxes. Basically two different strategies exist and can be combined. Positionally enriched substrates possess an uneven distribution of ^{13}C in the carbon backbone. These tracers are typically administered in the pure form, i.e. 100%, and are ideal to distinguish alternative pathways where only one branch losses or transfers the specifically labeled carbon (e.g. by decarboxylation). For example, [1-^{13}C]glucose is well suited to track fluxes in the oxidative branch of the pentose phosphate pathway (PPP) where the [1-^{13}C] atom is split to form

$^{13}CO_2$ and the resulting pentoses are label-free. In contrast, pentoses originating via the non-oxidative PPP are enriched in ^{13}C (Christensen et al. 2001). A shortcoming of positionally enriched tracers is that they are tailored for specific pathways and poorly suited for global fluxome estimates. Hence, they find wide application in networks that are highly constrained by stoichiometry and thus exhibit low degrees of freedom, or to determine the network structure in poorly characterized organisms (Cannizzaro et al. 2004; Fuhrer et al. 2005). On the other end, uniformly (fully) labeled substrates offer a larger scope in exchange for specificity. Uniformly labeled substrates are normally administered in combination with unlabeled isomers, e.g. as a 1:1 mix. When the tracer is metabolized, the carbon backbone of both labeled and unlabeled isoforms is broken and rearranged. Reactions that combine multiple carbon-containing intermediates will generate chimeric molecules with both ^{12}C and ^{13}C atoms, with characteristic labeling imprints. Examples are the transaldolase and transketolase in the non-oxidative PPP, anaplerotic reactions, or cyclic pathway such as the TCA cycle or the modular lipid biosynthesis. With uniformly labeled tracers, the essential information for pathway flux discrimination is not enclosed in the label that was lost during metabolic activity such as with positional enrichment, but in the presence of ^{13}C fine structures that reflect enzymatic scrambling specific for a pathway.

Compartments in higher cells complicate the problem in several ways: (i) additional reactions are necessary to model pathways independently for each compartment. Splitting of intermediate pools across distinct compartments considerably increases the degrees of freedom. (ii) The intracompartmental transport mechanisms are very relevant, in particular when coupled to sym- or antiport. (iii) Metabolites are measured as the sum of all compartments. When a metabolite is localized in two (or more) compartments with possibly different biosynthetic origin, the corresponding labeling patterns may differ and, thus, are typically discarded for flux calculation. Provided that the model of biochemical reactions is correct and complete, mathematical methods for the optimal selection of tracer and analytes exist (Möllney et al. 1999; Rantanen et al. 2006).

Experimental design is also influenced by the analytes that can be detected. The majority of ^{13}C-based flux studies published in the last decade was based on the labeling patterns of protein-bound amino acids because of their abundance (roughly half of total cell dry weight) that facilitates measurement of labeling pattern. High abundance is unfortunately coupled to lower turnover and, hence, short transients cannot be investigated. In the case of biomass macromolecules and the constituting monomers when turnover occurs, the inverse of the growth rate provides a rough estimate of the shortest interval that can be investigated with a stationary ^{13}C metabolic flux experiment (Wiechert and Nöh 2005).

4.2 From analytes to ^{13}C labeling patterns

Determination of carbon fluxes in isotopic steady state relies on macroscopic balances and ^{13}C labeling patterns: intermediates concentrations are superfluous unless isotopically non-stationary conditions are tackled (discussed in a later section). During sample harvest, it is important to quench metabolism rapidly enough to avoid post-sampling artifacts. The constraints are set by the turnover of the analytes. When the protein-bound fraction of amino acids is the target, handling operations in the range of minutes are safe. In contrast, when dealing with intracellular intermediates sub-second quenching and cooling is recommended because their pools are exchanged by orders of magnitude more rapidly. In sharp contrast to metabolome experiments, quantitative and reproducible extraction of intermediates from cells is not of relevance as long as detection is not compromised by poor recoveries.

Two techniques exist to distinguish and quantify isotopic distributions, namely nuclear magnetic resonance (NMR) spectroscopy and mass spectrometry (MS). Both platforms were equally successful in providing essential information from protein-bound amino acids for the estimation of fluxes in central carbon metabolism. In NMR, 2-dimensional heteronuclear [^{13}C,^{1}H] correlation spectroscopy resolves all relevant resonances in proteinogenic amino acids without prefractionation (Szyperski 1995). Labeling patterns are inferred from characteristic spin-spin couplings that arise when neighbor ^{13}C atoms are present, and uniformly labeled tracers are therefore typically utilized. With MS, amino acids mixtures have first to be resolved by chromatographic means. Gas chromatography – mass spectrometry (GC-MS) has become the principal workhorse for mainly two reasons. First, it combines robustness, fast measurements, fully baseline-resolved amino acids, and relative low instrument and running costs. Second, it delivers extensive fragment information to unravel central carbon fluxes (Christensen and Nielsen 1999; Dauner and Sauer 2000). The latter point is crucial for ^{13}C-experiments and an important digression must be made. Each carbon atom in a molecule can be either labeled (^{13}C) or unlabeled (^{12}C). A molecule with n carbon atoms possesses 2^n possible states, called *isotopomers* (from isotope isomers). MS discriminates only the mass and is not able to distinguish between all isotopomers: those with identical weight are detected as a lumped pool. This limits calculability of fluxes when alternative pathways lead to isotopomers with equal label content. The hurdle is often overcome by inducing analyte fragmentation in the MS: from intact molecules, smaller daughter ions are generated and their isotopic distribution is measured to yield the isotopic distribution of partial carbon backbone segments or even the enrichment of single atom positions. Fragmentation in GC-MS occurs spontaneously at the interface between GC and MS when a high energy electrons beam is used to ionize the analytes. The resulting fragments enable ^{13}C flux analysis in many organisms using various tracers, such as for example [1-^{13}C], [1,2-$^{13}C_2$], and [U-^{13}C]glucose, and have contributed to the diffusion of GC-MS as preferred platform.

Table 2. Summary of metabolic flux analysis methods for experiments with stable isotopic tracers.

Approach	**Pros**	**Cons**
NET FLUXES		
Isotopomer balancing	- integrates all available information - calculates net and exchange fluxes - universal framework is available that do not need adaptation for new networks or tracers	- requires correct network and physiological data - cumbersome troubleshooting
^{13}C-constrained metabolic flux analysis	- simple and fast computation	- exchange fluxes not calculated
RELATIVE FLUXES		
Flux ratios analysis	- direct evidence for pathway activity - independent from measured rates - fast, unsupervised	- tedious design of new equations - implicit assumptions on reversibility that might do not hold after severe genetic perturbations.
PROFILING		
Fluxome profiling	- independent from any model - suitable for complex media - applicable with any tracer (^{13}C, ^{15}N, ^{2}H and combinations)	- qualitative - large number of replicas/samples needed
SiDMAP	- optimized for mammalian cells and glucose	- qualitative - requires multiple experiments to obtain a complete analysis

For emerging applications based on free metabolites, MS is currently superseding NMR owing to its superior sensitivity, simpler hyphenation to chromatography, and optional fragmentation capabilities. MS methods are increasingly profiting from the continuous progresses made in liquid chromatography (LC) and capillary electrophoresis (CE) that bring about baseline separations of the majority of central carbon and other polar metabolites pivotal to unravel fluxes. Flux analyses can build directly upon MS-metabolomics with minor adjustments made to prioritize precise estimation of mass distributions before concentrations (cf. 5.2 and 5.3.1).

4.3 From ^{13}C labeling patterns to fluxes

A variety of computational approaches to interpret ^{13}C labeling blueprints have bloomed driven by the need to address well-defined questions or hypotheses in

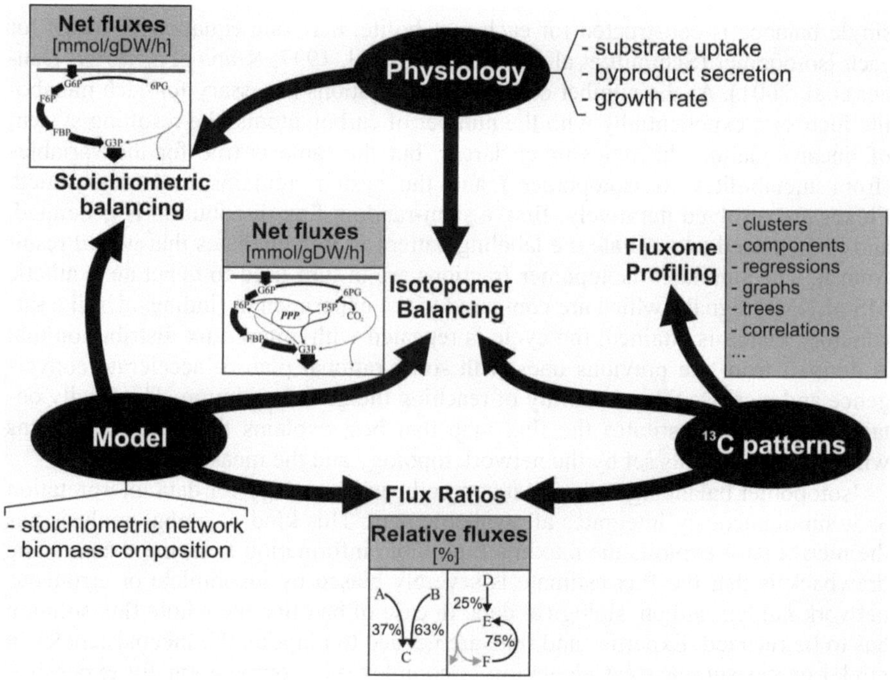

Fig. 2. Flow chart of data integration alternatives in ^{13}C metabolic flux analysis. Inputs and outputs are shown in black ellipsoids and grey boxes, respectively.

highly heterogeneous biological systems. Extensions and perhaps simplifications had to be introduced to face the sometimes scarce availability of measurements, ill-defined networks, and analytical imprecision. Three very different sets of information are utilized to estimate fluxes:

- Physiology: extracellular rates of substrate uptake and product formation, growth rate.
- Model of biochemical network: including - for each reaction in the system - stoichiometry, assumptions on the irreversibility, and the mapping of single atom positions between educts and products.
- ^{13}C labeling patterns: from NMR, MS, or both.

I define three major clusters of methods on the basis of which of the above information domains are utilized and combined to investigate metabolic fluxes (Fig. 2 and Table 2).

4.3.1 Isotopomer balancing

Isotopomer balancing is the natural extension of the stoichiometric balancing approach (cf. 2) to include ^{13}C data. It requires and concomitantly integrates extracellular fluxes, network model, and ^{13}C patterns. A network model is the basis for the balance equations. In contrast to simple stoichiometric balancing where a single balance is constructed for each metabolite, here one equation is drawn for each isotopomer (Schmidt et al. 1997; Zupke et al. 1997; Klapa et al. 1999; Dauner et al. 2001). As the number of additional equations necessary for each metabolite increases exponentially with the number of carbon atoms, the resulting system of linear equations becomes much larger, but the same is true for the variables (from metabolites to isotopomers) and the system remains underdetermined. Fluxes are resolved iteratively: first, a semi-random flux distribution is generated, and is then used to simulate the labeling pattern in intermediates that would result from it. The simulated isotopomer fractions are in turn used to generate synthetic MS or NMR signals, which are compared to the experimental findings. Until a satisfactory match is attained, the cycle is repeated with a new flux distribution that is derived from the previous ones with some rational plan to accelerate convergence and increase the probability of reaching the global optimum. The finally obtained solution constitutes the flux map that best explains the labeling patterns within the constraints set by the network topology and the measured rates.

Isotopomer balancing is the most comprehensive strategy for data interpretation as it simultaneously integrates all available data. This kind of global analysis has the merit that it exploits the maximum possible information from the dataset. The drawback is that the flux estimate is severely biased by incomplete or erroneous network models and physiological data. In case of bad fits, the whole flux solution has to be rejected. Expertise and time are needed to pinpoint the inconsistencies in model or measurements. Calculation is complex and computationally expensive, and special derivatives of isotopomer fractions such as cumomers (Wiechert et al. 1999) or bondomers (van Winden et al. 2002), were demonstrated to effectively improve the process. Antoniewicz et al. recently introduced a novel approach to reduce the number of systems variables by at least one order-of-magnitude while preserving a full description of the isotopomers. This decomposition in so-called *elementary metabolite units* dramatically simplifies the equation system and thus accelerates solving, and will most likely constitute a cornerstone for the rapid analysis of non-stationary experiments or of concomitant ^{2}H, ^{13}C, ^{18}O, and ^{15}N labeling in large networks (Antoniewicz et al. 2006). Notably, a detailed statistical analysis is crucial to correctly weight the outcomes (Antoniewicz et al. 2006).

To our knowledge, 13C-FLUX is currently the most complete and freely available software tool that offers rigorous ^{13}C-based balancing for generalized networks from both NMR or MS experiments (Wiechert et al. 2001). Alternatively, NMR2Flux computes fluxes in plants from 2D-NMR spectra of protein-bound amino acids (Sriram et al. 2004). Isotopomer balancing has been used to quantify fluxes for example from amino acids in microorganisms with NMR (Marx et al. 1996; Petersen et al. 2000; Emmerling et al. 2002; van Winden et al. 2003) and MS data (Fischer and Sauer 2003; Klapa et al. 2003), from free metabolites with MS (van Winden et al. 2005; Kleijn et al. 2006), or in plants with NMR of amino acids (Sriram et al. 2004).

4.3.2 Flux ratios

The isotopomer balancing approach outlined in the previous section sets strict requirements in terms of input data (Fig. 2). Initially driven by the need to analyze fluxes also in absence of physiological data, *metabolic flux ratio analysis* was developed to directly decipher ^{13}C labeling patterns (Szyperski 1995). Briefly, metabolic flux ratios quantify the relative fluxes of alternative pathways at the node (metabolite) of convergence. For this purpose, analytical equations are developed first for each branch point of interest. Each analytical equation is designed to take advantage of the labeling features that best discriminates between the theoretical ^{13}C blueprints of converging pathways. In central metabolism, about 10 independent flux ratios can be determined from amino acids for ^{13}C-glucose experiments with bacteria or yeast using either NMR (Szyperski 1995; Maaheimo et al. 2001) or MS data (Christensen et al. 2001; Fischer and Sauer 2003; Blank and Sauer 2004). For the broadly used flux ratios from ^{13}C experiments and GC-MS data, a detailed protocol is given in (Nanchen et al. 2006). Single flux ratios are calculated from the mass distributions of typically only 1-3 intermediates (or inferred from amino acids) and absolutely no kind of measured rate is required. The power of ratios lies in their local nature that renders them less susceptible to possibly erroneous models or measurements, and in the fact that they provide direct evidence for the operation of a particular pathway *in vivo*. In addition, the rapid and almost completely unsupervised computation of flux ratios enables high-throughput - and yet quantitative - flux studies. The major drawback is the initial time invested for development or adaptation of the analytical equations for new tracers or modified metabolic networks. Flux ratios were, for example, used to identify new pathways or unexpected cross-activity (Fischer and Sauer 2003; Zamboni et al. 2004), characterize unknown networks (Fuhrer et al. 2005), demonstrate metabolic robustness and suboptimal operation of *Bacillus* (Fischer and Sauer 2005), and to investigate adaptive evolution of metabolism (Hua et al. 2006).

In the so-called *^{13}C-constrained metabolic flux analysis*, flux ratios can be used to solve the problem of undetermined stoichiometric balances, because they provide additional, independent constraints to reduce the solution space (Fischer et al. 2004). If at least one flux ratio is available to fix each degree of freedom in the metabolic network, a unique flux map can be calculated by means of a linear system or least-square fit for fully and overdetermined systems, respectively. Results from ^{13}C-constrained metabolic flux analysis and isotopomer balancing are consistent (Fischer et al. 2004). Yet, the latter provides more detailed information with respect to the exchange fluxes in bidirectional reactions. These are neglected or implicitly assigned when developing the analytical equations to calculate flux ratios. In knockout mutants with severe growth defects, these tacit assumptions may not hold and lead to wrong ratio estimates and, in turn, erroneous net fluxes from ^{13}C-constrained metabolic flux analysis. Nevertheless, ratios-constrained net flux analyses are a robust tool for both large-scale (Blank et al. 2005) and detailed studies of cellular carbon, redox, and energy metabolism (Zamboni et al. 2003; Blank et al. 2005; Hua et al. 2006). For experiments on glucose minimal medium, software packages for metabolic flux ratio and ^{13}C-constrained metabolic balance analysis are freely available (Zamboni et al. 2005).

A related approach is the so-called *stable isotope based dynamic metabolic profiling* (SIDMAP), that - akin to metabolic flux ratios analysis - interprets ^{13}C-patterns according to a metabolic model without measured extracellular rates. It features a collection of analytical equations that were tailored to monitor specific changes in carbon metabolism of mammalian cells grown on [1,2-$^{13}C_2$]glucose and analyzed by GC-MS of biomass or secreted products. The complex composition of culture medium impairs large-scope fluxome quantitation. Nevertheless, this approach affords a specialized profiling tool to, for example, capture metabolic responses in tumoral cells or to lead targeted drug design (Boren et al. 2001; Boros et al. 2003; Marin et al. 2004).

4.3.3 Fluxome profiling

In analogy to data mining methods applied to other omics data, multivariate analysis can be used to explore large datasets of ^{13}C labeling patterns (Zamboni and Sauer 2005). This approach of fluxome profiling features the unique chance to infer structural and quantitative information from raw labeling data without any a priori knowledge of the biochemical reaction network.

What can be discovered in ^{13}C labeling patterns? A first proof-of-concept study with bacterial cultures and a variegated set of tracers and conditions was presented by our lab (Zamboni and Sauer 2004). The working hypothesis was that the absence or presence of pathway activity is reflected in the label fingerprints of metabolites. By purely unsupervised statistical techniques, this work (i) demonstrated that it is indeed possible to separate the overlapping signatures of independent pathways, (ii) proved that signatures are consistent with biosynthetic routes, (iii) showed that structural knowledge on biosynthesis of metabolites can be deduced from covariating patterns, (iv) showed that mutants can be clustered according to metabolic changes, and (v) mapped the effect of transcriptional regulators on metabolic activity. Current efforts aim at developing robust tools of machine learning and expertise to systematically scavenge all relevant features in large datasets. Albeit in progress, first results reveal that for each dataset the number of stable (not sensitive to algorithm parameters or to *in silico* superimposed noise) pathway signatures is well defined and sometimes exceeds the number of those calculable with the established metabolic flux ratio equations. This suggests that novel, still latent blueprints of metabolic activity are contained in the data in addition to those disclosed by today's metabolic flux ratio analysis.

Beyond the qualifiers obtained, for example, from hierarchical clustering or classification trees, it is obviously desirable to obtain quantitative insights on metabolic fluxes. In fact, quantitative estimators for flux partitioning ratios were successfully derived from unsupervised methods such as independent component analysis (Zamboni and Sauer 2004), but for some flux ratio no matching estimator could be identified. Supervised methods such as regressions or adaptive neural

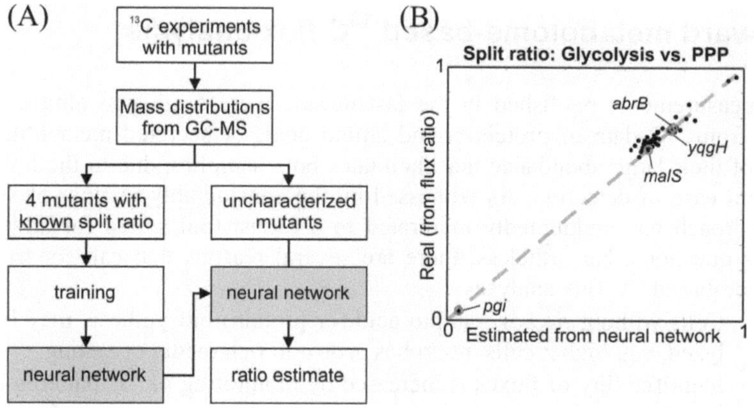

Fig. 3. Quantitative determination of glycolysis-to-PPP split ratio in *Bacillus subtilis* knockout using supervised machine learning and no a priori knowledge of metabolism. (A) Schematic representation of approach. Mutants were grown individually on a mix of 50% [U-^{13}C] and natural glucose. The mass distributions of 4 mutants were used to train an adaptive neural network to estimate the flux ratio. (B) The graph shows the validation of the trained neural network: for each mutant the estimated flux split estimated by the neural network is compared to the real value calculated from a model-based analytical equation. Circles and dots indicate the mutants used for training and validation, respectively. The dashed diagonal indicates perfect predictions.

networks can possibly fill this gap as shown exemplarily in Figure 3, but the general applicability and utility of supervised machine learning with ^{13}C labeling patterns is still questionable and has to be assessed in systematic studies.

Fluxome profiling, based on either supervised or unsupervised procedures, is still in its infancy and hence, in contrast to the well-established approaches of isotopomer balancing and flux ratio analysis, it is only possible to speculate on its practical applications. With this in mind, principally two advantages unique to fluxome profiling call for further development. First, fluxome profiling can handle labeling data from experiments with higher cells because it is compatible with virtually any network (unicellular – multicellular), isotopic tracer (^{13}C, ^{2}H, ^{18}O, ^{15}N, and combinations), and medium composition (Zamboni and Sauer 2004). Second, multivariate statistics afford a very simple basis for comparing different omics data. For example, if it is true that metabolic fluxes reflect the integration of all interactions between and within metabolites, proteins, RNA, etc., it can be expected that statistical correlations and anticorrelations between metabolic fluxes and concentration of species in the different layers will contribute to identify the loci were control is exerted and the mechanism how regulation occurs (Weckwerth et al. 2004; Morgenthal et al. 2006).

5 Toward metabolome-based ^{13}C flux analysis

Flux measurements published in the last decade were originated almost exclusively from ^{13}C data of protein-bound amino acids or secreted metabolites, because of their large abundance that facilitates both sampling due to the low turnover and ease of detection. As witnessed by the considerably number of studies, this approach has undoubtedly maturated to a robust tool suited for addressing various questions. Nevertheless, there are several reasons that call for true metabolome-based ^{13}C flux analyses:

- Cells without *de novo* amino acid (or protein) biosynthesis may be analyzed, e.g. higher cells, microbes grown in rich media or resting.
- Identifiability of fluxes is increased by monitoring of ^{13}C patterns in metabolites that are not precursors of proteinogenic amino acids. In addition, the risk of erroneous or ambiguous mapping of atoms between precursors and metabolic end products is circumvented.
- Labeling experiments are shortened because isotopic steady state is attained earlier. This leads to lower costs and enables the analysis of systems that cannot be kept long in metabolic steady state.
- Slow metabolic shifts (in the range of minutes to hours) become observable, as long as a metabolic steady state can be approximated throughout onset of the isotopic steady state in intracellular metabolites.

The full potential of metabolome-based ^{13}C flux analysis to tackle such conditions and questions can be unleashed only with direct measurements of intermediates in proximity of the pathway of interest.

5.1 Experimental proof-of-concept

Two landmark studies of cellular fluxes based on ^{13}C-patterns of primary metabolites have been published so far, both by van Winden and coworkers (van Winden et al. 2005; Kleijn et al. 2006). In the first one, baker's yeast was grown in glucose-limited continuous cultures, and at metabolic steady state the culture was fed with 100% [1-^{13}C]glucose. After 40 and 60 min of labeling, two cell aliquots were harvested rapidly, quenched, and central carbon metabolites were extracted and measured by liquid chromatography (LC)-MS. Isotopomer balancing (cf. 4.3.2) was successfully used to fit fluxes in glycolysis and PPP to the ^{13}C labeling pattern of ten intermediates. This study demonstrates the feasibility of metabolome-based flux analyses, and contributes further relevant observations. First, comparison of labeling pattern at the two time points of sampling confirm that already after 40 min the majority of metabolites is in isotopic steady state. Exceptions are discussed below. Second, the direct comparison of labeling patterns in reactants at both sides of every bidirectional reactions indicates which metabolite pools are equilibrated, and thus, which reversible enzymes operate in forward and backward direction at rates that are much higher than the apparent net metabolic flux, that is the difference of the two. Third, turnover of the storage carbohydrate glycogen was found to interfere with rapid onset of isotopic steady state in glucose-1-phosphate and glucose-6-phosphate, so that after 60 min isotopic steady state is not yet achieved. When a turnover reaction between glycogen and glucose-1-phosphate is introduced into the model, the result is a worse confidence interval for the flux split between glycolysis and the PPP. This is caused by the fact that both [1-^{13}C] label loss in the oxidative PPP and variable inflow of unlabeled hexose-phosphates from glycogen produce hardly distinguishable increases in unlabeled fractions of intermediate. Two solutions can obviate to the problem of large pools disturb onset of isotopic steady state. As anticipated by the authors in the above study, one option is to label for a longer period of time. The drawback is that extensive time is probably necessary to obtain isotopic equilibration of the large glycogen pool. Alternatively, differently labeled substrates can be adopted to experimentally assess the exchange of large reservoirs. For the aforementioned example, [1,2-$^{13}C_2$] or [U-^{13}C]glucose would have served to estimate more precisely the glycolysis-to-PPP split in the same span of time, because they enable concomitant quantitation of the collateral turnover of glycogen.

Indeed, the second and more recent metabolome-based ^{13}C flux study by the same lab affords determination of the flux split between oxidative PPP and glycolysis in filamentous fungi by an analytical equation that calculates the flux ratio from the isotopic mass distribution of tree intermediates close to the node (Kleijn et al. 2006). This study shows that the results obtained analytically are consistent with isotopomer balancing but more accurate, and demonstrates for the first time the potential of metabolome-based ^{13}C flux ratio analysis (cf. 4.3.2).

5.2 Analytics: lessons from metabolomics

The trivial analogy between metabolomics and metabolome-based ^{13}C flux analysis in terms of analytes is reflected by the similar experimental workflow in the steps from cells harvest to analysis. Hence, current best practices for accurate flux studies include the use of rapid sampling devices, immediate quenching of metabolism, tailored chromatographic separation to possibly reduce matrix effects, and highly-sensitive detection. MS is actually preferred to NMR in the detection of ^{13}C labeling in free metabolites due to the higher sensitivity. In addition, chromatographic separation becomes compulsory to capture the ^{13}C distributions of structurally similar metabolites as it often occur in the same pathway, for which MS is prioritized because on-line interfacing to GC, LC, or capillary electrophoresis (CE) is well established.

The topic of analytical separation introduces a relevant question: which of the MS-compatible platforms frequently used in metabolomics (i.e. GC, LC, and CE) is the most suited for metabolome-based ^{13}C metabolic flux analysis? For the specifics of the intermediates of interest, i.e. phosphorylated sugars and carboxylic acids in glycolysis, PPP, and TCA cycle, all three modes can be used for separation and subsequent MS detection. Here I survey these separation techniques, while the specifics of MS detection are addressed in the following sections.

For GC-MS acquisition, volatile derivatives of polar compounds are obtained after methoxymation and silylation and separated with simple protocols amenable to high-throughputs (Strelkov et al. 2004; Koek et al. 2006). The strength of this method is that it is generally suited to detect other classes of compounds such as alcohols, amines, amino acids, or purines. Although it suffers from derivatization efficiencies varying for the different classes (Koek et al. 2006), this does not affect the measurement of isotopic distributions because they do not depend on absolute concentrations. To increase the amount of sample introduced onto the column, temperature programmable injectors can be used to inject up to 1000x larger volumes. Notably, the benefits are marginal when low and highly concentrated analytes elute closely or overlapping, because overloading of the more abundant compound causes peak broadening and often detector saturation. Due to the extensive fragmentation that is normally caused by electron impact ionization, GC-MS spectra are very complex and identification of analytes relies on spectral databases of compound libraries (Schauer et al. 2005).

Analysis by LC-MS is slightly complicated by the ionic and polar character of central carbon metabolites because of the poor compatibility between MS ionization and the LC buffers commonly used for separating such anionic and hydrophilic compounds. Electrospray ionization is enhanced by solvents with high organic phase and low salt content, whereas chromatographic elution is controlled by concentrated sodium hydroxide gradients in water (van Dam et al. 2002). Interfacing to MS is then only possible with electrochemical exchangers of sodium cations-protons that are inserted in the liquid path between column and sprayer but comes at the cost of sensitivity and chromatographic resolution. Retention of ionic analytes in reverse phase LC can be mediated by hydrophobic ion pairing reagents (Huck et al. 2003). Although volatile counter-ions that are compatible with electrospray process can be used, particular care must be dedicated in instrument maintenance to loss of sensitivity and signal deterioration. A even more MS-friendly alternative is hydrophilic liquid interaction chromatography (HILIC), which exhibits improved separations of ions in high organic phases and is available in nanoscale systems, where maximum sensitivity is attained (Alpert et al. 1994; Tolstikov and Fiehn 2002; Bajad et al. 2006). In general, sensitivity in nano-LC can be further increased with preconcentration by loading large sample volumes to a short enrichment column that fully retains the analytes in a thin section. When the solvent gradient is started, a focused and highly concentrated analyte plug elutes from the enrichment column, and is separated on the analytical column. Unfortunately, the injection volumes of central carbon metabolites is still limited when their retention on commercially available phases is not sufficient to load large sample volumes without having a fraction already eluting from the enrichment column, e.g. with most HILIC material. In comparison to GC, the longer equilibration time and chromatographic separations of organic gradients reduce sample throughputs. In contrast, the milder ionization in LC-MS enables the detection of intact molecules, which produce less populated spectra and facilitates identification.

Among the three platforms, CE-MS features unsurpassed peak capacity, concomitant separation of anions and cations, and resolution of most isomers present in central metabolism within short runs (Soga et al. 2003; Harada et al. 2006). CE as well offers the possibility to focus the analytes in large volumes by sandwiched injection techniques (Britz-McKibbin and Terabe 2003). The drawback of CE-MS measurement lies in the expertise and time necessary to obtain reproducible measurements at high-throughputs. In addition, the narrow eluting peaks limit the number of different fragmentation cycles that can be performed over a peak.

Overall, all three systems provide access to key intermediates in central metabolism and can cope with large injection volumes which are used to enhance sensitivity. To date, GC-MS and LC-MS are the preferred platform to detect labeling patterns in amino acids and central carbon metabolism, respectively. CE-MS is superior in sensitivity and enables detection of both compound classes. Nevertheless, these advantages are apparently not yet sufficient to replace GC and LC.

5.3 Current developments

To fulfill the goals of metabolome-based ^{13}C flux analysis (cf. 5), further improvements are necessary. In the following sections I address three topics that are targets of current research. The first two are of experimental nature and aim at obtaining possibly detailed and accurate labeling information from free metabolites. Both aspects are pivotal in the quest of comprehensive flux analysis for cells grown in complex media. The third topic is the extension of metabolic flux analysis to cope with the frequently occurring isotopically non-stationary systems, which will promote metabolome-based flux analyses to a universally applicable tool.

5.3.1 How to measure precise isotopic mass distributions?

The analogies between fluxome and metabolome measurements stop upon subjecting metabolites to mass spectrometry, because measuring precise mass distributions differs from measuring concentrations, and MS instruments have to be set up accordingly. In quantitative concentrations measurements, MS/MS acquisitions are the mode of choice for best signal-to-noise and high scanning rate are employed to obtain more data points on a peak and reduce interpolation errors. In contrast, detection of isotopic mass distributions such as needed for ^{13}C flux analysis is generally done with full range MS acquisitions, because for each metabolite/fragment a range of 10-15 m/z has to be scanned (or fragmented) due to the overlapping presence of naturally occurring isotopes. In complex samples, where chromatographic coelution is frequent, or with in-source fragmentation (e.g. electron impact ionization in GC-MS), selected ion monitoring looses attractiveness because at least 50-100 m/z bins have to be scanned simultaneously and complicate acquisition programs must be prepared to ensure that the correct mass range is monitored at the elution time of each analyte.

As a rule of thumb, isotopic fractions of 1 mol% (better if lower) compared to the monoisotopic mass should be precisely quantifiable to obtain fluxes with good confidence. Hence, the limits of quantitation (LOQ) for mass distributions are at least 2 orders-of-magnitude higher than the LOQ for metabolite concentrations. Because of poor ion statistics, low abundant fractions are more prone to inaccuracy. Another consequence is that MS detectors must exhibit a wide linear dynamic range of >4 decades to effectively measure distributions in real samples where analytes are heterogeneously concentrated. If that is not the case (e.g. as in most ion traps) multiple injections of different amounts are necessary to characterize low and highly abundant species.

Low mass resolution is also detrimental for exact isotopic distributions, in particular when quadrupoles or ion traps are used for detection. High-resolving time-of-flight or Fourier Transform instruments are not affected. Resolution has to be increased to ensure that no overlap or crosstalk between neighbor m/z bins occur, also after slight calibration drifts. Unfortunately resolution comes always at the cost of sensitivity, but this drawback can be partly alleviated with slower scan speeds. As mentioned above, this is in conflict with the ideal settings for quantitative concentration measurements because less data points open for peak interpolation errors. In synthesis, detection of exact mass distributions depends on possibly high ion counts in full-range MS mode, good mass resolution, and an outstanding linear dynamic range. Due to the interdependency of these properties and the generally low abundance of free metabolites, sensitivity rapidly emerges as the major bottleneck in fluxome measurements.

5.3.2 Fragmentation: the key to obtain the labeling of single atoms

In metabolomics, fragmentation is extensively utilized for identification and selective detection. In fluxomics, fragmentation provides labeling imprints at sub-molecular level and eventually positional enrichment, i.e. the abundance of label at single atom positions (Fig. 4). Flux identifiability is subordinated to the ^{13}C patterns that are measurable and hence, in turn, to the fragments that can be generated. Novel fragments can enable more detailed analyses and more flexibility in the choice of the tracer. Also when equivalent fragments of the same metabolite are measured (e.g. those with the same carbon backbone), they lead to higher confidence in the flux estimation. As a general rule, it is thus desirable to obtain and detect the largest number of fragments possible.

Routine utilization of fragment data is, however, hindered by the overlap of three technical issues: (1) Fragmentation is inducible and happens when molecules collide at high energy with gas molecules or electrons, or when they are subject to strong electric fields. However, for each ion the break points can only be minimally controlled by the instrument settings. Increasing or decreasing of (collision) energy favor formation of low and high molecular weight daughter ions, respectively. It is, however, not possible to break every C-C bond at will, and some atoms are virtually not distinguishable (e.g. C_1 and C_2, or C_5 and C_6 in Fig. 4).

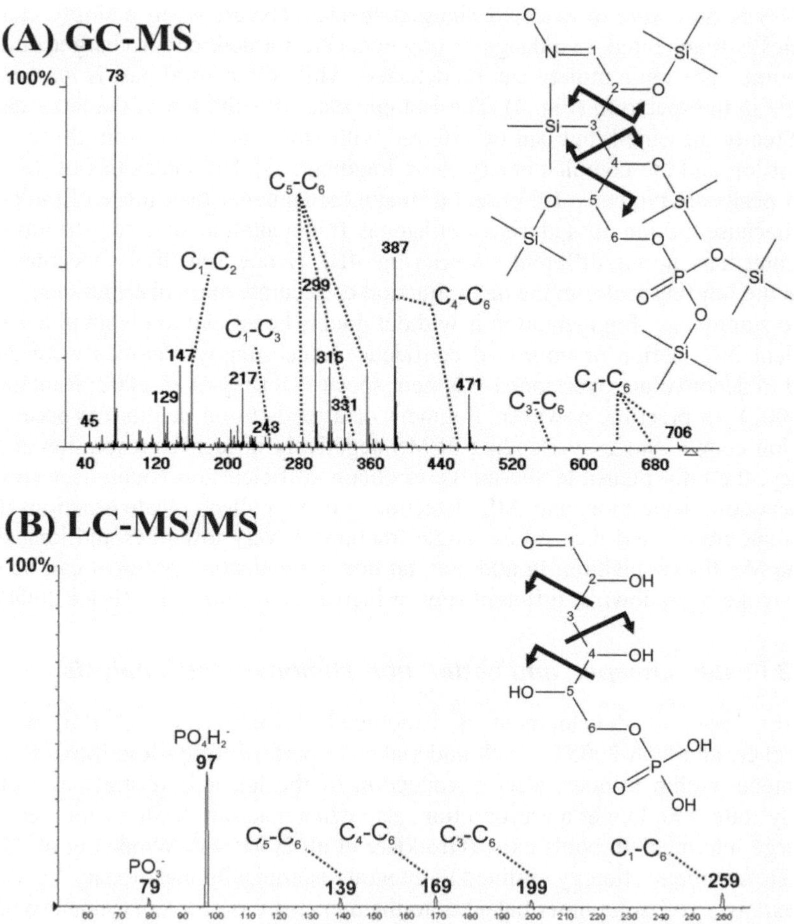

Fig. 4. MS spectra of fragmented glucose-6-phosphate. On each pane, the intact parent molecule is drawn with numbered carbon atoms. Directly measurable fragments are indicated by thick arrows. (A) Spectrum resulting by in-source fragmentation in GC-MS. The analyte was first methoximated and sylilated to obtain a volatile derivative. Many fragments are observable, but their intensity is too low to quantify isotopic distributions (e.g. C_3-C_6). (B) Spectrum provoked by collisional fragmentation in a LC-MS/MS experiment. 100% intensity corresponds to that of the parent ion (m/z -259) in absence of collisions. Sugar-phosphates are prone to break at the phosphoester bond, so that the carbon-containing fragments are underrepresented versus the non-informative phosphate ions (m/z -97 and -79). Since the charge is located on the phosphate group, only one daughter ion is observed when the carbon backbone is broken. Nevertheless, the mass distribution of the neutral complement can be calculated from that of the intact molecule. Hence, GC-MS and LC-MS/MS provide qualitatively equivalent information, and the true limitation is set by ion counts. Considering that for each fragment several m/z have to be measured, LC-MS/MS might be preferred here because no overlaps between fragments or unknown peaks occur. A decision must account for the expected mass shifts caused by ^{13}C enrichment.

(2) MS is only able to detected charged species. Hence, when a singly charged species is fragmented, two daughter fragments are formed: one is charged and one is neutral. The ionic moiety can be detected, while the neutral part is lost and invisible in the spectrum (Fig. 4). The isotopic mass distribution of the latter cannot be directly measured, but can be inferred with worse precision from those of the parent ion and the complementary ionic fragment. (3) The intensities of the fragment peaks are typically 1-2 orders of magnitude smaller than those of the parent ion because of ion loss during collisional fragmentation and redistribution of daughter ions among different masses (Fig. 4B). Hence, sensitivity becomes once more the limiting factor in the determination of accurate mass distributions.

To summarize, fragmentation is without doubts beneficial to obtain either independent information or improved confidence. Accordingly, theories were developed to deconvolute overlapped fragment spectra (Jeffrey et al. 2002; Rantanen et al. 2002). In practice, however, fragment data tends to be qualitative because of low ion counts. Since overloading of MS negatively influences resolution and accuracy, the only plausible alternative to obtain sufficient ion counts is seemingly to decouple separation and MS detection, i.e. to collect eluate fractions from chromatography and then infuse single fractions at very low rates and long times to the MS for acquisition. In addition, ad-hoc derivatization protocol can be used to provoke breakdown at different sites or increase the abundance (Price 2004).

5.3.3 Faster, cheaper, and better: non-stationary flux analysis

Another area of development is isotopically instationary ^{13}C flux analysis (Wiechert and Nöh 2005), which undertakes to perform fully-descriptive flux experiments within minutes after introduction of the labeled substrate as isotopic steady state is no longer a precondition, also when macromolecules turnover occur or large intermediate pools exist (Grotkjaer et al. 2004; van Winden et al. 2005). The so far unique strategy outlined to integrate isotopically instationary ^{13}C data is the extension of isotopomer balances to the dynamic case by replacement with ordinary differential equations. For this purpose, metabolite pool sizes are also newly introduced in the equations and fitted in an iterative procedure.

Time profiles of ^{13}C-patterns must be measured upon start of labeling to monitor the label propagation through the network. Conjoint measurement of metabolite concentrations is not strictly required. Omission, however, causes an increase in degrees of freedom, complicates the fitting procedure, and results in worse confidence intervals. Ideally, as many pool sizes as possible should be measured, and missing data can only be compensated by multiple labeling experiments (Nöh and Wiechert 2006). Notably, due to the metabolic steady state of the culture, the pool sizes are constant while the labeling pattern is still instationary. Thus, a single measurement fully describes concentrations throughout labeling. Solving the resulting highly non-linear system with thousands of ordinary differential equations is the most challenging and time-consuming step, although it can be speculated that implementation of elementary metabolite units decomposition would boost the calculation by a few orders of magnitude (Antoniewicz et al. 2006). Simulations done by Wiechert and coworkers demonstrate that the flux calculability is tightly connected to sampling time points, total labeling duration, and tracer choice. Optimal and detailed a priori design of experiments is therefore mandatory (Nöh and Wiechert 2006).

6 Conclusions

Metabolome-based ^{13}C metabolic flux analysis is on the track to become a universal tool to quantify metabolic activity in large networks, higher cells, and complex environments. Measuring metabolic fluxes under such conditions is a challenging task that demands conjoint experimental, analytical, and mathematical skills. Know-how on aspects such as experimental design, execution, and data integration can be transferred from existing ^{13}C metabolic flux methods developed for microbes, where expertise and computation tools were established over the last decade. Nevertheless, further technical improvements are still necessary in the domains of (i) analytics to increase sensitivity of MS detection, and (ii) mathematical algorithms to efficiently cope with isotopically non-stationary ^{13}C flux experiments.

These accomplishments will eventually enable to comprehensively estimate fluxes, help unravel the underlying control mechanisms that govern metabolic fluxes, discriminate genetic mutations, assess the effect of drugs and diet on metabolism, or monitor the metabolic response in health and disease in virtually any biochemical reaction network where intermediates are accessible.

Acknowledgements

I would like to thank Uwe Sauer and Matthias Heinemann for their critical comments on the manuscript.

References

Alpert AJ, Shukla M, Shukla AK, Zieske LR, Yuen SW, Ferguson MA, Mehlert A, Pauly M, Orlando R (1994) Hydrophilic-interaction chromatography of complex carbohydrates. J Chromatogr A 676:191-222

Antoniewicz MR, Kelleher JK, Stephanopoulos G (2006) Determination of confidence intervals of metabolic fluxes estimated from stable isotope measurements. Metab Eng 8:324-337

Antoniewicz MR, Kelleher JK, Stephanopoulos G (2007) Elementary metabolite units (EMU): A novel framework for modeling isotopic distributions. Metab Eng 9:68-86 (Epub: 2006 - Metab Eng: 2007)

Bajad SU, Lu W, Kimball EH, Yuan J, Peterson C, Rabinowitz JD (2006) Separation and quantitation of water soluble cellular metabolites by hydrophilic interaction chromatography-tandem mass spectrometry. J Chromatogr A 1125:76-88

Bequette BJ, Sunny NE, El-Kadi SW, Owens SL (2006) Application of stable isotopes and mass isotopomer distribution analysis to the study of intermediary metabolism of nutrients. J Anim Sci 84 Suppl:E50-59

Blank LM, Küpfer L, Sauer U (2005) Large-scale ^{13}C-flux analysis reveals mechanistic principles of metabolic network robustness to null mutations in yeast. Genome Biol 6:R49

Blank LM, Lehmbeck F, Sauer U (2005) Metabolic-flux and network analysis in fourteen hemiascomycetous yeasts. FEMS Yeast Res 5:545-558

Blank LM, Sauer U (2004) TCA cycle activity in *Saccharomyces cerevisiae* is a function of the environmentally determined specific growth and glucose uptake rates. Microbiology 150:1085-1093

Boren J, Cascante M, Marin S, Comin-Anduix B, Centelles JJ, Lim S, Bassilian S, Ahmed S, Lee WN, Boros LG (2001) Gleevec (STI571) influences metabolic enzyme activities and glucose carbon flow toward nucleic acid and fatty acid synthesis in myeloid tumor cells. J Biol Chem 276:37747-37753

Boros LG, Brackett DJ, Harrigan GG (2003) Metabolic biomarker and kinase drug target discovery in cancer using stable isotope-based dynamic metabolic profiling (SIDMAP). Curr Cancer Drug Targets 3:445-453

Britz-McKibbin P, Terabe S (2003) On-line preconcentration strategies for trace analysis of metabolites by capillary electrophoresis. J Chromatogr A 1000:917-934

Bulyk ML (2006) DNA microarray technologies for measuring protein-DNA interactions. Curr Opin Biotechnol 17:422-430

Cannizzaro C, Christensen B, Nielsen J, von Stockar U (2004) Metabolic network analysis on *Phaffia rhodozyma* yeast using ^{13}C-labeled glucose and gas chromatography-mass spectrometry. Metab Eng 6:340-351

Christensen B, Christiansen T, Gombert AK, Thykaer J, Nielsen J (2001) Simple and robust method for estimation of the split between the oxidative pentose phosphate pathway and the Embden-Meyerhof-Parnas pathway in microorganisms. Biotechnol Bioeng 74:517-523

Christensen B, Nielsen J (1999) Isotopomer analysis using GC-MS. Metab Eng 1:282-290

Cusick ME, Klitgord N, Vidal M, Hill DE (2005) Interactome: gateway into systems biology. Hum Mol Genet 14 Spec No. 2:R171-181

Dauner M, Bailey JE, Sauer U (2001) Metabolic flux analysis with a comprehensive isotopomer model in *Bacillus subtilis*. Biotechnol Bioeng 76:144-156

Dauner M, Sauer U (2000) GC-MS analysis of amino acids rapidly provides rich information for isotopomer balancing. Biotechnol Prog 16:642-649

Dauner M, Sonderegger M, Hochuli M, Szyperski T, Wüthrich K, Hohmann H-P, Sauer U, Bailey JE (2002) Intracellular carbon fluxes in riboflavin-producing *Bacillus subtilis* during growth on two-carbon substrate mixtures. Appl Environ Microbiol 68:1760-1771

Dauner M, Storni T, Sauer U (2001) *Bacillus subtilis* metabolism and energetics in carbon-limited and excess-carbon chemostat culture. J Bacteriol 183:7308-7317

Emmerling M, Dauner M, Ponti A, Fiaux J, Hochuli M, Szyperski T, Wüthrich K, Bailey JE, Sauer U (2002) Metabolic flux responses to pyruvate kinase knockout in *Escherichia coli*. J Bacteriol 184:152-164

Fischer E, Sauer U (2003) Metabolic flux profiling of *Escherichia coli* mutants in central carbon metabolism using GC-MS. Eur J Biochem 270:880-891

Fischer E, Sauer U (2003) A novel metabolic cycle catalyzes glucose oxidation and anaplerosis in hungry *Escherichia coli*. J Biol Chem 278:46446-46451

Fischer E, Sauer U (2005) Large-scale *in vivo* flux analysis shows rigidity and suboptimal performance of *Bacillus subtilis* metabolism. Nat Genet 37:636-640

Fischer E, Zamboni N, Sauer U (2004) High-throughput metabolic flux analysis based on gas chromatography-mass spectrometry derived ^{13}C constraints. Anal Biochem 325:308-316

Forbes NS, Meadows AL, Clark DS, Blanch HW (2006) Estradiol stimulates the biosynthetic pathways of breast cancer cells: detection by metabolic flux analysis. Metab Eng 8:639-652

Fuhrer T, Fischer E, Sauer U (2005) Experimental identification and quantification of glucose metabolism in seven bacterial species. J Bacteriol 187:1581-1590

Grotkjaer T, Akesson M, Christensen B, Gombert AK, Nielsen J (2004) Impact of transamination reactions and protein turnover on labeling dynamics in ^{13}C-labeling experiments. Biotechnol Bioeng 86:209-216

Harada K, Fukusaki E, Kobayashi A (2006) Pressure-assisted capillary electrophoresis mass spectrometry using combination of polarity reversion and electroosmotic flow for metabolomics anion analysis. J Biosci Bioeng 101:403-409

Hellerstein MK (2003) *In vivo* measurement of fluxes through metabolic pathways: the missing link in functional genomics and pharmaceutical research. Annu Rev Nutr 23:379-402

Hoglund A, Kohlbacher O (2004) From sequence to structure and back again: approaches for predicting protein-DNA binding. Proteome Sci 2:3

Hoon Yang T, Wittmann C, Heinzle E (2006) Respirometric ^{13}C flux analysis--Part II: *in vivo* flux estimation of lysine-producing *Corynebacterium glutamicum*. Metab Eng 8:432-446

Hua Q, Joyce AR, Fong SS, Palsson BO (2006) Metabolic analysis of adaptive evolution for *in silico* designed lactate-producing strains. Biotechnol Bioeng in press

Huck JH, Struys EA, Verhoeven NM, Jakobs C, van der Knaap MS (2003) Profiling of pentose phosphate pathway intermediates in blood spots by tandem mass spectrometry: application to transaldolase deficiency. Clin Chem 49:1375-1380

Isermann N, Wiechert W (2003) Metabolic isotopomer labeling systems. Part II: structural flux identifiability analysis. Math Biosci 183:175-214

Jeffrey FM, Roach JS, Storey CJ, Sherry AD, Malloy CR (2002) ^{13}C isotopomer analysis of glutamate by tandem mass spectrometry. Anal Biochem 300:192-205

Kauffman KJ, Prakash P, Edwards JS (2003) Advances in flux balance analysis. Curr Opin Biotechnol 14:491-496

Klamt S, Schuster S (2002) Calculating as many fluxes as possible in underdetermined metabolic networks. Mol Biol Rep 29:243-248

Klapa MI, Aon JC, Stephanopoulos G (2003) Systematic quantification of complex metabolic flux networks using stable isotopes and mass spectrometry. Eur J Biochem 270:3525-3542

Klapa MI, Park SM, Sinskey AJ, Stephanopoulos G (1999) Metabolite and isotopomer balancing in the analysis of metabolic cycles: I. Theory. Biotechnol Bioeng 62:375-391

Kleijn RJ, van Winden WA, Ras C, van Gulik WM, Schipper D, Heijnen JJ (2006) ^{13}C-labeled gluconate tracing as a direct and accurate method for determining the pentose phosphate pathway split ratio in *Penicillium chrysogenum*. Appl Environ Microbiol 72:4743-4754

Kleijn RJ, van Winden WA, van Gulik WM, Heijnen JJ (2005) Revisiting the ^{13}C-label distribution of the non-oxidative branch of the pentose phosphate pathway based upon kinetic and genetic evidence. FEBS J 272:4970-4982

Koek MM, Muilwijk B, van der Werf MJ, Hankemeier T (2006) Microbial metabolomics with gas chromatography/mass spectrometry. Anal Chem 78:1272-1281

Koffas M, Stephanopoulos G (2005) Strain improvement by metabolic engineering: lysine production as a case study for systems biology. Curr Opin Biotechnol 16:361-366

Krömer JO, Sorgenfrei O, Klopprogge K, Heinzle E, Wittmann C (2004) In-depth profiling of lysine-producing *Corynebacterium glutamicum* by combined analysis of the transcriptome, metabolome, and fluxome. J Bacteriol 186:1769-1784

Kümmel A, Panke S, Heinemann M (2006) Putative regulatory sites unraveled by network-embedded thermodynamic analysis of metabolome data. Mol Syst Biol 2:2006 0034

Küpfer L, Schütz R, Sauer U (2007) Predicting *in vivo* fluxes in *Escherichia coli* by constraint-based modeling. submitted

Maaheimo H, Fiaux J, Cakar ZP, Bailey JE, Sauer U, Szyperski T (2001) Central carbon metabolism of *Saccharomyces cerevisiae* explored by biosynthetic fractional ^{13}C labeling of common amino acids. Eur J Biochem 268:2464-2479

Marin S, Lee WN, Bassilian S, Lim S, Boros LG, Centelles JJ, FernAndez-Novell JM, Guinovart JJ, Cascante M (2004) Dynamic profiling of the glucose metabolic network in fasted rat hepatocytes using [1,2-$^{13}C_2$]glucose. Biochem J 381:287-294

Marx A, de Graaf AA, Wiechert W, Eggeling L, Sahm H (1996) Determination of the fluxes in the central metabolism of *Corynebacterium glutamicum* by nuclear magnetic resonance spectroscopy combined with metabolite balancing. Biotech Bioeng 49:111-129

McCabe BJ, Previs SF (2004) Using isotope tracers to study metabolism: application in mouse models. Metab Eng 6:25-35

Möllney M, Wiechert W, Kownatzki D, de Graaf AA (1999) Bidirectional reaction steps in metabolic networks: IV. Optimal design of isotopomer labeling experiments. Biotechnol Bioeng 66:86-103

Morgenthal K, Weckwerth W, Steuer R (2006) Metabolomic networks in plants: Transitions from pattern recognition to biological interpretation. Biosystems 83:108-117

Nanchen A, Fuhrer T, Sauer U (2006) Determination of metabolic flux ratios from ^{13}C-experiments and GC-MS data: protocols and principles. In: Weckwerth W (ed) Metabolomics. Humana Press

Nöh K, Wiechert W (2006) Experimental design principles for isotopically instationary ^{13}C labeling experiments. Biotechnol Bioeng 94:234-251

Petersen S, de Graaf AA, Eggeling L, Möllney M, Wiechert W, Sahm H (2000) *In vivo* quantification of parallel and bidirectional fluxes in the anaplerosis of *Corynebacterium glutamicum*. J Biol Chem 275:35932-35941

Price NP (2004) Acylic sugar derivatives for GC/MS analysis of ^{13}C-enrichment during carbohydrate metabolism. Anal Chem 76:6566-6574

Rantanen A, Mielikainen T, Rousu J, Maaheimo H, Ukkonen E (2006) Planning optimal measurements of isotopomer distributions for estimation of metabolic fluxes. Bioinformatics 22:1198-1206

Rantanen A, Rousu J, Kokkonen JT, Tarkiainen V, Ketola RA (2002) Computing positional isotopomer distributions from tandem mass spectrometric data. Metab Eng 4:285-294

Sauer U (2004) High-throughput phenomics: experimental methods for mapping fluxomes. Curr Opin Biotechnol 15:58-63

Sauer U, Canonaco F, Heri S, Perrenoud A, Fischer E (2004) The soluble and membrane-bound transhydrogenases UdhA and PntAB have divergent functions in NADPH metabolism of *Escherichia coli*. J Biol Chem 279:6613-6619

Schauer N, Steinhauser D, Strelkov S, Schomburg D, Allison G, Moritz T, Lundgren K, Roessner-Tunali U, Forbes MG, Willmitzer L, Fernie AR, Kopka J (2005) GC-MS libraries for the rapid identification of metabolites in complex biological samples. FEBS Lett 579:1332-1337

Schmidt K, Carlsen M, Nielsen J, Villadsen J (1997) Modeling isotopomer distributions in biochemical networks using isotopomer mapping matrices. Biotechnol Bioeng 55:831-840

Schwender J, Ohlrogge JB, Shachar-Hill Y (2003) A flux model of glycolysis and the oxidative pentosephosphate pathway in developing *Brassica napus* embryos. J Biol Chem 278:29442-29453

Segre D, Vitkup D, Church GM (2002) Analysis of optimality in natural and perturbed metabolic networks. Proc Natl Acad Sci USA 99:15112-15117

Shimizu K (2004) Metabolic flux analysis based on ^{13}C-labeling experiments and integration of the information with gene and protein expression patterns. Adv Biochem Eng Biotechnol 91:1-49

Soga T, Ohashi Y, Ueno Y, Naraoka H, Tomita M, Nishioka T (2003) Quantitative metabolome analysis using capillary electrophoresis mass spectrometry. J Proteome Res 2:488-494

Sriram G, Fulton DB, Iyer VV, Peterson JM, Zhou R, Westgate ME, Spalding MH, Shanks JV (2004) Quantification of compartmented metabolic fluxes in developing soybean embryos by employing biosynthetically directed fractional ^{13}C labeling, two-dimensional [^{13}C, ^{1}H] nuclear magnetic resonance, and comprehensive isotopomer balancing. Plant Physiol 136:3043-3057

Stephanopoulos G (1999) Metabolic fluxes and metabolic engineering. Metab Eng 1:1-11

Strelkov S, von Elstermann M, Schomburg D (2004) Comprehensive analysis of metabolites in *Corynebacterium glutamicum* by gas chromatography/mass spectrometry. Biol Chem 385:853-861

Szyperski T (1995) Biosynthetically directed fractional ^{13}C-labeling of proteinogenic amino acids. An efficient analytical tool to investigate intermediary metabolism. Eur J Biochem 232:433-448

Tolstikov VV, Fiehn O (2002) Analysis of highly polar compounds of plant origin: combination of hydrophilic interaction chromatography and electrospray ion trap mass spectrometry. Anal Biochem 301:298-307

Turner SM, Hellerstein MK (2005) Emerging applications of kinetic biomarkers in preclinical and clinical drug development. Curr Opin Drug Discov Devel 8:115-126

van Dam JC, Eman MR, Frank J, Lange HC, van Dedem GW, Heijnen JJ (2002) Analysis of glycolytic intermediates in *Saccharomyces cerevisiae* using anion exchange chromatography and electrospray ionization with tandem mass spectrometric detection. Anal Chim Acta 460:209-218

van Winden WA, Heijnen JJ, Verheijen PJ (2002) Cumulative bondomers: a new concept in flux analysis from 2D [^{13}C,^{1}H] COSY NMR data. Biotechnol Bioeng 80:731-745

van Winden WA, Heijnen JJ, Verheijen PJ, Grievink J (2001) A priori analysis of metabolic flux identifiability from ^{13}C-labeling data. Biotechnol Bioeng 74:505-516 van Winden WA, van Dam JC, Ras C, Kleijn RJ, Vinke JL, van Gulik WM, Heijnen JJ (2005) Metabolic-flux analysis of Saccharomyces cerevisiae CEN.PK113-7D based on mass isotopomer measurements of ^{13}C-labeled primary metabolites. FEMS Yeast Res 5:559-568

van Winden WA, van Gulik WM, Schipper D, Verheijen PJ, Krabben P, Vinke JL, Heijnen JJ (2003) Metabolic flux and metabolic network analysis of *Penicillium chrysogenum* using 2D [^{13}C, ^{1}H] COSY NMR measurements and cumulative bondomer simulation. Biotechnol Bioeng 83:75-92

Varma A, Palsson BO (1994) Stoichiometric flux balance models quantitatively predict growth and metabolic by-product secretion in wild-type *Escherichia coli* W3110. Appl Environ Microbiol 60:3724-3731

Vaseghi S, Baumeister A, Rizzi M, Reuss M (1999) *In vivo* dynamics of the pentose phosphate pathway in *Saccharomyces cerevisiae*. Metab Eng 1:128-140

Weckwerth W, Loureiro ME, Wenzel K, Fiehn O (2004) Differential metabolic networks unravel the effects of silent plant phenotypes. Proc Natl Acad Sci USA 101:7809-7814

Wiechert W, Möllney M, Isermann N, Wurzel M, de Graaf AA (1999) Bidirectional reaction steps in metabolic networks: III. Explicit solution and analysis of isotopomer labeling systems. Biotechnol Bioeng 66:69-85

Wiechert W, Möllney M, Petersen S, de Graaf AA (2001) A universal framework for ^{13}C metabolic flux analysis. Metab Eng 3:265-283

Wiechert W, Nöh K (2005) From stationary to instationary metabolic flux analysis. Adv Biochem Eng Biotechnol 92:145-172

Wittmann C, Heinzle E (2002) Genealogy profiling through strain improvement by using metabolic network analysis: metabolic flux genealogy of several generations of lysine-producing corynebacteria. Appl Environ Microbiol 68:5843-5859

Wu L, van Winden WA, van Gulik WM, Heijnen JJ (2005) Application of metabolome data in functional genomics: a conceptual strategy. Metab Eng 7:302-310

Zamboni N, Fischer E, Muffler A, Wyss M, Hohmann HP, Sauer U (2005) Transient expression and flux changes during a shift from high to low riboflavin production in continuous cultures of *Bacillus subtilis*. Biotechnol Bioeng 89:219-232

Zamboni N, Fischer E, Sauer U (2005) FiatFlux--a software for metabolic flux analysis from ^{13}C-glucose experiments. BMC Bioinformatics 6:209

Zamboni N, Maaheimo H, Szyperski T, Hohmann HP, Sauer U (2004) The PEP carboxykinase also catalyzes C3 carboxylation at the interface of glycolysis and TCA cycle in *Bacillus subtilis*. Metab Eng: in press

Zamboni N, Mouncey N, Hohmann HP, Sauer U (2003) Reducing maintenance metabolism by metabolic engineering of respiration improves riboflavin production by *Bacillus subtilis*. Metab Eng 5:49-55

Zamboni N, Sauer U (2004) Model-independent fluxome profiling from ^{2}H and ^{13}C experiments for high-throughput functional analyses. Genome Biol 5:R99

Zamboni N, Sauer U (2005) Fluxome profiling in microbes. In: Vaidyanathan S, Harrigan GG, Goodacre R (eds) Metabolome analyses: strategies for systems biology. Springer, New York, pp 307-322

Zupke C, Tompkins R, Yarmush D, Yarmush M (1997) Numerical isotopomer analysis: estimation of metabolic activity. Anal Biochem 247:287-293

Zamboni, Nicola
Institute of Molecular Systems Biology, ETH Zurich, Wolfgang-Pauli Strasse 16, 8093 Zurich, Switzerland
zamboni@imsb.biol.ethz.ch

List of abbreviations

CE: capillary electrophoresis
GC: gas chromatograph(y)
LC: liquid chromatograph(y)
MS: mass spectrometry
NMR: nuclear magnetic resonance
PPP: pentose phosphate pathways
TCA: tricarboxylic acid (cycle)

Data acquisition, analysis, and mining: Integrative tools for discerning metabolic function in *Saccharomyces cerevisiae*

Michael C. Jewett, Michael A.E. Hansen, and Jens Nielsen

Abstract

The well defined genetic architecture and metabolic network of *Saccharomyces cerevisiae* make this organism a cornerstone for metabolomics research. Recent efforts have focused on robust sample preparation techniques, analytical tools to quantitatively identify hundreds of metabolites at the same time, and elegant approaches for analyzing and interpreting the data. While equally important, we focus here on approaches for extracting useful information from the data itself. We outline several statistical and mathematical methods that can be used to digest and validate the most important features in the data. These multivariate approaches are from either the well established standard portfolio of statistical methods, or can be adapted from other areas where similar problems can be identified and where statistical and mathematical methods exist. Looking forward, we also describe approaches for fusing metabolome data with other cellular measurements and network structure to elucidate biosynthetic control mechanisms.

1 Yeast as a model system for metabolomics

Understanding and controlling complex biomolecular systems is critical for investigating natural biological phenomena, treating disease, and engineering cells with novel function (Goeddel et al. 1979; Cameron et al. 1998; Hood et al. 2004; Alper et al. 2005; Endy 2005; Fung et al. 2005; Isaacs et al. 2006; Ro et al. 2006). This task, however, is hampered by difficulties in accurately monitoring, understanding, and manipulating highly integrative metabolic pathways and multi-tiered regulatory circuits (Ideker et al. 2001; Alper et al. 2005; Jewett et al. 2006). As a major branch of systems biology efforts, metabolomics helps to address this limitation by quantitatively identifying cellular metabolites and understanding how their levels influence network topology and ultimately, control phenotype. Due to the highly connected nature of metabolic networks, we envision that the most powerful approach for using metabolomics data for systems biology is within the integrated web of complex interactions, cellular pathways, molecular participants, and environmental stimuli that they connect. This requires a model system with a

Topics in Current Genetics, Vol. 18
J. Nielsen, M. C. Jewett (Eds.): Metabolomics
DOI 10.1007/4735_2007_0222 / Published online: 14 April 2007

rich density of available biological information and one for which high-throughput data can be rapidly and robustly obtained.

The yeast, *S. cerevisiae*, is extremely well suited for this objective (Castrillo and Oliver 2004). First, as one of the most intensely studied eukaryotic cells, there is a large wealth of knowledge detailing its genetics, biochemistry, and physiology (Rose and Harrison 1987-1995). Second, curated genome-scale metabolic models that provide a roadmap of metabolites, their biochemical reactions, and gene products, which catalyze these reactions, are well established (Forster et al. 2003). Third, *S. cerevisiae* was the first eukaryotic organism for which the whole genome sequence was available (Goffeau et al. 1996) and this information gives us an inventory of parts. Fourth, yeast has well characterized genetics and facile techniques for genetic manipulation. Fifth, there is a comprehensive collection of knockout mutants in *S. cerevisiae* that provides an immense resource for exploring the impact of transcriptional regulation on metabolic phenotype (Ross-MacDonald et al. 1999; Winzeler et al. 1999; Giaever et al. 2002). Sixth, yeast has simple, inexpensive, and scalable methods for cultivation under well controlled conditions. Seventh, many of the high-throughput techniques cataloging gene expression (Spellman et al. 1998), protein levels (Zhu et al. 2001), metabolite levels (Castrillo at al. 2003), protein-protein interactions (Uetz et al. 2000), and protein-DNA interactions (Lee et al. 2002) have been developed and refined for *S. cerevisiae.*

In addition to the abundance of notable reasons, which promise to help in the analysis and integration of metabolomic data, there are also compelling application based motives for the use of yeast. For example, *S. cerevisiae* serves as an excellent model system for studying higher eukaryotic cells, including humans (Mager and Winderickx 2005), and has been used for rDNA protein expression in the pharmaceutical industry (Porro et al. 2005).

With so many incentives for using yeast as a model organism, a significant amount of attention has already been given to *S. cerevisiae* metabolomics. In general, recent work has been targeted towards three main categories: (a) sample preparation, (b) metabolite identification and quantification, and (c) data analysis. For example, in an effort to ensure robust, unbiased quantification of metabolites (which is still a major challenge in metabolome analysis) Villas-Bôas et al. (2005a) have examined the impact of different sample preparation protocols on targeted intracellular metabolite recovery. As a paradigm for functional genomics and mutant classification, metabolic fingerprints and footprints have been shown to reveal phenotypic insights from qualitative metabolite patterns (Raamsdonk et al. 2001; Allen et al. 2003). Because metabolic footprinting and fingerprinting techniques typically search for distinguishing patterns in data without quantitative metabolite levels, they often fail to provide insight into specific metabolic pathways. To address this limitation, targeted analyses obtain differential metabolite levels by absolute or at least semi-quantification and unambiguous metabolite detection of pre-defined metabolites. New developments in targeted analysis are enabling more and more metabolites to be detected and quantified in a single-shot (Mashego et al. 2006; Villas-Bôas et al. 2005b). This promises to aid in the development of models, which systematically bridge transcriptional regulation and

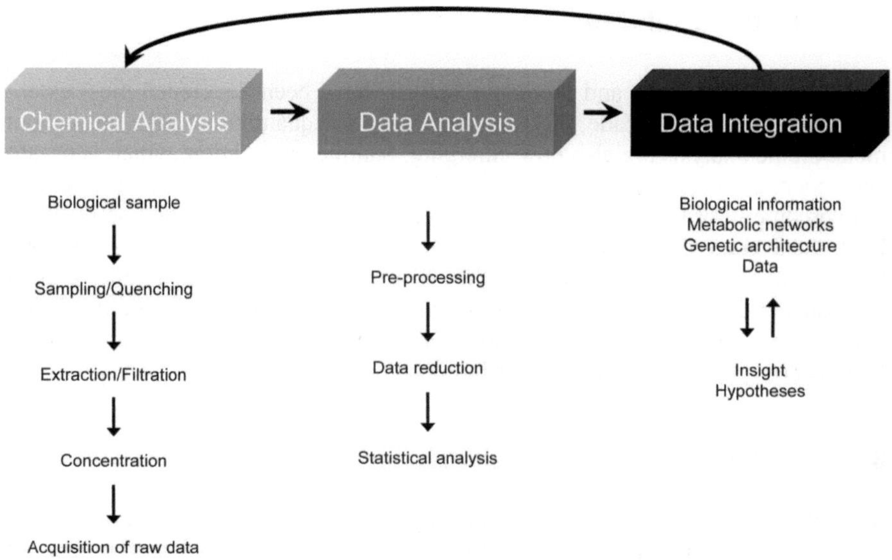

Fig. 1. Metabolite analysis workflow.

metabolic phenotypes (Kümmel et al. 2006; Çakir et al. 2006; Moxley et al. submitted). Since efforts in metabolomics motivate the need to track all potential biomarkers, identifying discriminatory features and removing noise from the data is crucial. A good example is the identification and study of plant defense metabolites (Kell et al. 2001).

2 Metabolite analysis workflow

What steps are taken to acquire a snapshot of metabolic composition? Typical metabolite analysis proceeds from chemical analysis to data analysis to data integration (Fig. 1). Upstream steps, particularly those involved in chemical analysis, have a significant impact on the final quality of the data. From experimental design to sample quenching, attention to every small detail is very important. Many experiments center on comparative profiling between genetic (e.g. wild type versus mutant) or environmental (e.g. growth on different carbon sources) perturbations. To obtain accurate differential data that minimize the influence of non-related factors, it is important that all conditions, except the variable of interest, are kept constant. Metabolite pools vary, for example, depending on growth rate, temperature, and media composition (which dynamically changes during batch cultures). Chemostat fermentations can be successfully used to eliminate variability arising from these factors (Hayes et al. 2002). We will now highlight major developments in the metabolite analysis workflow as they apply to yeast.

3 Chemical analysis

Although transcriptome and proteome analysis have been developed most extensively over the past decade, the tools necessary for quantitative high-throughput metabolome analysis are also now emerging. Sample preparation, which is organism specific, has been a key area of development. There are three main obstacles to overcome in order to acquire a reliable snapshot of the metabolic composition. First, the average lifetime of a typical intracellular metabolite is less than 1 second (Villas-Bôas et al. 2005c). Second, the chemical diversity of metabolite classes necessitates different extraction methods and different analytical methods. Third, current protocols are unable to resolve metabolites in different compartments, such as the cytosol, vacuole, or mitochondria.

3.1 Quenching

Since the average half-life of metabolites is so short, rapid inactivation of biological activity is required to prevent compositional changes. For *S. cerevisiae*, the method of choice was originally proposed by de Koning and van Dam (1992). Here, metabolism is stopped by rapidly spraying yeast cells into 60% (v/v) methanol kept at -40°C. Following quenching, the extracellular media is separated from the biomass by centrifugation. Washing the cell pellet with cold methanol (60% (v/v)) can be performed to ensure that there is no contamination from extracellular metabolites; however, leaking as a result of membrane weakening has been observed (Villas-Bôas et al. 2005a). Decreasing the time that the cells are exposed to the organic solution prior to extraction through faster centrifugations has been shown to significantly reduce leakage (Villas-Bôas et al. 2005a). Particularly crucial for kinetic experiments, techniques for rapid sampling have also been described (Lange et al. 2001; Mashego et al. 2006). While the time-scale is longer for most extracellular metabolites, the objective of instantaneously capturing an image of metabolism still requires rapid separation of the extracellular medium from the biomass and extracellular enzymes. In general, this is carried out by filtration.

3.2 Extraction

Due to their chemical diversity and location within the cell, extracting all cellular metabolites in their original state over a large dynamic range with a single or limited set of analytical techniques is impossible in practice. However, strategies attempting to cover a wide range of metabolites in a single step continue to evolve. For intracellular metabolites, the extraction method is critical and influences the kind of metabolites that are recovered and can be analyzed. The three most popular methods for *S. cerevisiae* are chloroform-methanol (de Koning and van Dam 1992), boiling ethanol (Gonzalez et al. 1997; Castrillo et al. 2003), and pure methanol (Villas-Bôas et al. 2005a). The chloroform – methanol method has a rich history of reproducibility and avoids instability of heat-labile metabolites; but, it is considered laborious and uses chloroform, a toxic and carcinogenic solvent. The recent workhorse in the field is the boiling ethanol method. This approach is reported as being simple, fast, and accurate. However, some metabolites show poor recoveries relative to the chloroform-methanol approach due to temperature instability (Villas-Bôas et al. 2005c). The pure methanol method has only recently been proposed. It appears to combine the advantages of both the chloroform-methanol and boiling ethanol strategies. Although the pure methanol method (Villas-Bôas et al. 2005a) shows similar recoveries of several metabolite classes relative to the chloroform-methanol method (amino and non-amino organic acids, nucleotides, among others), there are still questions surrounding whether or not the method completely inactivates all enzyme activities.

Villas-Bôas et al. have offered some insight into how extraction methods in yeast compare by investigating the recovery of a wide range of spiked metabolites covering several different classes (e.g. amino acids, sugar phosphates, etc.) from biological samples (Villas-Bôas et al. 2005a). Their results suggest that the pure methanol method offers perhaps the most reproducible, simple, and attractive approach for high-throughput metabolome coverage. One of the difficulties associated with comparing different extraction methods is that most laboratories have developed analytical techniques for only one specific class of metabolites most relevant to their work. For example, sugar phosphates, which were not observed in the comparison study by Villas-Bôas et al. (2005a), are routinely recovered by others when using the boiling ethanol method (Mashego et al. 2004, 2006; Wu et al. 2006). Hence, it is difficult to compare metabolome coverage between different laboratories. More rigorous comparative work, particularly among various laboratories with different expertise, is needed to characterize the best, or best set of, extraction protocols necessary to analyze a large number of metabolites spanning the metabolome.

3.3 Analytical methods

The choice of analytical technique for acquiring raw metabolome data is also important for metabolic state analysis. While NMR is sometimes used, mass spectrometry (MS) is the most widely used approach for recent methodologies developed for metabolomics. MS methods are highly sensitive, allow for identification of unknown compounds, and are high-throughput. Typical quantitative approaches couple an analytical separation technique (e.g. capillary electrophoresis (CE), liquid chromatography (LC), and gas chromatography (GC)) with MS based detection. Advantages and disadvantages to each approach are given in Table 1 and have been thoroughly described elsewhere (Villas-Bôas et al. 2005c).

To elucidate a quantitative image of yeast metabolism, LC-MS (van Dam et al. 2002) and GC-MS (Villas-Bôas et al. 2005b) methods have been mainly employed. Together, these methods cover a large fraction of the primary metabolites, a research area exploited in functional genomics efforts for elucidating general rules and descriptions of cellular behavior. The well-refined LC-MS platform

Table 1. Comparison of analytical techniques used in metabolomics (updated from Villas-Bôas et al. 2005c).

	Advantages	Disadvantages
1. GC-MS	- High separation efficiency - Easy interface between GC and MS - Simultaneously resolves different classes of metabolites - Reproducible	- Unable to analyze thermo-labile metabolites - Requires derivatization of non-volatile metabolites - Difficult to identify unknown compounds after derivatization
2. LC-MS	- High sensitivity - Enables analysis of thermo-labile metabolites - Average chromatographic resolution	- Matrix effects - Restrictions on LC eluents due to interface issues from LC to MS - De-salting may be necessary - More suitable for target analysis
3. CE-MS	- Uses small volumes - Average resolution - Fast separation of charged and uncharged species	- Difficult to interface CE with MS - Complex methodology and quantification - Least developed - Low sensitivity
4. MS	- Allows for rapid screening of metabolites (2-3 min per sample) - High sensitivity - Negligible sample clean-up for profiling	- Identification of metabolites generally requires tandem MS - Matrix effects - Requires elegant data deconvolution methods

offers an exceptionally sensitive and specific approach for studying the intermediates of the glycolytic pathway and some tricarboxylic acid cycle intermediates (van Dam et al. 2002). The GC-MS platform is designed to measure amino and non-amino organic acids (which comprise almost 40% of the yeast metabolome) that play crucial roles in central carbon metabolism, amino acid metabolism, and energy generation (Villas-Bôas et al. 2005b). A recent application of this GC-MS platform identified and quantified approximately 60 intracellular and extracellular metabolites per experimental condition (Villas-Bôas et al. 2005b).

3.4 Standardization

In addition to dynamic developments in refined analytical techniques and MS sensitivity, advances in internal standardization, one of the main challenges in quantitative metabolome analysis, are also paving the way for more robust measurements. Heijnen and coworkers have developed an approach which uses extracts from ^{13}C-saturated microbial cultivations to provide an internal standard for all intracellular metabolites to be quantified (Mashego et al. 2004; Wu et al. 2005). This work has created a platform that is independent of ion suppression effects, of metabolite modifications during extraction, and of variations in instrument response. Although less universal because limited to nitrogen-containing metabolites, ^{15}N-saturated cultivations have also shown a strong potential to impact metabolome study standardization (Lafaye et al. 2005).

4 Data analysis

After chemical analysis, it is important to draw conclusions based on latent structures hidden in the generated data (Fig. 1). Some of the first applications analyzing the metabolome were primarily focused on drawing taxonomical conclusions inferred from chemical information extracted manually from the analytical data files (Frisvad and Filtenborg 1983). Nowadays gene functions are studied through determination of intermediate and end product metabolites present in an organism at a given time (Sumner et al. 2003; Fiehn 2002; Kell 2004). In most of these cases, no *a priori* knowledge is available about where to look for changes. Whereas early applications analyzed a very limited part of the metabolome (quantitatively), studying gene functions and understanding the integrated nature of the cell requires analysis of whole metabolite profiles (qualitatively and quantitatively).

A number of reviews have in detail discussed the variety of modern analytical techniques and data collection and storage methods available (Fiehn 2002; Mendes 2002; Fiehn and Weckwerth 2003; Goodacre et al. 2004; Kell 2004). Brown et al. (2005) gives a comprehensive overview of the different methods and addresses the need for a streamlined pipeline, describing the wide-ranging methods and approaches that are used in metabolomics at different levels. Brown et al. also discuss issues that have to be considered when analyzing the metabolome, from data generation to the analysis into useful knowledge.

In general the analysis of metabolome data is done in *at least* three stages (Hansen and Smedsgaard 2006; Brown et al. 2005): pre-processing, data reduction, and statistical (inference) analysis. In the following, we briefly describe some of the popular techniques used for analyzing the metabolome.

4.1 Pre-processing

Pre-processing is an important part of data analysis since it primarily deals with tasks concerning the improvement and the enhancement of the parts in the data regarded as “signal” in relation to the parts of the data regarded as “noise”.. In some cases, noise is defined as the true random variations which can be seen as high frequent changes in a MS spectrum. In other cases “noise” can be defined in a much broader sense; as an “artifact” which is caused by internal drift in instrument parameters, etc. An example of this would be the drift in baseline that can be seen for many types of chromatographic data where the “true” peak signal seems to be superimposed on a slowly varying surface.

4.1.1 Noise reduction

For any given profile (spectrum, chromatographic trace, etc.), one of the common ways of removing noise is based on a so-called “moving window” filter (Antoniou 1993; Mitra 1998). The filter can be imagined as a window of a certain size moving along the profile, one profile element at a time. The middle element of the window is replaced with the weighted average of all elements in the window. The weights chosen in the window are important for the properties of the filter (Hansen and Smedsgaard 2006). Figure 2 illustrates the principle of the filter. The illustration shows a window with Gaussian weights, $N(\mu=0, \sigma=1)$.

If all weights would have been chosen to be one over the length of the filter (i.e. 1/7 in this example) the window would have calculated the mean of the overlaid cells in the profile, and hence the filter would have been called a “mean filter”. Other types of filters are based on this moving principle have different properties (Savitzky and Golay 1964; Eilers 2003). Some calculate the dot-product between the window weights and the profile values, as illustrated here, whereas others estimate the median, a quantile or other (weighted) measure derived from the profile values within the window.

It is important to remember the value of new elements and not make the replacement until the window has passed. This must be done since all calculations shall be based on the original data in the array. When the ends of the profile are filtered and parts of the window are outside the spectrum, the calculation must be done on fewer elements than when the entire window is inside the array. This implementation leaves the ends of the array unfiltered. For a 7-point filter, this means that when n elements are filtered, elements 1, 2, 3, and n-2, n-1, n remain unchanged when filtering is complete. For many applications – long profiles and short filters – this is no problem. Alternatively, the profiles can be padded with the values found at the end or padded with zeros.

Unfortunately, smoothing with these fixed filters does not preserve the height and width (i.e. the area) of a peak and the (centroid) position if the peak is skewed. Some of the existing filter algorithms can be made adaptive based on measured peak properties, such as intensity or width. To accommodate for this problem, the spectrum can be approximated locally by a higher order polynomial (of some order) within a moving window. This filtering method is closely related to the so

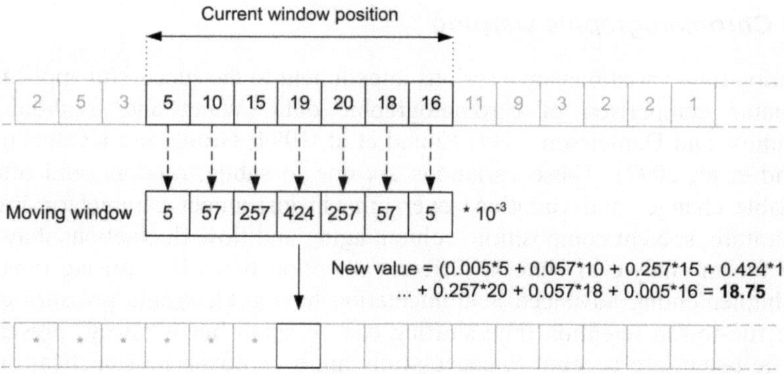

Fig. 2. Illustration of the moving window filter. The values marked with * are values that have been calculated although not shown in the figure. The weights used in the filter are chosen to be based on the normal distribution with mean zero and standard deviation equal to one.

called Savitsky-Golay filter (Savitzky and Golay 1964) available in most of the instrumental software packages (e.g. MassLynx and HP Chemstation).

4.1.2 Baseline correction

In Section 4.1.1 the noise was regarded as stochastic, but in some cases it has to be considered deterministic. As described above most of the chromatographic data can be regarded as "real information" added on to a "noisy" background or baseline (Goehner 1978). In the case of chromatographic data the baseline is defined as the part recorded when only carrier gas or solvent elute from the column (from the IUPAC compendium of technical terminology).

The baseline can be either flat, linear with a positive or negative slope, curved or a combination of all three. In most correction algorithms it is the goal to estimate the baseline $g(t)$, which then is subtracted from the original chromatogram. The methods available for estimating the background vary a lot in complexity (Mazet 2005). Some of the methods work on the whole chromatogram at once and some are local methods that break the chromatogram into relevant smaller sections where the background function is estimated better. In addition, some methods are parametric methods that try to estimate a function (piecewise or global) to the background. The simplest are the (locally) linear correction methods (Nielsen et al. 1998) followed by the more complex nonlinear methods. In most software packages like Origin (by OriginLab) or PeakFit (by Systat Software), as well as for many published papers (Vickers et al. 2001; Torres-Lapasió et al. 1997), the background is estimated by a least-squares polynomial fitting performed on a user defined subset of points, which should belong to the background. In 1997, Depczynski et al. introduced the wavelet transform as a new tool for removing the background from chromatographic data (Shao et al. 1999; Cai 2001; Tan and Brown 2002; Liu et al. 2003).

4.1.3 Chromatographic warping

Retention time variations are a serious impediment to the successful application of automated comparison of chromatographic data (Wang and Isenhour 1987; Malmquist and Danielsson 1994; Round et al. 1994; Grung and Kvalheim 1995; Bylund et al. 2002). These variations are due to subtle, random, and often unavoidable changes and variations over time in instrument parameters. Pressure, temperature, solvent composition, column aging and flow fluctuations may be the cause for an analyte to elute at different retention times in replicate runs. Even with implementing advanced instrumentation having electronic pressure control, subtle run-to-run retention time shifting can be small but is always present, and must be taken into account to successfully apply multivariate statistical methods (known as chemometric methods, when being applied on chemical data). Matrix effects and stationary phase decomposition may also cause variation in retention time. The main reason is that most pattern recognition techniques and chemometric methods are based on point to point comparison for successful analysis.

Many of the available alignment algorithms do not require knowledge or identification of peaks. The main goal for all warping algorithms is the same: to align chromatograms so that they correct for the random differences from run to run. These algorithms typically search to find the optimal "warp" according to some criterion. Warping is the process of stretching or shrinking one profile of peaks in order to make it match to another profile of peaks. The "simple" warp here would be stretching or shrinking the profiles in a linear manner just by moving the ends (like an elastic band).

Unfortunately, this simple approach can't be used since the difference between the profiles might vary along the retention time, for example. Therefore, all of the warping algorithms try to warp the profiles to each other in such a way that they are fitted warped locally. And the criteria used for what can be regarded as a local good fit is how well the pieces correlate to each other.

Since the first publications (Wang and Isenhour 1987), many attempts have been made to increase speed and performance of warping algorithms for finding the best match between two chemical profiles. Pravdova et al. (2002) and Tomasi et al. (2004) both review and evaluate two competing state of the art warping methods: Dynamic Time Warping (DTW) (Kassidas et al. 1998) and Correlation Optimized Warping (COW) (Nielsen et al. 1998). In 2003, Forshed et al. suggested a method called Peak Alignment with Genetic Algorithm (PAGA) and Johnson et al. (2003) came up with a method called Local Warping (LW). Some of the latest attempts have been made by Eilers (2004) Parametric Time Warping (PTW), Tibshirani et al. 2004 applying a hierarchical clustering method to construct a dendrogram of all peaks from multiple samples, and recently, a Hidden Morkov Model-based approach was proposed by Listgarten et al. (2004) to align multiple time series data. Finally, the year after Walczaka and Wub (2005) described a method called Fuzzy warping (FW).

One of the conclusions that can be made based on all existing literature is that no matter what method one chooses, it will be based on a tradeoff between performance and speed. Some methods, like DTW, are relatively fast, but have been found to perform not as well as the COW algorithm (Pravdova et al. 2002; Tomasi et al. 2004).

4.1.4 Deconvolution

Chromatographic techniques often give rise to situations where reaching complete resolution is not possible. Deconvolution is the process of separating the signals from overlapping peaks (Dromey et al. 1976; Stein 1999). Together with the development of high-performance instruments, the number of algorithms and tools for deconvolution of chromatographic data has been increasing steadily. As a result, compounds hidden within a peak cluster can now be quantified with relatively small errors.

Deconvolution methods can be divided into two fundamental categories: those that are use single (one-dimensional) profiles (Vivó-Truyols et al. 2002, 2005b) and those that are based on higher dimensional chromatographic data (Kong et al. 2005; Maleknia and Downard 2005). Most one-dimensional approaches (e.g. only retention time profile) rely on fitting a sum of peak-models to the profile (Li 2002; Vivó-Truyols et al. 2005a) whereas some utilize the unique information available along the other dimensions (e.g. spectral dimension is added to the retention time dimension). Adding more dimensions to the deconvolution algorithm makes the task easier. For example, two peaks eluting at almost the same time might differ in MS spectrum or UV absorbance spectrum.

Deconvolution can be either achieved in an automated fashion by the software packages provided with most GC-MS instruments (Pegasus, Leco, St. Josephs, USA) or separate software can be applied, such as AMDIS (http://chemdata.nist.gov/mass-spc/amdis; National Institute of Standards and Technology, Gaithersburg, USA). The AMDIS software is originally based on the algorithm developed by Stein (1999).

4.1.5 Data normalization/standardization

In some cases, it is interesting to look at the relative amounts of different compounds between samples. In these cases it is necessary to remove the effect of the total amount from the analysis. This type of correction is commonly known as normalization, standardization, and sometimes, multiplicative correction of the data. Data standardization is the process of making all data comparable to an established convention or procedure to ensure consistency and comparability across different types of variables.

4.2 Statistical analysis

In the following sections, some of the tools used for statistical analysis will be described. Before we go into detail with some of the more common multivariate methods, we would like to say a few words about univariate statistics. As emphasized by Brown et al. (2005) it is always useful to look at data with the "simple to understand" methods before continuing to the more complex ones. Simple statistics, like the mean and standard deviation, might detect outliers that have to be removed from the dataset. As an example, by looking at the mean value of the Total Ion Count (TIC) from direct ESI-MS data, one can easily detect samples that for some reason or another might be an outlier due to an injection problem.

4.2.1 Data reduction

One of the main challenges of analyzing the metabolome is the huge dimensionality of the data. For most analytical instruments, the amount of variables returned for each sample can be extremely high, and caution has to be taken. The performance of most multivariate statistical algorithms depends highly on the interrelationship between both the sample size and the number of dimensions.

In cases with many more dimensions than observations, it can be necessary to reduce the effective dimension to employ some of the more efficient methods that work best at lower dimensions. Based on redundant information in the data, observations can well be approximated by "projections" into a lower-dimensionality space – more or less removing the redundancy from the data (explained below). Many of the techniques used for data reduction and visualization of multivariate data are based on a so-called decomposition of **X** followed by a projection of the data onto the axes defined by the extracted factors.

Figure 3 illustrates the principle behind transforming data in a simple way. Here data are distributed as an ellipse in the (x_1,x_2) coordinate system. The purpose of any (linear) transformation is to place a new coordinate system in the original one based on some criterion. It is the **criterion** that makes the difference between the different methods such as, for example, Principal Component Analysis (PCA) (Mardia et al. 1979) and Fisher Discriminant Analysis (FDA) (Fisher 1936). Since PCA and FDA can be considered as the two most popular transformations, we will provide a brief description of the principles behind these methods. Other dimensionality reduction methods can also be applied to data, including non-linear transformations. Factor Analysis (Bartlett 1937), Projection Pursuit (Friedman and Stuetzle 1981), Wavelet Transforms (Percival and Walden 2000), and Independent Component Analysis (Karhunen and Joutsensalo 1995) are some of the additional methods that can be found in the literature. Common for all methods is their property allowing characterization of a low-dimensional subspace from the original data.

Principal component analysis - first described by Pearson (1901), PCA is probably the most popular technique for simplifying a dataset by reducing dimensionality. More formally PCA is a linear transformation (rotation of data) that chooses a new coordinate system (p_1,p_2) for the data set in such a way, that the greatest variance by any projection of the data is found along the first axis (p_1) – called the first principal component (PC) – the second largest variance on the second axis (p_2), and so on. PCA can be used to reduce the dimension of data while retaining those characteristics of the dataset that contribute mostly to the variance

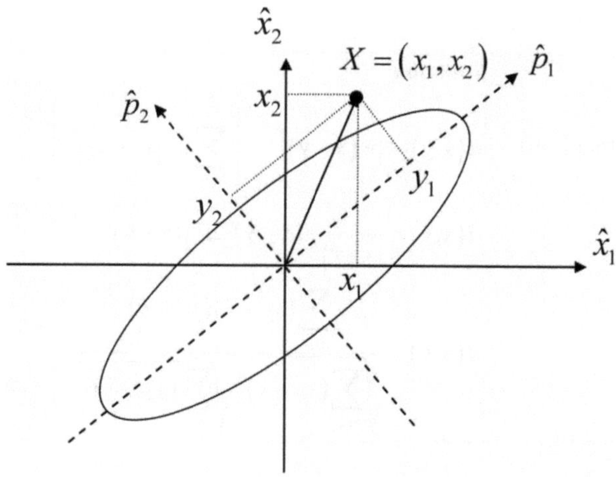

Fig. 3. Illustration of the principle of transforming data (see text for further details).

by eliminating the higher principal components, by a more or less heuristic decision. Characteristics retained may be the "most important", but this is not necessarily the case and depends on the application. In the following, we describe in detail the mathematics behind PCA.

Figure 3 illustrates the basic principles. The data are distributed in an ellipsoid along x_1 and x_2. From the plot we conclude that the metabolites x_1 and x_2 are correlated.

PCA places a new coordinate system (p_1,p_2) in (x_1,x_2) with the origin in the center of the data, and having the same number of axes. The axes are placed in such a way that the first axis is pointing along the direction with most variance, and the second is placed orthogonal in the direction of second most variance (Fig. 3). As one can see on the figure, p_1 now contains most of the variation in the data. In the case where the main variation in the data was caused by differences in the metabolite concentrations produced between two or more species, p_1 would enhance (capture) these differences.

It is possible to calculate the amount of variance captured by each PC. Often the amount of PCs chosen is defined as a percentage of the total variance present in the dataset. In the above case x_1 and x_2 each describe 50% of the total variance whereas p_1 describes approximately 90%. Often a number of PCs is chosen so that >90% of the total variance is captured. Hopefully, this will reduce the number of dimensions to a fraction of the original.

Fisher discriminant analysis - understanding the principle of PCA makes it easy to get a brief understanding of most of the other available data transformation techniques, such as Fisher Discriminant Analysis (FDA) (Fisher 1936). Whereas PCA finds directions in the data that contain the most variance, FDA finds the directions in which data segregate. More formally it is a linear transformation (rotation of data) that chooses a new coordinate system (f_1,f_2) for the data set in such a way, that the largest segregation between groups is found along the first axis (f_1),

Table 2. List of distance functions

Name	Function
Weighted L_p-norm ($\|\cdot\|_p$)	$d(\mathbf{x},\mathbf{y}) = \left\|\mathbf{w}(\mathbf{x}-\mathbf{y})\right\|_p = \left[\sum_{\forall k} w_k \left|x_k - y_k\right|^p\right]^{1/p}$
Mahalanobis	$d(\mathbf{x},\mathbf{y}) = \frac{1}{\left|\det \boldsymbol{\Sigma}\right|^p}(\mathbf{x}-\mathbf{y})^t \boldsymbol{\Sigma}^{-1}(\mathbf{x}-\mathbf{y})$
Correlation	$d(\mathbf{x},\mathbf{y}) = \frac{\sum_{\forall k}(x_k-\bar{x})(y_k-\bar{y})}{\sqrt{\sum_{\forall k}(x_k-\bar{x})^2}\sqrt{\sum_{\forall k}(y_k-\bar{y})^2}}$

the second largest segregation between groups is found along the second axis (f_2), and so on.

Whereas PCA is said to be unsupervised, FDA requires group information in order to calculate the projection. As for PCA, the number of dimensions that are included in FDA can be chosen so that the FDA coordinates describe >90% of variation within and between groups.

Further details about the technical aspects of PCA and FDA can be found in a huge amount of literature available on the topic, both in the literature as well as on the Internet.

An example to illustrate the power of FDA was recently reported by Mas et al. (2006). They compared two distinct analytical approaches based on mass spectrometry for their potential in revealing specific metabolic footprints of yeast single-deletion mutants. In the study, filtered fermentation broth samples were analyzed both by GC-MS and direct infusion ESI-MS. The mutants evaluated were *cat8*, *gln3*, *ino2*, *opi1*, and *nil1*, all with deletion of genes involved in nutrient sensing and regulation. Using FDA, they found that it is possible to discriminate the mutants in both the exponential and stationary growth phase, but the data from the exponential growth phase provide more physiological relevant information.

Another recent example exploiting FDA is given by Villas-Bôas et al. (2005b), presenting a novel derivatization method for metabolome analysis of yeast. Their sample workup method enables simultaneous metabolite measurements throughout central carbon metabolism and amino acid biosynthesis, using a standard GC-MS platform that was optimized for this purpose. As an implementation proof-of-concept, Villas-Bôas et al. (2005b) assayed metabolite levels in two yeast strains and two different environmental conditions in the context of a metabolic network. They demonstrated that these differential metabolite level data distinguish among sample types, such as typical metabolic fingerprinting or footprinting. More importantly, they showed that this differential metabolite level data provides insight into specific metabolic pathways and lays the groundwork for integrated transcription–metabolism studies of yeasts.

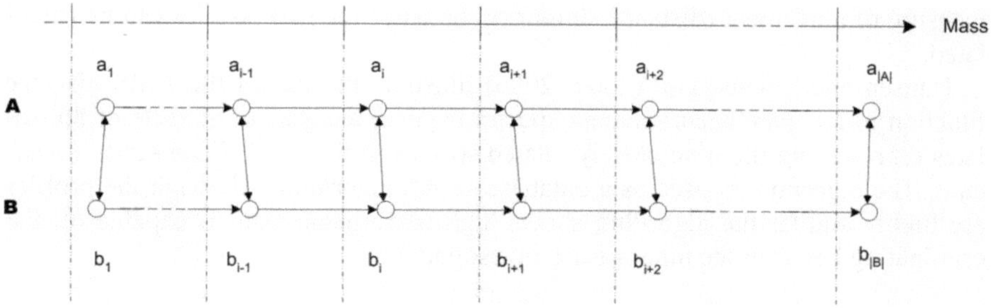

Fig. 4. Illustration of the linkage principle. The figure shows two profiles, **A** and **B**, (e.g. by direct ESI-MS) and the peaks found in both profiles, $a_1,\ldots,a_{|A|}$, and $b_1,\ldots,b_{|B|}$. In the figure |A| and |B| means the length (number of peaks) of both profiles.

4.2.2 Similarity (distance) based comparison

Both PCA and FDA see data as points in a hyper dimensional space – before and after the transformations. When data are represented as points in an appropriate space, it is possible to compare these observations by how far away they are located from each other. A distance function comparing the distances between two vectors, $\mathbf{x}_i$ and $\mathbf{x}_j$, of data (i.e. observations) is defined as a so-called distance function, $d_{ij} = d\left(\mathbf{x}_i, \mathbf{x}_j\right)$.

Distance functions are generally divided into two categories: continuous and binary functions. Table 2 shows a list of some of the most commonly used continuous distance functions. In the table **x** and **y** are two profiles and $d(\mathbf{x},\mathbf{y})$ is the distance between them. x_i and y_i are the i'th element in each of the profiles. Based on the choice of p, the most widely used are the 1-norm, 2-norm, and ∞-norm ($\left\|(\mathbf{x}-\mathbf{y})\right\|_\infty = \max\left|x_k - y_k\right|, \forall k$) referred to as the city-block or Manhattan distance, the Euclidian, and the Chebychev distances.

In some cases, as when comparing spectra, special distance functions can be developed utilizing special properties in the data (Stein 1995). Examples like this include library search methods, where reference spectra are stored in a database, and subsequent searches are done based on comparing a query spectrum with each of the references in the database based on a distance function. Some of the suggestions proposed include the Probability Based Matching (McLafferty 1974) algorithm and the Hertz similarity index (Hertz 1971).

Hansen and Smedsgaard (2004) have presented a new matching algorithm designed to compare high-resolution spectra. Whereas all of the existing distance functions methods are bound to compare fixed intervals of ion masses, the Accurate Mass Spectrum (AMS) distance method is independent of any alignment. The method takes into account that there may be differences in resolution of the spectra, and it is independent of any variable alignment procedures or binning. Figure 4 illustrates the principle of the algorithm. Given two profiles, the algorithm first detects the peaks followed by a matching of the peaks between them. Based on the number of peaks that could be matched, and how well they are matched, an overall measure of agreement (distance/similarity) between the two profiles can be calculated.

Hansen and Smedsgaard (2004, 2006) illustrate the use of the AMS distance function to compare accurate mass spectra from an analysis of extracts of 80 isolates representing the nine closely related species in the *Penicillium* series *Viridicata*. The algorithm is used as a database search algorithm. Although the profiles are highly similar the algorithm shows high performance and is capable of discriminating between the nine species investigated.

4.2.3 Clustering

Clustering algorithms can be hierarchical or partitional. Hierarchical algorithms find successive clusters using previously established clusters, whereas partitional algorithms determine all clusters at once.

Hierarchical Clustering - hierarchical algorithms can be agglomerative (bottom-up) or divisive (top-down) (Anderberg 1973; Hartigan 1975; Kaufman and Rousseeuw 1990). Agglomerative algorithms begin with each element as a separate cluster and merge them in successively larger clusters. Divisive algorithms begin with the whole set and proceed to divide it into successively smaller clusters. A key step in a hierarchical clustering is to select a distance measure. A simple measure is the Manhattan distance, equal to the sum of absolute distances for each variable. The name comes from the fact that in a two-variable case, the variables can be plotted on a grid that can be compared to city streets, and the distance between two points is the number of blocks a person would walk.

A more common measure is Euclidean distance, computed by finding the square of the distance between each variable, summing the squares, and finding the square root of that sum. In the two-variable case, the distance is analogous to finding the length of the hypotenuse in a triangle; that is, it is the distance "as the crow flies."

Given a distance measure, elements can be combined. Hierarchical clustering builds (agglomerative), or breaks up (divisive), a hierarchy of clusters. The traditional representation of this hierarchy is a tree data structure (called a dendrogram), with individual elements at one end and a single cluster with every element at the other. Agglomerative algorithms begin at the top of the tree, whereas divisive algorithms begin at the bottom. Cutting the tree at a given height will give a clustering at a selected precision. Mas et al. (2006) used hierarchical clustering after applying FDA. The cluster analysis was used to gain further insight into metabolic similarities in different mutants with deletions in transcription factors.

***k*-means Clustering** - the *k*-means algorithm assigns each point to the cluster whose center (also called centroid) it is nearest. The center of the cluster is calculated as the arithmetic mean (of each dimension) of all the points regarded to be in the cluster. The algorithm is roughly (MacQueen 1967):

1. Randomly generate k clusters and determine the cluster centers, or directly generate k seed points as cluster centers.
2. Assign each point to the nearest cluster center.

3. Recompute the new cluster centers.
4. Repeat 1-3 until some convergence criterion is met (usually that the assignment hasn't changed).

The main advantages of this algorithm are its simplicity and speed which allows it to run on large datasets. Its disadvantage is that it does not yield the same result with each run, since the resulting clusters depend on the initial random assignments. The algorithm maximizes inter-cluster (or minimizes intra-cluster) variance, but does not ensure that the result has a global minimum of variance.

4.3 Classification

Classification is a procedure in which individual items are placed into groups based on quantitative information on one or more characteristics based on a training set of previously labeled items.

One of the most known methods for classification is the k-nearest neighbor. Nearest neighbor methods in general are based on a measure of distance between observations, e.g. the Euclidean distance or one minus the correlation between two metabolite profiles. The *k*–nearest neighbor rule, *k*–NN (Fix and Hodges 1951), classifies an observation **x** as follows:

1. Find the *k* observations in the learning set that are closest to **x**.
2. Predict the class of **x** by majority vote, i.e., chooses the class that is most common among those *k* observations.

Other methods for classification can be found in Duda et al. (2001) and Herbrich (2001).

4.4 Genetic programming

As an alternative to the traditional (classical) statistical methods described above, a technique for the analysis of multivariate data by genetic programming (GP) has been described by Gilbert *et al.* (1997) (also see Kell 2002). Genetic programming (GP) is an evolutionary technique, which uses the concepts of Darwinian selection to generate and optimize a desired computational function or mathematical expression (Koza 1992). An initial random population of individuals, each encoding a function or expression, is generated and their fitness to reproduce the desired output is assessed. New individuals are generated either by mutation (the introduction of one or more random changes to a single parent individual) or by crossover (randomly rearranging functional components between two or more parent individuals). The fitness of the new individuals is assessed, and the best individuals from the total population become the parents of the next generation. This process is repeated until either the desired result is achieved or the rate of improvement in the population becomes zero. It has been shown that if the parent individuals are chosen according to their fitness values, the genetic method can approach the theoretical optimum efficiency for a search algorithm (Koza 1994).

An excellent example of the use of genetic programming is Kell et al. (2001). This was a "transgene discovery" problem in which they measured a series of metabolites via HPLC and used these as the inputs to a Genetic Program designed to find a rule which would tell from the metabolome data whether the transgene of interest was present or absent. The experiment was also aimed at investigating the biosynthesis and function of salicylic acid in plant defense by the expression of a salicylate hydroxylase enzyme (SH-L) to block accumulation.

A total of 48 peaks ($V_1 - V_{48}$) from the HPLC traces were digitized and integrated using standard software provided with the instrument, and a total of 36 samples studied. The metabolite peak values were used as inputs to the Genomic Computing software Gmax-bio (Aber Genomic Computing, Unit 8, Science Park, Aberystwyth SY23 3AH, UK), with the presence or absence of SH-L in the genotype being encoded 1 or 0. One of many rules which evolved could be written as follows:

```
SCORE = Sqrt((V37/V24)) + Sqrt(V30/(V24+V42))
```

Probability that plant contains the transgene:

```
1 / (1 + Exp(-(-8.046777 + SCORE * 1.872833)))
```

This rule had an accuracy of more than 95%. A power of genomic computing is that it ranks variables in order of their utility in successful rules. The top 3 variables were peaks 24, 30, and 42, and peak 24 was indeed salicylate, known to play a key role in defense mechanisms in many plants. Thus, the GP discovered not only what differences there were but also which were important to the biological pathway of interest, and turned metabolomic data into biochemical knowledge.

4.5 SpectConnect

Styczynski et al. (2007) have presented a new method, SpectConnect, to automatically catalogue and track otherwise unidentifiable conserved metabolite peaks across sample replicates and different sample condition groups without use of reference spectra. SpectConnect compares every spectrum in each sample to the spectra in every other sample within a user specified time window. By doing so, it is capable of determining which components are conserved across replicate samples. SpectConnect also determines which of these components differentiate one sample condition (e.g. time or treatment) from another, whether by changes in concentrations or merely by their presence/absence. The only requirement of the experimental measurements is that each sample condition must have replicates. In a sense, SpectConnect relies on an increase in signal relative to noise that is created by this requirement of replicates. While injection ("technical") replicates are the easiest way to provide the required replicates, it is also desirable to include biological replicates in a group of samples. This is due to systematic error in peak detection and deconvolution software that may consistently find a noise peak in technical replicates. Though this approach adds more biological variability to a group in terms of metabolite concentrations, it should have significantly less impact on the presence/absence of a metabolite. Ultimately, the authors hypothesize that these "true", important, spectra will be conserved across most or all replicates of a sample, while spectra that are artifacts of noise will not.

With SpectConnect, it is possible to find three distinct types of information. First, it identifies all of the components that are conserved across a single group of samples or replicates. Second, for each metabolite peak conserved in at least one group, it can determine all other groups in which it occurs. Finally, for any given pair of groups, it can find the likelihood, to some degree of statistical significance, that each metabolite peak is present in unequal amounts in the groups.

Styczynski et al. (2007) illustrates the use of SpectConnect on biological samples. The samples are time-course data from fermentation runs conducted with three different strains of *E. coli* analyzed with GC-MS. They found that across five time points over 30 hours, there were a total of 544 metabolite peaks (chromatogram peaks) that occurred in at least one of the strains in at least one time point, while 184 of those occurred in all of the strains in at least one time point. Qualitatively, this indicates that the genetic differences of these strains have caused significant differences in their respective metabolisms. This result is to be expected for mutants with deletions of metabolic enzymes: some subsets of metabolites are rendered inaccessible, so a significant metabolic adjustment is necessary to compensate for such changes.

Using this cumulative library of 544 metabolite peaks, they analyzed the metabolomic profile of one mutant strain relative to that of the reference strain in the course of the fed-batch cultivation. While a few compounds are identified using the known library, significantly more spectral signatures are detected with SpectConnect.

5 Data integration

Data analysis strategies are turning raw data into metabolite knowledge, but the true power of metabolome analysis is in using observed metabolite profiles to understand the interaction of components in biomolecular systems more completely. As mentioned, metabolome data, which comprise qualitative metabolite patterns, are useful for characterizing mutants and exploring metabolic chemodiversity (see Data analysis section). However, these data are difficult to integrate with other measurements of cellular molecules (e.g. mRNA and/or protein levels) because metabolite identities and levels are unknown. Metabolite identities and levels are required for systems biology efforts that seek to map different layers of regulation and understand the connectivity between genes, mRNAs, proteins, fluxes, and metabolites. For these efforts, chemical analysis and data analysis are just the starting point for inferring novel insight (Fig. 1).

Although the ultimate goal of unraveling regulatory schemes and communication mechanisms that dictate cellular function is clear, exactly how best to use metabolome data for achieving this goal is still a work in progress. One of the main reasons is that discerning metabolic heritage is a formidable task. Because metabolite levels are determined by changes in fluxes and the activities of enzymes (in general, 2 or more), metabolites represent the integration of signals broadcast from several functional levels (genes, mRNAs, and proteins) (Nielsen 2003). There is not a one-to-one link between genes and metabolites (Nielsen and Oliver 2005). Therefore, knowing that the level of metabolite X increases fourfold between condition 1 and condition 2 does not, in general, directly tell us what is happening to gene Y. Another challenge to bridging the gap between genetic structure and metabolic phenotype is that current analyses miss many metabolites that may be critical to the particular system of study (e.g. amino acids may be measured but not nucleotides). Hence, we may have quantitative metabolite levels for amino acid metabolism, but lack information detailing glycolysis. To address these issues, recent efforts have focused on imposing thermodynamic and/or network constraints (Kümmel et al. 2006; Çakir et al. 2006) or linear modeling (Pir et al. 2006) for upgrading the information content obtained from metabolome measurements (Fig. 5).

Network-embedded thermodynamic (NET) analysis is a new method for model-based interpretation of quantitative metabolite data (Kümmel et al. 2006). By using metabolome data, estimated or known flux directions that reflect the metabolic network operation, standard Gibbs free energies of formation, and the second law of thermodynamics, NET analysis offers a tool to calculate network constrained and thermodynamically possible ranges of metabolite concentrations. The power of this systems-level approach lies in determining metabolite concentrations that are feasible within the network rather than for just a single reaction (van Dien and Schilling 2006). As a result, fundamental principles of the metabolic 'system' can be discriminated. For example, NET analysis has already shown utility for determining the thermodynamic consistency of metabolome data (i.e. does reported data make sense in the context of the whole cell?). Strikingly, the authors show that of seven published *Escherichia coli* data sets, three were not consistent with the assumed flux distribution. This result emphasizes the need for improving our ability to accurately measure intracellular metabolite concentrations and promises to be a simple strategy for checking experimental errors. Another application of NET analysis is to resolve compartmentalized and/or pooled metabolites. As previously noted, a major bottleneck for endometabolome analysis is that compartmental metabolite concentration differences, between the cytosol and mitochondria for example, cannot be resolved. Kümmel et al. (2006) demonstrate that their computational framework was able to resolve compartmental differences for malate and oxaloacetate using *S. cerevisiae* data. One of the most compelling applications of NET analysis is to identify reactions that lie near or far from equilibrium conditions. This can, as the authors highlight, be used to gain insight into putative regulatory sites, with reactions far from equilibrium being more likely to impose regulatory control. NET analysis is currently limited by our inability to accurately measure a wide range of intracellular metabolite concentrations, by the requirement for knowing the direction of metabolite flux through the metabolic network, and by the need for standard Gibbs energies of formation. It should also be noted that while grounded in known network topology, NET analysis may suffer from thermodynamic assumptions.

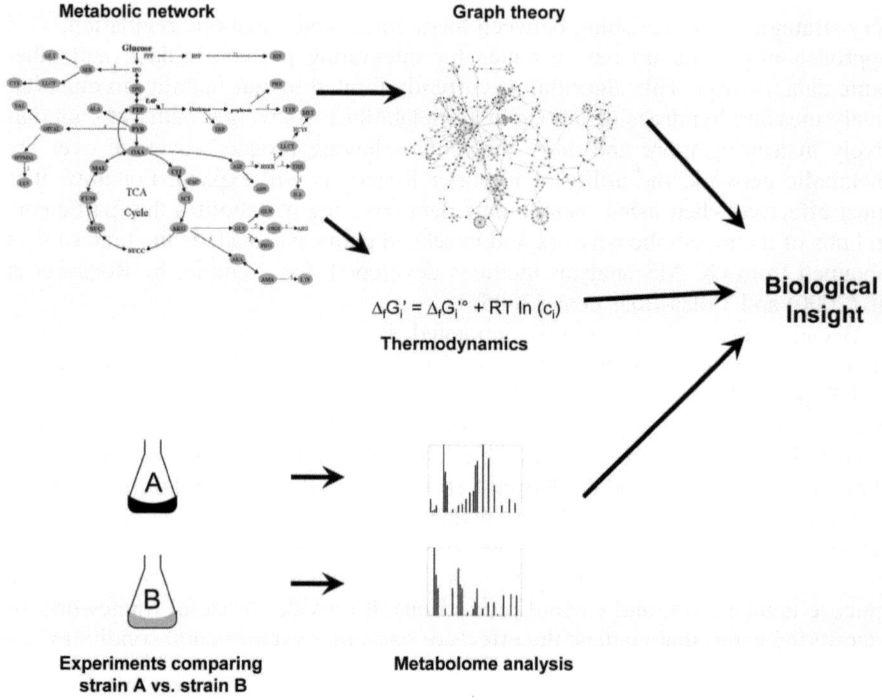

Fig. 5. New strategies to upgrade the information content in metabolome data for gaining biological insight are employing metabolic networks, graph theory, and thermodynamics.

By using graph theory to integrate metabolome data with metabolic network topology, Çakir et al. (2006) have proposed a general framework for identifying biochemical reactions around which the most significant coordinated metabolite changes are observed, termed 'Reporter Reactions'. Their approach is based, in part, on an earlier report used for discovering co-regulated sub-networks and reporter metabolites from transcriptome data (Patil and Nielsen 2005). The Reporter Reaction algorithm uses the significance of change in metabolite levels between conditions of interest to calculate a normalized Z-score for each reaction in the network. The calculated Z-score is ranked and used to determine Reporter Reactions. The major hurdle that the authors had to overcome was how to deal with only having measurements for a small fraction of metabolites present in the genome-scale metabolic model (84 metabolites were measured and the full genome-scale model consists of 844 metabolites). To address the lack of quantitative data, the authors used several pre-processing steps, including flux balance analysis, to judiciously obtain a reduced metabolic model containing 178 metabolites. Using several case stories that looked at redox metabolism, oxygen availability, and high gravity fermentation effects, the authors demonstrate that their algorithm enables identification of key reactions in the yeast metabolism affected by genetic and environmental perturbations. Equally important, they show that their 'reporter' platform can be integrated with transcriptomic data to map different cellular regulatory strategies, distinguishing between hierarchical and metabolic regulation. This approach may open up new avenues for integrating proteomic data, and other omic data, as well. This algorithm is currently limited by our inability to quantitatively measure hundreds of intracellular metabolites. As we get better at quantitatively measuring more and more metabolites having broader coverage over the metabolic network, the utility of Reporter Reactions will expand. For now, it is most effective when using metabolome data covering metabolites that participate in hubs of the metabolic network and in related pathways, such as the kind of data obtained from GC-MS analysis methods developed, for example, by Roessner et al. (2000) and Villas-Bôas et al. (2005b).

Because metabolism is highly annotated, it offers an extraordinary steppingstone for using metabolome data to understand cellular behavior. NET analysis and Reporter Reactions are excellent examples of this. However, other analysis methods, that do not impose network structure constraints, are also being developed to integrate metabolic data with additional cellular response measurements. Pir et al. (2006) have used the Partial Least Squares (PLS) method to linear model and integrate the transcriptomic and metabolomic response of *S. cerevisiae* to different media composition, growth rate, and specific gene deletion. While their approach falls short of using many measured metabolites (they only used biomass, glucose uptake rates, and ethanol production), it provides a useful framework for identifying genes that mediate the effects of particular experimental conditions.

6 Future outlook

Looking forward, we anticipate that new standards in chemical analysis, novel analytical strategies for quantitatively identifying increasing numbers of metabolites, and integration strategies for connecting metabolome data with molecular measurements from other functional levels of the cell will play an ever important role in systems biology efforts.

Acknowledgements

M.C. Jewett, M.A.E. Hansen, and J. Nielsen are most grateful to the NSF International Research Fellowship Program and the Danish Research Council for Technology and Production Sciences for supporting our work.

References

Allen J, Davey HM, Broadhurst D, Heald JK, Rowland JJ, Oliver SG, Kell DB (2003) High-throughput classification of yeast mutants for functional genomics using metabolic footprinting. Nat Biotechnol 21:692-696

Alper H, Miyaoku K, Stephanopoulos G (2005) Construction of lycopene-overproducing *E. coli* strains by combining systematic and combinatorial gene knockout targets. Nat Biotechnol 23:612-616

Anderberg MR (1973) Cluster Analysis for Applications. Academic Press: New York, NY

Antoniou A (1993) Digital Filters: Analysis, Design, and Applications. McGraw-Hill: New York, NY

Bartlett MS (1937) The statistical conception of mental factors. Br J Psychol 28:97-104

Brown M, Dunn WB, Ellis DI, Goodacre R, Handl J, Knowles JD, O'Hagan S, Spasić I, Kell DB (2005) A metabolome pipeline: from concept to data to knowledge. Metabolomics 1:39-51

Bylund D, Danielsson R, Malmquist G, Markides KE (2002) Chromatographic alignment by warping and dynamic programming as a pre-processing tool for PARAFAC modelling of liquid chromatography-mass spectrometry data. J Chromatogr A 961:237-244

Cai T, Zhang D, Ben-Amotz D (2001) Enhanced chemical classification of Raman images using multiresolution wavelet transformation. Appl Spectrosc 55:1124-1130

Çakir T, Patil KR, Onsan ZI, Ulgen KO, Kirdar B, Nielsen J (2006) Integration of metabolome data with metabolic networks reveals reporter reactions Mol Sys Biol 2:0050 doi:10.1038/msb4100085

Cameron DC, Altaras NE, Hoffman ML, Shaw AJ (1998) Metabolic engineering of propanediol pathways. Biotechnol Prog 14:116-125

Castrillo JI, Hayes A, Mohammed S, Gaskell SJ, Oliver SG (2003) An optimized protocol for metabolome analysis in yeast using direct infusion electrospray mass spectrometry. Phytochemistry 62:929-937

Castrillo JI, Oliver SG (2004) Yeast as a touchstone in post-genomic research: strategies for integrative analysis in functional genomics. J Biochem Mol Biol 37:93-106

Depczynski U, Jetter K, Molt K, Niemöller A (1997) The fast wavelet transform on compact intervals as a tool in chemometrics. I. Mathematical background. Chemom Intell Lab Syst 39:19-27

Dromey RG, Stefik MJ, Rindfleisch TC, Duffield AM (1976) Extraction of mass spectra free of background and neighboring component contributions from gas chromatography/mass spectrometry data. Anal Chem 48:1368-1375

Duda RO, Hart PE, Stork DG (2001) Pattern Classification. Wiley: ISBN 0471056693

Eilers PHC (2003) A perfect smoother. Anal Chem 75:3631-3636

Eilers PHC (2004) Parametric time warping. Anal Chem 76:404-411

Endy D (2005) Foundations for engineering biology. Nature 438:449-453

Fiehn O (2002) Metabolomics: the link between genotypes and phenotypes. Plant Mol Biol 48:155-171

Fiehn O, Weckwerth W (2003) Deciphering metabolic networks. Eur J Biochem 270:579-588

Fisher RA (1936) The use of multiple measures in taxonomic problems. Ann Eugenics 7:179-188

Fix E, Hodges JL (1951) Discriminatory analysis, nonparametric discrimination: Consistency properties. Technical Report Technical Report 4. USAF School of Aviation Medicine, Randolph Field, TX

Forshed J, Schuppe-Koistinen I, Jacobsson SP (2003) Peak alignment of NMR signals by means of a genetic algorithm. Analytica Chimica Acta 487:189-199

Forster J, Famili I, Fu P, Palsson BO, Nielsen J (2003) Genome-scale reconstruction of the *Saccharomyces cerevisiae* metabolic network. Genome Res 13:244-253

Friedman JH, Stuetzle W (1981) Projection pursuit regression. J American Stat Assoc 76:817-823

Frisvad JC, Filtenborg O (1983) Classification of terverticillate *penicillia* based on profiles of mycotoxins and other secondary metabolites. Appl Environ Microbiol 46:1301-1310

Fung E, Wong WW, Suen JK, Bulter T, Lee SG, Liao JC (2005) A synthetic gene-metabolic oscillator. Nature 435:118-122

Giaever G, Chu AM, Ni L, Connelly C, Riles L, Veronneau S, Dow S, Lucau-Danila A, Anderson K, Andre B, Arkin AP, Astromoff A, El Bakkoury M, Bangham R, Benito R, Brachat S, Campanaro S, Curtiss M, Davis K, Deutschbauer A, Entian KD, Flaherty P, Foury F, Garfinkel DJ, Gerstein M, Gotte D, Guldener U, Hegemann JH, Hempel S, Herman Z, Jaramillo DF, Kelly DE, Kelly SL, Kotter P, LaBonte D, Lamb DC, Lan N, Liang H, Liao H, Liu L, Luo CY, Lussier M, Mao R, Menard P, Ooi SL, Revuelta JL, Roberts CJ, Rose M, Ross-Macdonald P, Scherens B, Schimmack G, Shafer B, Shoemaker DD, Sookhai-Mahadeo S, Storms RK, Strathern JN, Valle G, Voet M, Volckaert G, Wang CY, Ward TR, Wilhelmy J, Winzeler EA, Yang YH, Yen G, Youngman E, Yu KX, Bussey H, Boeke JD, Snyder M, Philippsen P, Davis RW, Johnston M (2002) Functional profiling of the *Saccharomyces cerevisiae* genome. Nature 418:387-391

Gilbert RJ, Goodacre R, Woodward AM, Kell DB (1997) Genetic programming: A novel method for the quantitative analysis of pyrolysis mass spectral data. Anal Chem 69:4381-4389

Goeddel DV, Kleid DG, Bolivar F, Heyneker HL, Yansura DG, Crea R, Hirose T, Kraszewski A, Itakura K, Riggs AD (1979) Expression in *Escherichia coli* of chemically synthesized genes for human insulin. Proc Natl Acad Sci USA 76:106-110

Goehner R (1978) Background subtract subroutine for spectral data. Anal Chem 50:1223-1225

Goffeau A, Barrell BG, Bussey H, Davis RW, Dujon B, Feldmann H, Galibert F, Hoheisel JD, Jacq C, Johnston M, Louis EJ, Mewes HW, Murakami Y, Philippsen P, Tettelin H, Oliver SG (1996) Life with 6000 genes. Science 274:546, 563-547

Gonzalez B, Francois J, Renaud M (1997) A Rapid and reliable method for metabolite extraction in yeast using boiling buffered ethanol. Yeast 13:1347-1356

Goodacre R, Vaidyanathan S, Dunn WB, Harrigan GG, Kell DB (2004) Metabolomics by numbers: acquiring and understanding global metabolite data. Trends Biotechnol 22:245-252

Grung B, Kvalheim OM (1995) Retention time adjustments of two-way chromatograms using Bessel's inequality. Anal Chim Acta 304:57-66

Hansen ME, Smedsgaard J (2004) A new matching algorithm for high resolution mass spectra. J Am Soc Mass Spectrom 15:1173-1180

Hansen ME, Smedsgaard J (2006) Automated work-flow for processing of high-resolution mass spectra from direct infusion fingerprinting. Metabolomics (Submitted)

Hartigan J (1975) Clustering Algorithms. Wiley: New York, NY

Hayes A, Zhang N, Wu J, Butler PR, Hauser NC, Hoheisel JD, Lim FL, Sharrocks AD, Oliver SG (2002) Hybridization array technology coupled with chemostat culture: Tools to interrogate gene expression in *Saccharomyces cerevisiae*. Methods 26:281-290

Herbrich R (2001) Learning Kernel Classifiers: Theory and Algorithms. MIT Press: ISBN 026208306X

Hertz HS, Hites RA, Biemann K (1971) Identification of mass spectra by computer searching a file of known spectra. Anal Chem 43:681-691

Hood L, Heath JR, Phelps ME, Lin B (2004) Systems biology and new technologies enable predictive and preventative medicine. Science 306:640-643

Isaacs FJ, Dwyer DJ, Collins JJ (2006) RNA synthetic biology. Nat Biotechnol 24:545-554

Ideker T, Thorsson V, Ranish JA, Christmas R, Buhler J, Eng JK, Bumgarner R, Goodlett DR, Aebersold R, Hood L (2001) Integrated genomic and proteomic analyses of a systematically perturbed metabolic network. Science 292:929-934

Jewett MC, Hofmann G, Nielsen J (2006) Fungal metabolite analysis in genomics and phenomics. Curr Opin Biotechnol 17:191-197

Johnson KJ, Wright BW, Jarman KH, Synoveca RE (2003) High-speed peak matching algorithm for retention time alignment of gas chromatographic data for chemometric analysis. J Chromatogr A 996:141-155

Karhunen J, Joutsensalo J (1995) Generalization of principal component analysis, optimization problems, and neural networks. Neural Netw 8:549-562

Kassidas A, MacGregor JF, Taylor PA (1998) Synchronization of batch trajectories using dynamic time warping. Amer Inst Chem Eng J 44:864-873

Kaufman L, Rousseeuw PJ (1990) Finding Groups in Data. John Wiley & Sons

Kell DB, Darby RM, Draper J (2001) Genomic Computing. Explanatory Analysis of Plant Expression Profiling Data Using Machine Learning. Plant Phys 126:943–951

Kell DB (2002) Metabolomics and machine learning: explanatory analysis of complex metabolome data using genetic programming to produce simple, robust rules. Mol Biol Rep 29:237-241

Kell DB (2004) Metabolomics and systems biology: making sense of the soup. Curr Opin Microbiol 7:296-307

Kong H, Ye F, Lu X, Guo L, Tian J, Xu G (2005) Deconvolution of overlapped peaks based on the exponentially modified Gaussian model in comprehensive two-dimensional gas chromatography. Journal of Chromatography A 1086:160-164

de Koning W, van Dam K (1992) A method for the determination of changes of glycolytic metabolites in yeast on a subsecond time scale using extraction at neutral pH. Anal Biochem 204:118-123

Koza JR (1992) Genetic Programming: On the Programming of Computers by Means of Natural Selection; MIT Press: Cambridge, MA

Koza JR (1994) Genetic Programming II: Automatic Discovery of Reusable Programs; MIT Press: Cambridge, MA

Kümmel A, Panke S, Heinemann M (2006) Putative regulatory sites unraveled by network-embedded thermodynamic analysis of metabolome data. Mol Syst Biol 2:0034 doi:10.1038/msb4100074

Lafaye A, Labarre J, Tabet JC, Ezan E, Junot C (2005) Liquid chromatography-mass spectrometry and 15N metabolic labeling for quantitative metabolic profiling. Anal Chem 77:2026-2033

Lange HC, Eman M, van Zuijlen G, Visser D, van Dam JC, Frank J, de Mattos MJ, Heijnen JJ (2001) Improved rapid sampling for *in vivo* kinetics of intracellular metabolites in *Saccharomyces cerevisiae*. Biotechnol Bioeng 75:406-415

Lee TI, Rinaldi NJ, Robert F, Odom DT, Bar-Joseph Z, Gerber GK, Hannett NM, Harbison CT, Thompson CM, Simon I, Zeitlinger J, Jennings EG, Murray HL, Gordon DB, Ren B, Wyrick JJ, Tagne JB, Volkert TL, Fraenkel E, Gifford DK, Young RA (2002)

Transcriptional regulatory networks in *Saccharomyces cerevisiae*. Science 298:799-804

Li J (2002) Comparison of the capability of peak functions in describing real chromatographic peaks. Journal of Chromatography A 952:63-70

Listgarten J, Nealy RM, Roweis ST, Emili A (2004) Multiple Alignment of Continuous Time Series. Advances in Neural Information Processing Systems 17. MIT Press: Cambridge, MA

Liu B, Sera Y, Matsubara N, Otsuka K, Terabe S (2003) Signal denoising and baseline correction by discrete wavelet transform for microchip capillary electrophoresis. Electrophoresis 24:3260-3265

MacQueen J (1966) Some methods for classification and analysis of multivariate observations. Proc Fifth Berkeley Symp University of California Press 1:281-297

Mager WH, Winderickx J (2005) Yeast as a model for medical and medicinal research. Trends Pharmacol Sci 26:265-273

Maleknia SD, Downard KM (2005) Charge ratio analysis method: approach for the deconvolution of electrospray mass spectra. Anal Chem 77:111-119

Malmquist G, Danielsson R (1994) Alignment of chromatographic profiles for principal component analysis: a prerequisite for fingerprinting methods. J Chromatogr A 687:71-88

Mardia KV, Kent JT, Bibby JM (1979) Multivariate Analysis. Academic Press: London

Mas S, Villas-Bôas SG, Hansen ME, Åkesson M, Nielsen J (2006) A comparison of direct infusion MS and GC-MS for metabolic footprinting of yeast mutants. Biotechnol Bioeng: in press

Mashego MR, Wu L, Van Dam JC, Ras C, Vinke JL, Van Winden WA, Van Gulik WM, Heijnen JJ (2004) MIRACLE: mass isotopomer ratio analysis of U-13C-labeled extracts. A new method for accurate quantification of changes in concentrations of intracellular metabolites. Biotechnol Bioeng 85:620-628

Mashego MR, van Gulik WM, Vinke JL, Visser D, Heijnen JJ (2006) *In vivo* kinetics with rapid perturbation experiments in *Saccharomyces cerevisiae* using a second-generation BioScope. Metab Eng 8:370-383

Mazet V, Carteret C, Briea D, Idier J, Humbert B (2005). Background removal from spectra by designing and minimizing a non-quadratic cost function. Chemom Intell Lab Syst 76:121-133

McLafferty FW (1974) Probability based matching of mass spectra. Org Mass Spectrom 9:690-702

Mendes P (2002) Emerging bioinformatics for the metabolome. Brief Bioinform 3:134-145

Mitra SK (1998) Digital Signal Processing: A Computer-Based Approach. McGraw-Hill: New York, NY

Nielsen J (2003) It is all about metabolic fluxes. J Bacteriol 185:7031-7035

Nielsen J, Oliver S (2005) The next wave in metabolome analysis. Trends Biotechnol 23:544-546

Nielsen NV, Carstensen JM, Smedsgaard J (1998) Aligning of single and multiple wavelength chromatographic profiles for chemometric data analysis using correlation optimized warping. J Chromatogr A 805:17-35

Patil KR, Nielsen J (2005) Uncovering transcriptional regulation of metabolism by using metabolic network topology. Proc Natl Acad Sci USA 102:2685-2689

Pearson K (1901) On Lines and Planes of Closest Fit to Systems of Points in Space. Phil Mag 2, 559-572

Percival DB, Walden AT (2000) Wavelet Methods for Time Series Analysis. Cambridge University Press: Cambridge, MA

Pir P, Kirdar B, Hayes A, Onsan ZY, Ulgen KO, Oliver SG (2006) Integrative investigation of metabolic and transcriptomic data. BMC Bioinformatics 7:203

Porro D, Sauer M, Branduardi P, Mattanovich D (2005) Recombinant protein production in yeasts. Mol Biotechnol 31:245-259

Pravdova V, Walczak B, Massart DL (2002) A comparison of two algorithms for warping of analytical signals. Anal Chim Acta 456:77-92

Raamsdonk LM, Teusink B, Broadhurst D, Zhang N, Hayes A, Walsh MC, Berden JA, Brindle KM, Kell DB, Rowland JJ, Westerhoff HV, Dam KV, Oliver SG (2001) A functional genomics strategy that uses metabolome data to reveal the phenotype of silent mutations. Nat Biotechnol 19:45-50

Ro DK, Paradise EM, Ouellet M, Fisher KJ, Newman KL, Ndungu JM, Ho KA, Eachus RA, Ham TS, Kirby J, Chang MC, Withers ST, Shiba Y, Sarpong R, Keasling JD (2006) Production of the antimalarial drug precursor artemisinic acid in engineered yeast. Nature 440:940-943

Roessner U, Wagner C, Kopka J, Trethewey RN, Willmitzer L (2000) Simultaneous analysis of metabolites in potato tuber by gas chromatography–mass spectrometry. Plant J 23:131-142

Rose AH, Harrison JS (1987-1995) The yeasts. Vol. 1-6. Academic Press: London, UK

Ross-Macdonald P, Coelho PS, Roemer T, Agarwal S, Kumar A, Jansen R, Cheung KH, Sheehan A, Symoniatis D, Umansky L, Heidtman M, Nelson FK, Iwasaki H, Hager K, Gerstein M, Miller P, Roeder GS, Snyder M (1999) Large-scale analysis of the yeast genome by transposon tagging and gene disruption. Nature 402:413-418

Round AJ, Aguilar MI, Hearn MTW (1994) High-performance liquid chromatography of amino acids, peptides and proteins: CXXXIII. Peak tracking of peptides in reversed-phase high-performance liquid chromatography. J Chromatogr A 661:61-75

Savitzky A, Golay MJE (1964) Smoothing and differentiation of data by simplified least squares procedures. Anal Chem 36:1627-1639

Shao X, Cai W, Pan Z (1999) Wavelet transform and its applications in high performance liquid chromatography (HPLC) analysis. Chemom Intell Lab Syst 45:249–256

Spellman PT, Sherlock G, Zhang MQ, Iyer VR, Anders K, Eisen MB, Brown PO, Botstein D, Futcher B (1998) Comprehensive identification of cell cycle-regulated genes of the yeast *Saccharomyces cerevisiae* by microarray hybridization. Mol Biol Cell 9:3273-3297

Stein SE (1995) Chemical Substructure Identification by Mass Spectral Library Searching. J Am Soc Mass Spectrom 6(8):644-655

Stein SE (1999) An Integrated Method for Spectrum Extraction and Compound Identification from Gas Chromatography/Mass Spectrometry Data. J Am Soc Mass Spectrom 10:770–781

Styczynski MP, Moxley JF, Tong LV, Walther JL, Jensen KL, Stephanopoulos GN (2007) Systematic identification of conserved metabolites in GC/MS data for metabolomics and biomarker discovery. Anal Chem *79*:966-973

Sumner LW, Mendes P, Dixon RA (2003) Plant metabolomics: large-scale phytochemistry in the functional genomics area. Phytochemistry 62:817-836

Tan HW, Brown S (2002) Wavelet analysis applied to removing nonconstant, varying spectroscopic background in multivariate calibration. J Chemom 16:228-240

Tibshirani R, Hastie T, Narasimhan B, Soltys S, Shi G, Koong A, Le Q (2004) Sample Classification from Protein Mass Spectrometry, by 'Peak Probability Contrasts'. Bioinformatics 20(17):3034-3044

Tomasi G, van den Bergand F, Andersson C (2004) Correlation optimized warping and dynamic time warping as preprocessing methods for chromatographic data. J Chemom 18:231-241

Torres-Lapasió JR, Baeza-Baeza JJ, García-Alvarez-Coque MC (1997) A Model for the Description, Simulation, and Deconvolution of Skewed Chromatographic Peaks. Anal Chem 69:3822-3831

Uetz P, Giot L, Cagney G, Mansfield TA, Judson RS, Knight JR, Lockshon D, Narayan V, Srinivasan M, Pochart P, Qureshi-Emili A, Li Y, Godwin B, Conover D, Kalbfleisch T, Vijayadamodar G, Yang M, Johnston M, Fields S, Rothberg JM (2000) A comprehensive analysis of protein-protein interactions in *Saccharomyces cerevisiae*. Nature 403:623-627

van Dam JC, Eman MR, Frank J, Lange HC, van Dedem GWK, Heijnen SJ (2002) Analysis of glycolytic intermediates in *Saccharomyces cerevisiae* using anion exchange chromatography and electrospray ionization with tandem mass spectrometric detection. Anal Chimica Acta 460:209-218

Van Dien S, Schilling CH (2006) Bringing metabolomics data into the forefront of systems biology. Mol Syst Biol 2:0035 doi:10.1038/msb4100078

Vickers T, Wambles R, Mann C (2001) Curve fitting and linearity: data processing in Raman spectroscopy. Appl Spectrosc 55:389-393

Villas-Bôas SG, Hojer-Pedersen J, Akesson M, Smedsgaard J, Nielsen J (2005a) Global metabolite analysis of yeast: evaluation of sample preparation methods. Yeast 22:1155-1169

Villas-Bôas SG, Moxley JF, Akesson M, Stephanopoulos G, Nielsen J (2005b) High-throughput metabolic state analysis: the missing link in integrated functional genomics of yeasts. Biochem J 388:669-677

Villas-Bôas SG, Mas S, Akesson M, Smedsgaard J, Nielsen J (2005c) Mass spectrometry in metabolome analysis. Mass Spectrom Rev 24:613-646

Vivó-Truyols G, Torres-Lapasió JR, Caballero RD, García-Alvarez-Coque MC (2002) Peak deconvolution in one-dimensional chromatography using a two-way data approach. J Chromatogr A 958:35-49

Vivó-Truyols G, Torres-Lapasió JR, van Nederkassel AM, Heyden YV, Massart DL (2005a) Automatic program for peak detection and deconvolution of multi-overlapped chromatographic signals Part I: Peak detection. J Chromatogr A 1096:133-145

Vivó-Truyols G, Torres-Lapasió JR, van Nederkassel AM, Heyden YV, Massart DL (2005b) Automatic program for peak detection and deconvolution of multi-overlapped chromatographic signals Part II: Peak model and deconvolution algorithms. Journal of Chromatography A 1096:146-155

Walczaka B, Wub W (2005) Fuzzy warping of chromatograms. Chemom Intell Lab Syst 77:173-180

Wang CP, Isenhour TL (1987). Time-warping algorithm applied to chromatographic peak matching gas-chromatography Fourier-transform infrared mass-spectrometry. Anal Chem 59: 649-654

Winzeler EA, Shoemaker DD, Astromoff A, Liang H, Anderson K, Andre B, Bangham R, Benito R, Boeke JD, Bussey H, Chu AM, Connelly C, Davis K, Dietrich F, Dow SW, El Bakkoury M, Foury F, Friend SH, Gentalen E, Giaever G, Hegemann JH, Jones T,

Laub M, Liao H, Davis RW (1999) Functional characterization of the *S. cerevisiae* genome by gene deletion and parallel analysis. Science 285:901-906

Wu L, Mashego MR, van Dam JC, Proell AM, Vinke JL, Ras C, van Winden WA, van Gulik WM, Heijnen JJ (2005) Quantitative analysis of the microbial metabolome by isotope dilution mass spectrometry using uniformly 13C-labeled cell extracts as internal standards. Anal Biochem 336:164-171

Wu L, Mashego MR, Proell AM, Vinke JL, Ras C, van Dam J, van Winden WA, van Gulik WM, Heijnen JJ (2006) *In vivo* kinetics of primary metabolism in *Saccharomyces cerevisiae* studied through prolonged chemostat cultivation. Metab Eng 8:160-171

Zhu H, Bilgin M, Bangham R, Hall D, Casamayor A, Bertone P, Lan N, Jansen R, Bidlingmaier S, Houfek T, Mitchell T, Miller P, Dean RA, Gerstein M, Snyder M (2001) Global analysis of protein activities using proteome chips. Science 293:2101-2105

MC Jewett and MAE Hansen contributed equally to this work

Hansen, Michael A. E.
Center for Microbial Biotechnology, BioCentrum-DTU, Technical University of Denmark, Søltofts Plads, DK-2800 Kgs. Lyngby, Denmark

Jewett, Michael C.
Center for Microbial Biotechnology, BioCentrum-DTU, Technical University of Denmark, Søltofts Plads, DK-2800 Kgs. Lyngby, Denmark

Nielsen, Jens
Center for Microbial Biotechnology, BioCentrum-DTU, Technical University of Denmark, Søltofts Plads, DK-2800 Kgs. Lyngby, Denmark
jn@biocentrum.dtu.dk

E. coli metabolomics: capturing the complexity of a "simple" model

Martin Robert, Tomoyoshi Soga and Masaru Tomita

Abstract

As the workhorse of early studies on metabolism, the metabolic pathways of *E. coli* are arguably the best characterized. The richness of information available about its pathways is broader than for any other model. However, in spite of decades of descriptive work, only recently can a significant number of *E. coli* metabolic network constituents be analyzed simultaneously. The advent of metabolomic methods that allow to capture qualitative as well as quantitative information about the intracellular and extracellular metabolite profiles is starting to shed light on the remaining complexity of this simpler model. Here we describe important findings about the physiology of *E. coli* resulting from emerging metabolomic studies. While a vast number of intracellular metabolites in *E. coli* still remain to be characterized, the information obtained from those studies can provide an unprecedented amount of information about metabolic pathways including their functional elucidation, enzyme activity, metabolic fluxes, network robustness, or even the discovery of completely novel reactions or pathways. These results are also being used to populate rich databases and to develop computational models of *E. coli* metabolism that have already proven effective to predict cellular states and will shed light on complex and until now still elusive regulatory principles.

1 Introduction

Most scientists (biochemists) will remember their introductory biochemistry class as an introduction to the intricacies and inherent network structure of biochemical pathways. The related biochemical pathway maps are both surprisingly complex and puzzling and therefore command respect. We also know that most of these pathways were elucidated by brave, and often isolated, scientists working with this popular model organism, *Escherichia coli (E. coli)*. In fact, much of the original biochemical knowledge regarding enzyme activities originates from work derived from this organism. Now that we have firmly stepped into the metabolomic era, the sheer complexity of the metabolome, as far as we can see it (only the tip of the iceberg?), definitely commands respect even from the most active and prolific research groups. *E. coli* remains a friendly ally in this continuing quest for the ex-

Topics in Current Genetics, Vol. 18
J. Nielsen, M. C. Jewett (Eds.): Metabolomics
DOI 10.1007/4735_2007_0221 / Published online: 14 April 2007

haustive (both qualitative and quantitative) elucidation of the biochemical intermediates (metabolites) of a single unicellular organism.

The metabolome can be defined as the complement of low molecular weight metabolites in the cell under particular physiological conditions (Kell et al. 2005). It is thus a very dynamic molecular ensemble whose profile changes rapidly under environmental conditions. It is currently still not possible to obtain a complete picture of the metabolome with any single experimental method, and at present, only a subset of the metabolome is actually being sampled, surveyed, and consequently used to infer new biological insight. However, biochemical reactions are often conserved across species and by definition the chemical species that make them up are exactly the same, in contrast to proteins and genes that are usually only partially conserved during evolution. In spite of the remaining difficulties in developing analytical methods that can sample a more substantial proportion of the metabolome, metabolomics, the qualitative and quantitative analysis and characterization of the metabolome, benefits from the fact that effective methods can be applicable directly to other organisms since compounds are conserved.

For all the secrets that *E. coli* has shed, sometimes reluctantly, about its functions, it has yet to reveal a considerable number. These are the subject of intense research efforts and the speed and ease at which *E. coli* can be grown and manipulated together with the availability of genome-wide resources such as the Keio collection (Baba et al. 2006) and ASKA libraries (Kitagawa et al. 2005) make it a leading player for the limelight among other model organisms such as yeast. As a unicellular organism, *E. coli* presents obvious advantages; the necessary biomass for extraction of metabolites can be easily and rapidly obtained and it is easier to analyze with its well-known and simpler genetics. At the same time, the challenges of preserving and then breaking through a double membrane, and the presence of culture medium can complicate some analytical tasks.

Reliable experimental data is the cornerstone for building useful models of *E. coli* and metabolomic data is one of the most desirable data type toward this goal (Arita et al. 2005; Ishii et al. 2005). In this chapter, we introduce the basic methods that are making possible the analysis of the *E. coli* metabolome and the main studies that have recently emerged. Applications of *E. coli* metabolomics as a tool for functional genomics and metabolic engineering are later introduced. Finally, the initial efforts reporting the construction of large-scale models of its metabolism and various bioinformatics resources for *E. coli* metabolomics are described. As a general review of multiple facets of *E. coli* metabolomics this chapter is not meant to be exhaustive. Hopefully the reader will be inspired to seek additional information among the described references and resources.

2 Experimental methods

Prior to extraction of metabolites, *E. coli* cells obviously need to be grown. Bacterial cells such as *E. coli* usually reproduce rapidly so that cultures times are minimal and a large number of cells can be obtained rapidly. Culture conditions can vary between standard batch culture and steady-state continuous cultures where substrate (usually glucose) is limiting and is gradually provided while spent media is removed and collected. This allows *E. coli* to be grown at a constant growth rate and avoids the growth-reducing effects of accumulation of metabolites. Since biological variation is usually larger than variation due to analytical methods, a reproducible system such as continuous culture can thus be highly desirable.

2.1 Quenching of metabolism and metabolite extraction

Current methods of metabolomic analysis generally make use of so-called invasive procedures where tissues or cells are sampled, sometimes pretreated, and then disrupted to produce a soup of metabolites that can be analyzed using standard or emerging analytical methods. The extraction of metabolites from *E. coli* and other microorganisms requires special care since numerous compounds are also usually present in the culture medium and must thus usually be removed to facilitate the analysis of intracellular metabolites. Since metabolic reactions occur very quickly, rapid quenching of biochemical reactions is important to obtain a metabolite mixture that is representative of the *in vivo* situation at sampling time. Unfortunately, in practice, this is often difficult to do due to the necessity to remove the extracellular culture medium. When seeking accurate quantification of intracellular metabolites one must thus decide on a trade-off between cleaner samples whose profile might have changed since sampling versus more rapidly quenched cellular samples that likely contain extracellular metabolites and from which intracellular metabolites likely have leaked. The extracellular medium can first be removed, to avoid contamination with medium components, by filtration or centrifugation but these processes happen over time scales much longer than the turnover time of most metabolites in the cell. Some of the basic difficulties regarding metabolism quenching and metabolite extraction have recently been discussed (Villas-Boas et al. 2005a). While this issue remains mostly unsolved, comparative analysis of samples originating from different strains, for example, are still possible and useful although acknowledging uncertainties about the accurate quantification of certain metabolites is necessary.

The rapid and efficient quenching of cellular metabolism and extraction of metabolites is thus crucial and, as for earlier work done in yeast (de Koning and van Dam 1992), rapidly spraying *E. coli* cells directly into cold methanol has been a commonly used quenching method (Buchholz et al. 2001, 2002). Alternately, Bhattacharya used boiling water to both quench and extract metabolites (Bhattacharya et al. 1995). While this can effectively inactivate most enzymes, heat labile metabolites will obviously be lost in such a way. A recent study however reports that heat treatment (< 95°C) may not be so problematic for most metabolites and can be used for both very rapid quenching and extraction of metabolites (Schaub et al. 2006). Perchloric acid extraction of intracellular metabolites has also been used successfully (Larsson and Tornkvist 1996; Emmerling et al. 1999; Buchholz et al. 2001, 2002) though it is also limited by the stability of metabolites under

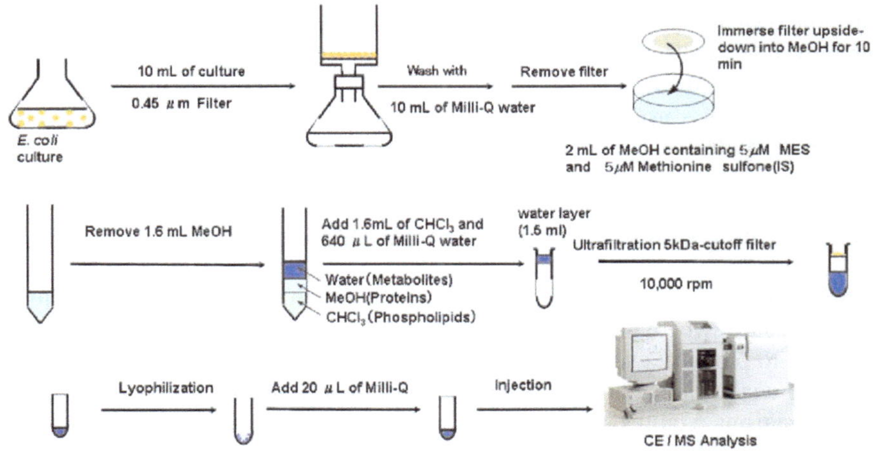

Fig. 1. Basic protocol for extraction of polar metabolites from *E. coli* for analysis by CE-MS. An aliquot from a culture of *E. coli* is collected, rapidly filtered, and washed with water to remove extracellular components. Quenching of metabolism is performed by immersing the filter directly in 100% methanol. Liquid phase extraction is performed by adding water and chloroform. The water layer, containing mostly polar metabolites, is isolated, ultrafiltered to remove proteins and other large molecules, lyophilized, and stored until analysis. Prior to analysis by CE-MS, the metabolites are redissolved in 20 μl of water. Reproduced and modified with permission from (Tomita and Nishioka 2005).

such conditions. One study compared six different methods for extracting metabolites from *E. coli* (Maharjan and Ferenci 2003) including hot ethanol, hot or cold methanol, perchloric acid, alkaline (potassium hydroxide), and methanol/chloroform extraction. It is obvious that any specific extraction solution is bound to result in the loss of specific metabolites that are unstable in its presence. Cold methanol was found to be the most promising with respect to the total number of metabolites extracted, providing acceptable recovery and being also simple and rapid. The results possibly reflect both the extracting power of methanol, for a broad range of metabolites, as well as the relative stability of many metabolites in this solvent. Probably because of its ease and versatility, the cold methanol extraction procedure has thus continued to be widely used for *E. coli* (Bajad et al. 2006; Koek et al. 2006). At the Institute for Advanced Biosciences (IAB), Keio University, as originally developed for *B. subtilis* sample preparation (Soga et al. 2003), *E. coli* culture samples are sampled, rapidly filtered, and the filter rinsed with water (Fig. 1). Metabolites are then extracted in methanol at room temperature. This procedure seems to produce the best results in terms of recovery, reproducibility, and overall diversity of observable metabolites (T. Soga, unpublished). In addition, it allows the elimination of medium-derived metabolites and the maintenance of membrane integrity, something that cannot be achieved by plunging cells directly into cold methanol (Wittmann et al. 2004).

An important thing to keep in mind is that as with any other organism, it is clear that no single method allows to quantitatively extract all compounds efficiently due to their wide variety of physico-chemical properties. However, a combination of methods can potentially be used for particular subgroups of metabolites. In addition, as mentioned above, removal of culture medium requires precious time that may result in post-sampling changes in metabolite content, whereas plunging culture samples directly into cold methanol can result in metabolite leakage and sampling-induced profile changes (Wittmann et al. 2004). A promising recent advance that addresses this two-fold issue makes use of *E. coli* cells grown on filters instead of liquid culture, to minimize extracellular medium contamination and for rapid and easy quenching/extraction by plunging the filters directly in cold methanol (Brauer et al. 2006).

Note: The following two sections, together, provide a look at existing and emerging studies of the *E. coli* metabolome. Section 2.2 is mostly method-oriented while Section 3 describes mostly applications-oriented metabolomic studies. However, the two sections partially overlap in studies and content and the somewhat artificial separation should be considered non-rigid since many of the studies reviewed are both development- and applications-oriented. It is only for the purpose of illustrating methods and results separately that studies have been grouped this way. Here, emphasis is placed on methods that have already been used or shown to be applicable to the analysis of *E. coli* metabolites.

2.2 Main analytical methods tested with *E. coli*

Traditionally, the elucidation of metabolic intermediates and pathways progressed by the analysis of a single or relatively few chemical species at a time. This approach has been widely successful and most existing qualitative descriptions originate from such efforts. However, emerging tools for more comprehensive analysis of metabolites based on hyphenated mass spectrometry methods and nuclear magnetic resonance (NMR) are powerful for measuring in parallel the concentration of multiple metabolites, making possible the determination of pathway functions in a systems fashion. While there are still only relatively few applications in *E. coli*, some of the main methods used to interrogate the *E. coli* metabolome in both focused and large-scale studies will be described and are displayed in Figure 2. As mentioned above, the analytical platforms described are universal and can in theory be used similarly for any species with similar results since the structure of metabolites is shared for all organisms. It is nonetheless interesting to examine some of the major analytical methods that have been used specifically to analyze *E. coli* metabolites. As mentioned in Section 2.1, the main challenges for this bacterium, as for most other microorganisms, are to quantitatively extract most metabolites and avoid contamination and interference from culture medium-derived components.

2.2.1 Enzyme assays

For any analytical method, sensitivity is an important issue since under typical culture conditions, for microorganisms, the volume of cells may represent about only 0.1% of the culture volume (Kaderbhai et al. 2003) and metabolites represent only a minor fraction of *E. coli* dry cell weight (approximately 3%) (Neidhardt et al. 1990). Traditional assays making use of enzymatic reactions to quantify metabolites have been and continue to be widely used in targeted approaches. Enzyme assays have proven useful to quantify several *E. coli* metabolites due to their high specificity (Lowry et al. 1971; Emmerling et al. 1999; Schaefer et al. 1999; Chassagnole et al. 2002) and usually do not require pre-fractionation of samples. Most such assays are based on ultra-violet (UV), visible, bioluminescence or fluorescence spectroscopy for the detection of reaction intermediates. In most cases, direct or indirect quantification of the production or conversion of a metabolite is used to evaluate the concentration of the target metabolite in the sample. Some of the disadvantages of using enzymatic assays are that they usually require large amounts of sample, can be time-consuming and costly, and are sometimes difficult to conduct in parallel from the same sample. They are also prone to unpredictable effects due to modulators of enzyme activity that may be present in the sample matrix. Moreover, enzymatic assays are usually limited to the measurement of a relatively limited number of metabolites due to difficulties in multiplexing.

2.2.2 Chromatography

A common way to characterize components of a complex sample is usually to use one or more separation method(s) usually based on chromatography, making use of the differential interaction of analytes with a stationary phase. As such, thin-layer chromatography (TLC) has been used successfully to monitor a limited number of metabolites in *E. coli* (Tweeddale et al. 1998, 1999; Liu et al. 2000). The method is based on the differential migration of metabolites along a solid support using specific solvents. More recently up to 95 different metabolites could be observed in *E. coli* extracts using two-dimensional TLC (Maharjan and Ferenci 2003). In addition, high-performance liquid chromatography with conductivity and UV detection has also been used to monitor and identify multiple metabolites in *E. coli* (Bhattacharya et al. 1995; Buchholz et al. 2002).

2.2.3 Mass spectrometry-based methods

The advent of methods based on mass spectrometry (MS) and their inherent high sensitivity and selectivity has been an important step for the emergence of metabolomics. Mass spectrometry draws its power from the ability to precisely measure the mass (or rather the mass to charge ratio) of a molecule thereby providing high selectivity of detection even in complex matrices and also potentially allowing to unambiguously identify a metabolite when its mass is unique. Mass spectrometric methods, including direct infusion, have been used in yeast for intracellular (Castrillo et al. 2003) and extracellular (Allen et al. 2003) measurements of a large number of metabolites and the approach has been used to functionally characterize *E. coli* mutants based on metabolite footprints (Kaderbhai et al. 2003) (see Section 6.1). While infusion methods are useful for surveying the complexity of the metabolome and for functional genomics using metabolic footprinting - defined as the profiling of secreted metabolites - they are not really quantitative. Hyphenated MS methods, where pre-separation of complex samples is performed on-line prior to MS analysis, are preferable for quantitative work.

Recently, hyphenated MS methods have made possible the detection, and sometimes identification, of an increasing variety of metabolites in *E. coli*. These methods combine the power of metabolite physical separation with the additional level of selectivity provided by mass spectrometry to considerably increase the number of co-eluting metabolites that can be determined.

Recently developed capillary electrophoresis-mass spectrometry (CE-MS) methods allow the reliable and quantitative analysis of complex mixtures of anions, cations, and nucleotides (Soga and Heiger 2000; Soga et al. 2002a, 2002b, 2004). Using these methods, more than 1600 compounds could be detected in *B. subtilis* (Soga et al. 2003) including most amino acids, a large number of nucleotides and most intermediates of central carbon metabolism. Though not using *E. coli* as a model, a more recently developed but very similar CE-time-of-flight-MS method, with improved sensitivity for both cations and anions, was recently reported (Soga et al. 2006) and is currently being used to characterize various *E. coli* strains (T. Soga, unpublished). In addition, a modified method to facilitate the analysis of anions by using a reversed electroosmotic flow and pressure-assisted CE-MS has recently been reported (Harada et al. 2006). Resolution seems excellent and the use of a standard fused-silica capillary for the separation of anions is a significant advantage. However, the overall reproducibility of the method and its quantitative potential have not yet been clearly demonstrated.

Gas-chromatography mass spectrometry (GC-MS) and liquid-chromatography mass spectrometry (LC-MS) have been used successfully to characterize *E. coli* metabolites and some of this work will also be introduced in the next section. GC-MS is a powerful method with very high resolution, good quantification ability, and often simplifies compound identification due to the use of electron impact ionization (together with the availability of standard databases) (Kopka et al. 2005). However, it requires sample derivatization, which can introduce errors and is time-consuming, and is more problematic for thermally unstable compounds such as phosphorylated intermediates of central carbon metabolism. Nevertheless, the power of GC-MS-based bacterial metabolomics was demonstrated through the quantification of 121 metabolites in *Corynebacterium glutamicum* (Strelkov et al. 2004). Recently, using an improved derivatization method,the detection of over 200 metabolites in *E. coli*, 60 of which could be reliably identified was reported (Koek et al. 2006). With this method most classes of organic molecules could be analyzed satisfactorily including carboxylic acids, phosphates, and amines. On the other hand, the linearity and detection limits for metabolites containing amides, thiols, or sulfonic acid were not as good. Emerging two-dimensional GC X GC-MS methods promise even better results due to the increased separation efficiency though there are yet no reports of using such methods for *E. coli*.

Since in general polar metabolites are poorly retained on standard reverse-phase chromatography columns, GC-MS together with CE-MS have been the tools of choice to analyze the large number of polar and charged metabolic intermediates. However, alternative LC-MS methods are appearing that are likely to have a significant impact in the field. To improve retention, peak shape, and recovery of polar metabolites, an optimized system using hexylamine as the ion-pairing agent and a pH gradient was recently developed (Coulier et al. 2006). The method allows separating and quantifying multiple *E. coli* metabolites using standard reverse phase columns (IP-LC-ESI-MS). Only limited (<10%) ion-suppression effects were observed, at least for the few metabolites tested. Overall, while a total of more than 150 commercially available standards could be analyzed, 68 different nucleotides and 24 coenzyme A esters were determined in *E. coli*. However, in contrast to the CE-MS methods, IP-LC-ESI-MS still does not perform as well for highly polar metabolites (sugars, amino acids, etc.). Alternately, an LC-MS system based on separation by hydrophilic-interaction chromatography with an amino column, followed by detection using tandem MS on a triple-quadruple system looks promising (Bajad et al. 2006). This system allowed the identification and relative quantification of 69 mostly polar *E. coli* metabolites. To achieve good sensitivity and specificity, mass spectrometry is performed in single reaction monitoring (SRM) mode and thus the approach is targeted to selected metabolites of interest. The relative standard deviation (RSD) median was a respectable 13 and 31% for intra- and inter-sample measurements, respectively.

The reader interested in more complete details about CE-MS, GC-MS, and LC-MS methods for metabolomics is referred to other publications (Fiehn 2002; Villas-Boas et al. 2005b). In addition, emerging MS-based methods such as desorption electrospray ionization (DESI) MS and direct analysis in real time (DART) (Chen et al. 2006; Cooks et al. 2006) promise to increase the throughput by minimizing the sample preparation steps. They are highly sensitive, selective and may find useful applications in microbial metabolomics. Ion-mobility based methods that can separate isomers and conformers should also find applications in metabolome analysis (Ochoa and Harrington 2005; Koeniger et al. 2006).

The use of isotopically labeled internal standards for each monitored metabolite in hyphenated-MS methods would be ideal for quantitative work since analyses of samples with complex matrices are prone to suffer from ion-suppression effects. However, this is impractical and at first appears prohibitively expensive for metabolome-wide studies. To circumvent these difficulties, some recent studies have reported the use of saturating *in vivo* labeling of biomass with stable isotopes. This generates labeled internal standards that can be used for more reliable absolute quantification by isotope ratio, and such methods could find broad applications (Birkemeyer et al. 2005; Wu et al. 2005; Bajad et al. 2006).

2.2.4 Nuclear magnetic resonance

To resolve a large number of compounds, NMR is another powerful method for metabolomics that can even be performed in a non-invasive manner, a very desirable property. However, the sensitivity of NRM-based methods is typically orders of magnitude lower than MS-based methods. The potential for high-throughput analysis has yet to be exploited fully although the methods are quantitative and reproducible, factors that can make up for the reduced sensitivity (Pan and Raftery 2006). The complexity of metabolomic samples and the resulting overlap and redundancy in chemical shifts for numerous metabolites can also limit to some extent the power of the method. However, samples can also be fractionated and NMR has also been linked to off-line LC systems to address these issues (Noteborn et al. 2000). The existing need for non-invasive methods can thus possibly be filled by a combination of NMR-based methods, metabolic footprinting (Section 6.1), and emerging methods that have yet to be used to probe the *E. coli* metabolome.

As is the case for most organisms or sample source, MS-based methods currently dominate the landscape of analytical methods for microbial metabolome analysis. However, as it has been in yeast (Raamsdonk et al. 2001), NMR is bound to be useful for *E. coli* metabolomics too. The inherent generic nature of these two methods make them particularly well suited for the analysis of metabolite pools of considerable physico-chemical diversity and the combination of MS and NMR methods can prove to be even more powerful (Crockford et al. 2006). Overall, as is the case for extraction methods, it should be apparent that only a combination of methods can possibly allow to exhaustively survey the metabolome since each method is somewhat biased for a specific subset of metabolites.

While the above description is meant as a brief overview, more details about analytical methods previously used for microbial metabolomics have been reviewed elsewhere (van der Werf et al. 2005; Villas-Boas et al. 2005b; Wang et al. 2006) and the interested reader is invited to consult these publications. The main steps and methods involved in metabolome analysis in *E. coli* can be found in Figure 2. While the issue is not addressed here, all analytical methods are prone to variation and errors due to multiple factors, including sample preparation, instruments errors and data post processing and these are discussed in more details in Chapter 2. In addition, the analysis of the typically very large and complex, high dimensionality metabolomic datasets derived from hyphenated mass spectrometry and NMR methods is non-trivial. Multiple multivariate statistical data analysis tools can be used (e.g. principal component analysis, partial least squares-discriminant analysis, etc.) and some of the common and emerging methods are introduced in Chapters 9 and 12.

3 Existing E. coli metabolomic studies

While *E. coli* has been biochemically scrutinized and dismantled for decades, still only a relatively small number of studies have examined a significant number of metabolites simultaneously in this organism. So far, very few large-scale metabolite measurements in *E. coli* are available. Thus, while some studies may not exactly qualify as metabolomic studies *per se* due to the still limited number of metabolites measured, some of the work investigating the levels of multiple metabolites using standard biochemical assays will be introduced in this section. The reader should keep in mind that large metabolomic datasets in *E. coli* have yet

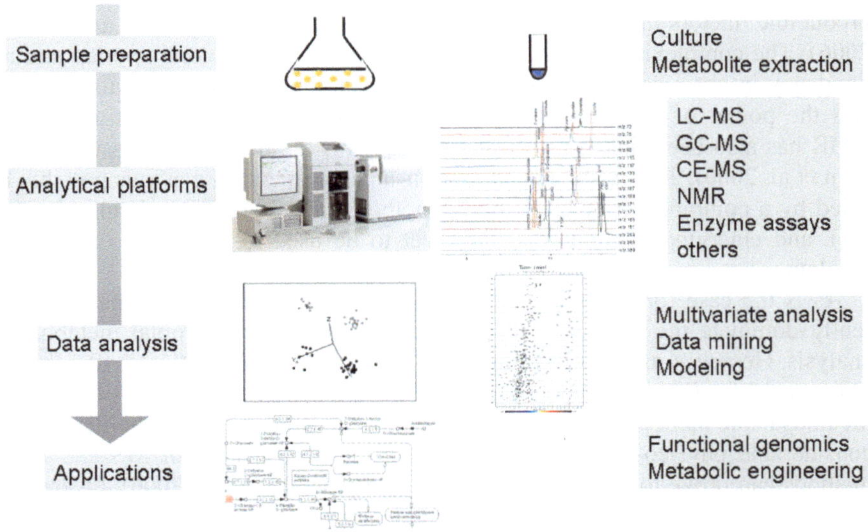

Fig. 2. Overall view of a typical *E. coli* metabolome analysis pipeline and possible applications. Sample preparation, analytical methods, data analysis and applications form the basic steps in most *E. coli* metabolomic experiments. There are numerous possibilities at each stage, each having specific advantages and disadvantages (see text).

to emerge due in part to some of the technical difficulties mentioned above, but emerging technologies promise to significantly expand the measurable metabolome in the near future. A summary of the main available *E. coli* studies described below that are both methods- or application-oriented can be found in Table 1.

3.1 Groundwork

In what might arguably be considered as one of the earliest *E. coli* metabolomic study, the effect of different carbon and nitrogen sources on the intracellular levels of several metabolites in glycolysis, the tricarboxylic acid (TCA) cycle, and amino acids was assessed (Lowry et al. 1971). Using only enzymatic assays for quantification, the levels of glycolytic intermediates were found to be in agreement with the known enzyme regulation occurring during gluconeogenesis. Decreased adenosine triphosphate (ATP) levels in slowly growing cells were observed while cells grown in acetate showed increased levels of glycolytic intermediates suggesting new regulatory factors for isocitrate lyase activity. More recently, Bhattacharya et al. (Bhattacharya et al. 1995) measured about 16 metabolites of the

Table 1. Main *E. coli* metabolomic studies

Subject(s)	Method(s)	Reference(s)
Effect of carbon and nitrogen source on metabolite levels	Enzyme assays (fluorometry)	(Lowry et al. 1971)
Method development	HPLC-UV	(Bhattacharya et al. 1995)
Effect of slow growth and ROS on metabolome	TLC	(Tweeddale et al. 1998, 1999)
Metabolite dynamics	Enzyme assays and HPLC	(Schaefer et al. 1999)
Glucose metabolism upon overexpression of glycolytic enzymes	Enzyme assays	(Emmerling et al. 1999)
Adaptation to high density growth	TLC	(Liu et al. 2000)
Dynamics of intracellular metabolites	HPLC and LC-MS	(Buchholz et al. 2001, 2002)
Dynamic model of central metabolism	Enzyme assays	(Chassagnole et al. 2002)
Metabolic footprinting for functional genomics	FT-IR and infusion-MS	(Kaderbhai et al. 2003)
Effect of extraction method on metabolome	2D-TLC	(Maharjan and Ferenci 2003)
Effect of central metabolism enzyme deletion on metabolic network	Enzyme assays, flux analysis	(Peng et al. 2004; Siddiquee et al. 2004; Li et al. 2006; Rahman et al. 2006)
Metabolite dynamics following substrate pulse	Enzyme assays	(Hoque et al. 2005)
Method development and metabolite changes during growth	GC-MS	(Koek et al. 2006)
Method development	LC-MS/*in vivo* isotopic labeling	(Coulier et al. 2006) (Bajad et al. 2006)
Effect of carbon or nitrogen starvation on metabolomic response	LC-MS	(Brauer et al. 2006)

Glycolytic, pentose phosphate, and TCA cycle pathways by HPLC using standard UV detection. In addition, Liu et al. (2000) measured several metabolites in *E. coli* to explore adaptation to high density culture and found that large changes in the level of trehalose reflected the known role of the sigma factor RpoS (σ^S) in the control of trehalose biosynthesis genes (Liu et al. 2000).

In order to perform a dynamic study in *E. coli* taking into account the rapid turnover of metabolites *in vivo*, an automated and rapid sampling system was developed to collect samples in about 0.2 seconds. Such a system allowed to observe oscillations in glycolytic intermediates using standard enzyme assays to measure metabolite concentrations (Schaefer et al. 1999). A shift in the capacity to analyze a larger number of *E. coli* K-12 metabolites was later reported using LC-MS for the measurements of intracellular metabolites (Buchholz et al. 2001). The authors could achieve the quantification of 15 different intermediates and showed that the method gave results comparable to those obtained with enzymatic assays and the metabolite limit of detection varied between 0.02 and 0.5 mM. Continuing the dynamic analysis of metabolites using of a limited substrate pulse, the same group looked at over 30 different *E. coli* metabolites using a combination of enzymatic assays, UV-HPLC, and LC-MS that provided the type of high-resolution time-course data required for dynamic modeling of metabolic networks (Buchholz et al. 2002). Another early study compared the effect of slow growth as well as mutation of rpoS on more than 70 ^{14}C-labeled metabolites using 2D-TLC and standard amino acid analysis (Tweeddale et al. 1998). By revealing numerous changes in intracellular sugar and amino acid levels, the results highlighted the power of metabolomic studies to clarify global regulation of metabolism under various experimental conditions. The same group then examined the effect of superoxide stress on *E. coli* using similar methods and observed changes in the level of several antioxydants (Tweeddale et al. 1999).

In one of the first combined experimental and computational analysis of metabolic regulation, researchers using enzymatic assays and HPLC measured the concentrations of multiple intermediates of glycolysis and pentose phosphate shunt (Chassagnole et al. 2002). Stopped-flow sampling of continuous cultures of *E. coli* was used to measure the dynamics of intracellular intermediates following a glucose pulse. Using mass balance equations to derive kinetic rate equations, the experimental results allowed to obtain the enzyme kinetic parameters necessary to develop a dynamic model of *E. coli* metabolism, that could reproduce many experimentally observed phenomena (see Section 10). Among these, the link between glycolysis intermediate levels and the phosphotransferase system (PTS) was established and previously observed metabolite oscillations (Schaefer et al. 1999) could be described.

A similar study looked at the metabolite dynamics of the response of *E. coli* to pulse addition of substrate, under both glucose- and ammonia-limited steady-state continuous cultures (Hoque et al. 2005). In addition, the differences between the wildtype and pykA mutant were evaluated and quantified immediately after substrate addition using a specially designed rapid sampling device to collect several samples in the first few seconds after substrate pulse followed by sampling at increasingly longer intervals. Specifically, many intermediates of glycolysis were found to rapidly accumulate after glucose pulse addition. Several pentose phosphate pathway intermediates accumulated in the pykA mutant while the intracellular NADPH concentration decreased in ammonia-limited conditions, presumably because of its depletion through glutamate biosynthetic pathway. The results of such experiments demonstrate how the measurements of multiple intracellular metabolites can facilitate the elucidation of functional differences between strains under different growth conditions.

Several other recent and rather targeted *E. coli* metabolite studies originate from applications in metabolic engineering. For this purpose, the monitoring of several key metabolites by enzyme assays can provide, in a simplified way, an integrated view of intracellular metabolism and is often important to phenotypically evaluate the overall effects of engineering the cell. In one example, the effect of overexpressing endogenous and exogenous glycolytic enzymes (PfkA and PykF) in non-growing *E. coli* strains engineered for ethanol fermentation -as a way of synthetically "simplifying" the metabolic network- was examined (Emmerling et al. 1999). Using metabolite concentrations obtained for about 15 different key extracellular and intracellular metabolites by enzymatic assays, a considerable increase in flux through glycolysis in a PfkA overexpressor strain and a shift from ethanol to lactate production that seems independent of any concomitant increase in glucose transport were observed. The results demonstrate that a single enzyme overexpression can thus potentially increase flux to a specific metabolite, a desirable fate for metabolic engineers.

3.2 Combining concentration data with enzyme activity and flux measurements

Several studies combined the power of metabolic flux analysis (see Section 8) together with assays of enzyme activities and the measurement of concentrations of key intracellular metabolites to evaluate the phenotypic effects of deletion of genes encoding central carbon metabolism enzymes. Enzymatic assays were used to measure intracellular metabolite concentrations. In a pykF mutant (deficient in phosphoenol pyruvate (PEP) to pyruvate conversion), decreased transcript levels for most glycolytic enzymes concurrently with increased levels of pentose phosphate pathway enzymes were observed (Siddiquee et al. 2004). On the other hand, whereas enzyme transcript levels usually correlated well with enzyme activity, fluxes through specific parts of the pathways did not correlate as well with enzyme activity but rather appeared related to changes in metabolite concentrations. Some of the regulatory intricacies of central carbon metabolism were thus more clearly revealed by using metabolite concentration data in combination with fluxes and enzyme activity data than they would have been using single layer datasets. Similarly, in a ppc mutant (deficient in PEP to oxaloacetate conversion), the activities of multiple glycolytic and pentose phosphate shunt pathways were found to decrease, in agreement with the observed slower growth and glucose utilization (Peng et al. 2004). Increased utilization of the glyoxylate shunt was suggested by increased enzyme activity levels of AceA and several TCA cycle enzymes, in accordance with the accumulation of glycolytic metabolites and a decrease of acetyl-coA and oxaloacetate. The reduced PckA flux in the ppc mutant was associated with the observed increase in PEP concentration, an allosteric inhibitor of PckA. By integrating these observations, the authors suggest that the need for TCA cycle function for growth maintenance was achieved by increased glyoxylate shunt activity thus replenishing oxaloacetate. Similarly, the main role of lpdA in pyruvate metabolism was clarified by measuring intracellular metabolite concentrations in combination with flux and enzyme activity measurements (Li et al. 2006). Finally, the same group also analyzed the effects of sigma factor rpoS gene deletion on gene expression, enzyme activity, and the level of a few key metabolites during exponential and early stationary growth phases (Rahman et al. 2006). Broad changes were observed in both gene expression and enzyme activity and acetyl-coA levels were found to be considerably lower in the mutant. The reduced ability of rpoS mutant to utilize acetate seems to explain the premature entry into the stationary phase.

3.3 Emerging metabolomic studies in *E. coli*

The only *E. coli* studies that might be considered as genuine metabolomic analyses in terms of numbers of measured metabolites, are just starting to appear (Bajad et al. 2006; Coulier et al. 2006; Koek et al. 2006). These datasets have been collected using GC-MS and novel LC-MS methods that are described in more detail in Section 2.2.3. While these studies were aimed at method developments, experimental validation of the platforms with *E. coli* samples allowed partial survey of the *E. coli* metabolome during different growth phases (Bajad et al. 2006). As expected, the levels of most monitored metabolites decreased in the stationary phase with the largest decreases occurring in fructose-1,6-biphosphate and inosine monophosphate (IMP) levels, while cyclic adenosine monophosphate (cAMP), phenylalanine, and histidine dramatically increased during the stationary phase. The authors pointed out that some of the results might be specific to the sampling method used. Others observed several metabolite-specific patterns of change in concentration during different growth phases of *E. coli* (Koek et al. 2006). Overall, these new methods currently allow the simultaneous quantification of several dozen metabolites in *E. coli* and were also shown to be useful to monitor the cellular energy charge (Coulier et al. 2006). Finally, using the above LC-MS method (Bajad et al. 2006) the same group then evaluated the consequences of carbon or nitrogen starvation on 68 metabolites of *E. coli* (Brauer et al. 2006). A general response common to both treatments (depletion of biosynthetic intermediates) and a more specific response could be observed and interestingly the main features of this response appear conserved in *S. cerevisiae.*

Together, the above studies demonstrate how metabolomic data can provide functional information about *E. coli* physiology in response to environmental or genetic perturbation and how this information can complement that obtained with other -omic methods. Importantly, because of the numerous remaining technical obstacles mentioned earlier, the validity of quantitative metabolite measurements obtained by emerging analytical platforms will require further evaluation. This issue may not be easy to address directly but the recently described "NET analysis" approach may turn out to be very useful for this (Kummel et al. 2006) (see Section 10). In addition, while all these studies provide useful information, the breadth of the sampled metabolome still remains deceivingly small (see next section) and other methods are likely to be necessary to improve the coverage. Some MS-based methods also face reliability issues due to ion-suppression effects in complex matrices, detector saturation, and a still somewhat limited dynamic range. Moreover, the challenge of identifying most detected metabolites remains. While these represent current trade-offs to be able to analyze a large number of metabolites in parallel, expected gains in resolution and throughput justify the efforts to improve quantitative methods based on MS.

4 Evaluating the size of the *E. coli* metabolome

One can wonder what the total number of possible *E. coli* metabolites is. While no definitive answer exists yet and there is still a lack of large-scale metabolomic data sets that could shed light on this issue, estimates can nonetheless be derived from experimental data in the literature (usually small-scale) and from genome-based reconstruction of the *E. coli* metabolic network based on literature data (Reed et al. 2003) or database annotations (Arakawa et al. 2006).

4.1 Hints from genome-based models

The latest publicly available version of the *E. coli* genome-based model (iJR904) contains 904 genes and 931 unique biochemical reactions involving 625 metabolites. The total number of metabolites is likely larger than this, as not all pathways or reactions are currently included in the model, let alone discovered. The list of all biochemical components included in the iJR904 model is available on-line (see Section 11) and eventual newer versions can be expected to contain even more reactions and a significantly increased number of metabolites. In contrast, the genome-based modeling (GEM) system is an automated-model construction tool (Arakawa et al. 2006) that was used to generate a list of 1195 potential distinct metabolites in *E. coli*, using data compiled from multiple databases. The model constituents were found to cover more than 90% of *E. coli* metabolism data in KEGG and EcoCyc databases (see Section 11) and the iJR904 model. While a few false-positives and false-negatives were reported during model building (Arakawa et al. 2006), these seem to be mostly associated with discrepancies such as lack of, or ambiguous EC numbers. The most significant difference in terms of total metabolite number appears to be related to the use of whole genome information (every possible enzyme) in the GEM-based model construction, in contrast to a growing but yet incomplete selection of pathways in iJR904.

4.2 Experimental clues

From an experimental point of view, global analysis of *B. subtilis* extracts by CE-MS, revealed the presence more than 1600 compounds (Soga et al. 2003). Among these, only about 150 could be unambiguously identified using standards and another 83 more based on prediction of migration times in CE-MS (Sugimoto et al. 2005). While an hypothetical figure of 576 metabolites has been derived from *B. subtilis* genome-based predictions, estimates based on indirect experimental evidence suggest about 1200-1400 metabolites (van der Werf et al. 2005). Together these studies thus suggest the existence of more than 1200 metabolites in *B. subtilis*. Since the genome size of *B. subtilis* and *E. coli* is comparable, assuming an overall similar distribution of functional protein classes between the two organisms, the *E. coli* metabolome can be expected to be roughly the same size (>1200). Our preliminary CE-MS analysis of the polar metabolite (methanol) fraction of *E.*

coli suggests that more than 680 different compounds can be observed in such *E. coli* extracts. As chromatographic peaks, all these may not necessarily represent distinct metabolites since some may represent redundant forms of the same metabolite such as metal or other adducts, or alternately charged form of the same compound. However, it is fair to say that there are probably more than 500 detectable compounds in this polar extract alone, among which we can currently identify and quantify about 130. This CE-MS derived estimate might at first suggest an *E. coli* metabolome significantly smaller than that of *B. subtilis*. However, this may be due to significant differences in extraction efficiency between these different types of bacteria and also to the physico-chemical nature of the metabolites so that only a fraction of the metabolome is actually surveyed with the current methods. In addition, there might be a more extensive array of secondary metabolites produced by *B. subtilis*. To reconcile the smaller *E. coli* metabolome estimates obtained from experimental data with those of genome-based predictions, one must also consider that some metabolites are easily degraded or lost during isolation, not measurable by CE-MS because of neutral or non-polar character, or display poor ionization efficiency. Finally, many metabolites are simply not expressed under specific experimental conditions (rich medium, batch culture, aerobic conditions, etc.) or their concentrations are below the current detection levels. Sampling all possible metabolites at once is thus not possible due to both the inherent dynamic nature of the metabolome as well as the physical limits of extraction and analytical methods. Only once all possible reactions are confirmed experimentally will an accurate estimate be possible. A graphical representation of the metabolome estimates and possible gross distribution in various data sets is shown in Figure 3.

4.3 Improving metabolite identification

To experimentally measure and identify most *E. coli* metabolites, there is a pressing need for greater availability of a large collection of chemical standards and more detailed structural analysis. The availability of mass spectral databases will also be important. Because of conservation of the chemical structure of metabolites across species, mass spectral databases collected from any other organisms can be used in theory (Wagner et al. 2003; Kopka et al. 2005). However, some of them, particularly tandem mass spectrometry databases are sometimes instrument- or technology-specific and there is also a need to address this issue. Chapter 4 of this book describes in more detail a large GC-MS database developed to facilitate the identification of metabolites. In addition, the identity of many metabolites can already be predicted reasonably well for some analytical methods. Using an artificial neural network used to extract descriptors from about 500 standards whose CE-MS migration time was experimentally measured, a good prediction of the migration time of cations was possible (Sugimoto et al. 2005). The system generates a list of candidate metabolites whose masses closely match that of an unknown compound. Evaluation of the results showed that the correct metabolite is

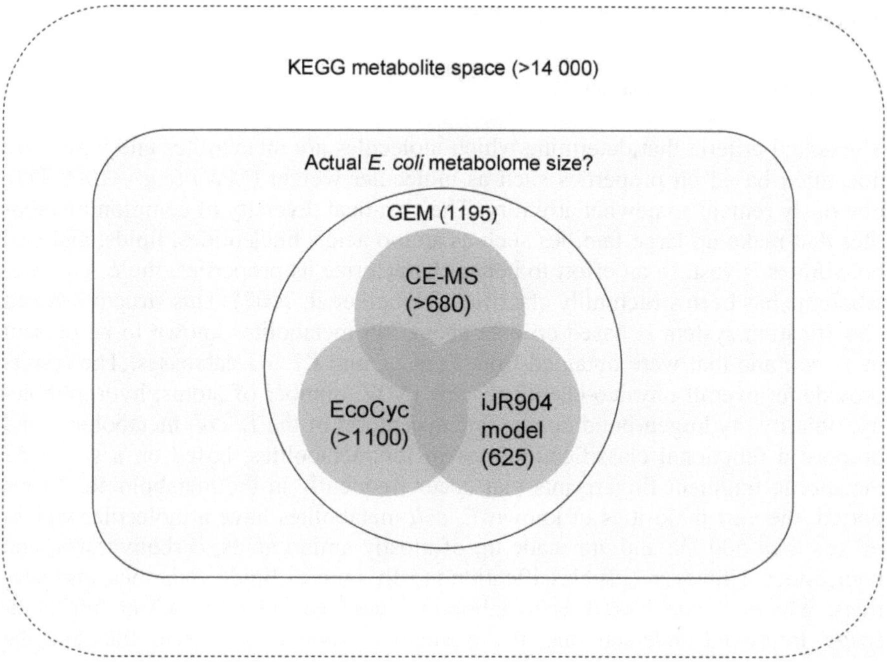

Fig. 3. Diagram of size estimates and distribution of the *E. coli* metabolome. The included information comes from experimental datasets, databases and metabolic models (see text). The size of the different compartments is not meant to be to scale and the overlapping proportion of each compartment is only symbolic. While the outer area is used to illustrate the larger number of known metabolites (KEGG metabolite space), not all *E. coli* metabolites are necessarily yet in the KEGG COMPOUND database. Figures in parenthesis indicate the number of metabolites as of February 2007.

within the top three candidates, 78% of the time. Therefore, for global identification of metabolites in complex samples, in addition to the obvious utility of tandem mass spectrometry, a robust prediction algorithm can prove to be both time-saving and very powerful.

Overall, the above estimates most likely represent rather conservative and underestimated values for the size of the *E. coli* metabolome. The reported broad substrate specificity, or promiscuity, of many metabolic enzymes can potentially lead to a much larger actual number of metabolites than estimates derived from gene number may lead to conclude (Schwab 2003; Kuznetsova et al. 2006). In addition, the combinatorial composition of sugar and lipid metabolites can also in theory greatly inflate metabolome space.

5 Architecture/anatomy of the *E. coli* metabolome

5.1 Metabolite architecture

The actual criteria that determine which molecules are metabolites and which are not, often based on properties such as molecular weight (MW) (e.g. <1000 Da), obviously remain somewhat arbitrary. The chemical diversity of common metabolites that make up large families such as amino acids, nucleotides, lipids, and carbohydrates is vast. In an effort to better characterize its properties, the *E. coli* metabolome has been structurally classified (Nobeli et al. 2003). This structure-based classification system is based on data about 745 metabolites known to be present in *E. coli* and that were obtained from EcoCyc and KEGG databases. The results provide an overall physico-chemical view (MW, number of atoms, hydrophobicity, polarity, hydrogen-bond donors and acceptors) of the *E. coli* metabolome and propose a functional classification system for metabolites, based on a set of 57 metabolite fragment fingerprints that occur frequently in the metabolome. As reported, the vast majorities of known *E. coli* metabolites have a molecular weight of less than 500 Da and are made up of mostly amino acids, carbohydrates, and nucleotides. However, this classification mostly ignores lipids, quinones, and peptides, which, if considered as metabolites, could considerably affect this landscape. Improved understanding of the chemical diversity of metabolites and the classification system thus developed can have repercussions on the characterization of enzyme-substrate interaction specificity (or promiscuity). This original characterization of the metabolome constitutes an interesting example of interfacing bioinformatic tools and metabolite databases to make possible the interpretation of metabolomic datasets. This, in turn, can facilitate the understanding of metabolic pathways and predicting, through simulation, the effect of environmental changes on the cell (Nobeli and Thornton 2006).

In a related study, the clustering of metabolites based on deconvolution of structural elements using graph comparison methods revealed that the largest group of metabolite represents carbohydrates (Hattori et al. 2003). In addition, the known structural link, albeit of limited scale, between metabolic pathway and genome architecture could be observed.

5.2 Pathway architecture

At a different architectural level, the determination of routes or pathways linking two metabolites can now also be derived from computational tools (Arita 2003). The method uses atomic-scale tracing of atoms within molecules to decipher the relevant paths. The approach is unique in that it captures information regarding links between parts of molecules that are otherwise not obtainable from classic maps. The analysis provides a list of the most commonly used metabolic intermediates in the reconstituted pathways of *E. coli*, namely pyruvate, acetyl coA, ATP, glucose, glutamate as well as other co-factors and nucleotides (Arita 2003). A software tool that allows to compute such relationships in pathway intermediates is also available.

Network architecture terminology defines small world network as those characterized by a small distance (pathway length) separating any pair of network nodes on average. However, metabolite moieties are not necessarily structurally conserved along traditional pathway maps that simply graphically connect metabolites (nodes) through enzymes (edges). At least according to the criteria of structure conservation in biosynthetic and degradation pathways, the *E. coli* metabolic network is therefore not "small" (Arita 2004) but displays pathway lengths much longer than previously expected. Earlier views that concluded otherwise may have failed to account for the lack of structural conservation in adjacent enzymatic reactions, leading to a structurally incorrect overview of some metabolic pathways.

Regarding the network structure of metabolic pathways, pathway hubs are usually suspected to be essential for an organism. A genome-scale model of *E. coli* metabolism together with experimental data about the effect of enzyme deletion on survival (essentiality) revealed essential roles for several metabolites and conversely suggested the non-essentiality of specific compounds previously suspected of being essential (Imielinski et al. 2005). Not necessarily contradictory to the work of Arita (2004), studies of enzyme essentiality in *E. coli* demonstrate the presence of a minority (9%) of enzymes (hubs) whose essentiality was previously experimentally demonstrated (Lemke et al. 2004; Baba et al. 2006). The rationale is that well connected compounds are produced by multiple reactions (enzymes) and the enzymes producing them are thus not essential, while essential enzymes produce poorly connected metabolites in important metabolic routes. However, it should be kept in mind that metabolite essentiality, while obviously linked to synthesizability through pathway connectivity, is also eventually linked to gene and protein expression, as well as enzyme activity. Specific genetic changes or environmental conditions that modify this balance may result in "indirect" essentiality not linked to physical synthesizability as inferred from network structure alone (Baba et al. 2006). Finally, a general overview of gene essentiality in the metabolic pathways of *E. coli*, drawing attention to the distinction between essentiality for survival and essentiality for fitness, was recently published (Gerdes et al. 2006).

6 *E. coli* metabolomics as a powerful tool for functional genomics

Despite the availability of a complete genome sequence, as of 2004, nearly half the *E. coli* proteins did not have any experimentally confirmed function (Serres et al. 2004). For the characterized portion of the genome, most of the functions were assigned in the classic way using small targeted experiments based on hypothesis testing. To improve the capacity to assign function to uncharacterized proteins, unbiased analyses harnessing the power of metabolomics as a tool to provide detailed functional information about metabolic phenotypes are emerging. The metabolome, being the end product of the concerted action of the genome, transcriptome, proteome, and interactome, integrates the final measurable fingerprints of cellular activity and is particularly well-suited for functional genomics and systems biology discovery (Kell 2006). Another important aspect is that metabolomics, also integrates both the cellular and environmental factors, a fate which may be difficult to achieve at other levels (Clayton et al. 2006).

6.1 Metabolic footprinting

The effect of gene deletions, which may not reveal any apparent growth defect or other easily tractable phenotype, may become clearer following profiling of intracellular or extracellular metabolites. In this regard, a method referred to as metabolic footprinting has been shown to be a useful functional tool (Allen et al. 2003; Kell et al. 2005). The approach is based on comparisons of extracellular metabolite profiles obtained from wildtype and single gene deletion mutant cells instead of analyzing the more complex intracellular profiles. While the profile of extracellular metabolites may not reveal as easily or directly the functional inner workings of the cell it has the main advantage of looking at a smaller subset of secreted metabolites and also bypasses the difficulties of efficiently extracting the intracellular metabolites (Hollywood et al. 2006). Moreover, the method does not require the identification of metabolites. The analysis and comparison of mutants of genes of both known and unknown function using advanced clustering and chemometrics analysis based on machine learning (metabolic footprint), can allow to associate uncharacterized genes to a particular pathway or function (Kell et al. 2005). The whole process is reminiscent of the functional analysis by co-responses in yeast (FANCY) and the profiling of intracellular metabolites by NMR (Raamsdonk et al. 2001).

Metabolic footprinting has been used for analyzing secreted metabolites in *E coli* using Fourier transform infrared spectroscopy and infusion MS. Its discriminating potential was studied by analyzing disruptants of genes in the tryptophan biosynthetic pathway (Kaderbhai et al. 2003). The relative ease, low cost, speed, and reproducibility with which such experiments can be performed thus make them attractive tools for *E. coli* functional genomics using single gene disruptant libraries such as the Keio collection (Baba et al. 2006).

6.2 Enzyme discovery using non-targeted metabolomics

Among all gene products in *E. coli*, more than a third are enzymes, making it the single largest functional protein group in the genome. Based on recent annotation data, about 500 gene products are listed as putative enzymes, meaning that their activity has been inferred from sequence similarity but has not been confirmed experimentally (Serres et al. 2004). In addition, another group of close to 600 members bears no annotated function at all. Many of them can also be expected to be novel enzymes having completely novel activities and mechanisms. The fact that there is still a lot to discover in terms of possible new metabolic pathways, even in the well-characterized *E. coli*, was recently demonstrated. To the surprise of some (Osterman 2006), components of a known operon were found to be enzymes constituting a whole new pathway for pyrimidine utilization (Loh et al. 2006). Thus, the potential of metabolomics for functional discovery in *E. coli* remains important.

Traditionally, enzymes have been characterized by purifying the proteins having activity from a cell or tissue extract. With the availability of complete genome sequences and large experimental resources it can be more efficient to use the reverse approach of taking easily purified recombinant proteins and trying to associate them to specific activities. At the same time, to discover completely novel activities in an unbiased manner, it is desirable to use a system that allows assaying for reactions for which a priori no information is available about either the substrate or the product. Toward this objective, a generic assay system, combining the power of libraries of recombinant proteins (candidate enzymes) and mass spectrometry-based metabolite profiling as an unbiased and generic assay read-out, was developed and tested for the discovery of novel enzyme activities (Saito et al. 2006). Mass spectrometry has previously been successfully used for monitoring enzyme activity and kinetics (Jankowski et al. 2001; Zea and Pohl 2004). However, previous attempts were made using specific and selected enzymes and by monitoring only a few molecules by MS. More recently, approaches to monitor any type of reaction where a mass change occurs have also been reported (Yu et al. 2004; Zea and Pohl 2004; Pohl 2005) for different classes of enzymes, demonstrating the feasibility of generic MS assays.

Using a recently developed system (Saito et al. 2006), *in vitro* screening for enzyme activity can be performed by incubating purified recombinant proteins (selected as likely enzymes using tools such as those mentioned in Section 6.3) with a metabolite soup as substrate source, followed by analysis of the resulting reaction mixture by capillary electrophoresis mass spectrometry (CE-MS) to detect possible activities (Fig. 4). The deconvolution of CE-MS data obtained from reaction mixtures in the presence or absence of an enzyme can yield direct evidence of the presence of an enzymatic activity and moreover directly pinpoint the substrate(s) and product(s) of the reaction. To facilitate the comparison of large datasets containing subtle differences, software tools to facilitate the identification of specific metabolite whose level vary between two or more samples were also developed (Baran et al. 2006). The originality of the screening method lies in the fact that it is completely generic, i.e., the same method can potentially be used for any target. It also means that activities can be discovered without any prior information about the activity or the reaction type, a task difficult to accomplish with more standard approaches. The method was used first to demonstrate the direct detection of activities of several known enzymes belonging to different classes (Saito et al. 2006). As a proof-of-concept the same system was then used to screen uncharacterized enzyme candidates and assign novel sugar phosphatase activity to

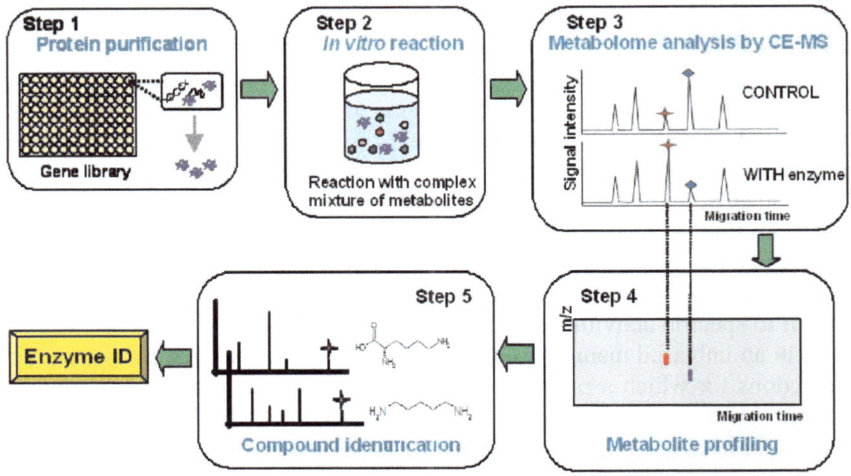

Fig. 4. Overview of a metabolomic approach for enzyme discovery. A few candidate proteins from a library of *E. coli* purified proteins (obtained from ASKA collection (Kitagawa et al. 2005)) are pooled and used for an *in vitro* assay using a metabolite soup, in the presence or absence of proteins. Profiling of the assay mixtures is performed by CE-MS and the identification of specific differences in profiles is assisted by software tools. Then, the identity of the compound(s) can be verified using chemical standards and/or by tandem mass spectrometry. Finally, the newly discovered enzyme can be given an appropriate name, and its activity further characterized. Reprinted with permission from (Saito et al. 2006), Copyright 2006 American Chemical Society.

YbhA and YbiV (Fig. 5). An important issue is to confirm that the detected activity is really linked with the original protein target. This can be accomplished by analyzing extracts of cells where the candidate enzyme gene has been either deleted or overexpressed.

Complementary assays that make use of class-specific but generic synthetic substrates to screen unknown proteins for enzymatic activities have recently been reported and were originally meant to experimentally classify activities within a subclass (phosphatase, protease, esterase, etc.) (Proudfoot et al. 2004; Kuznetsova et al. 2005). In this manner a large family of novel sugar phosphatases was discovered. Recent results show that many members of the *E. coli* haloacid dehydrogenase-like phosphatase family are relatively promiscuous, catalyzing the conversion of several structurally similar metabolites, suggesting that some of this flexibility may represent a reservoir for the evolution of novel phosphatases (Kuznetsova et al. 2006). More definite assignment of these activities to a specific pathway *in vivo* might eventually come from the differential analysis of metabolite extracts obtained from mutant strains by metabolite profiling, where changes in the levels of specific metabolites may reveal the actual *in vivo* substrate(s). Such an approach of using whole-cell extract metabolite profiling for non-targeted enzyme activity discovery has already been reported (Saghatelian et al. 2004; Saghatelian and Cravatt 2005). In this case, the metabolite profiles of brain tissue

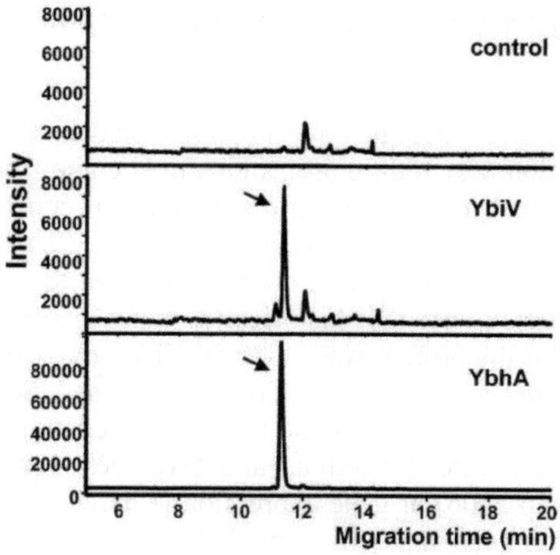

Fig. 5. Example of CE-MS-based monitoring of metabolites for the discovery of enzymatic activities. Changes in a complex metabolite mixture incubated with uncharacterized *E. coli* proteins are monitored. Selected ion electropherograms (m/z 171) for control (without protein; upper panel), YbiV (middle), and YbhA (lower) samples. Peaks indicated by arrows correspond to the compounds specifically produced in the presence of YbiV and YbhA and are products of the reaction. Reprinted with permission from (Saito et al. 2006), Copyright 2006 American Chemical Society.

obtained from wildtype and enzyme mutant mice were analyzed to reveal the origin of the reaction. The method has been termed discovery metabolite profiling and has been successfully used to assign substrates to a mammalian fatty acid amide hydrolase (Saghatelian et al. 2004). It will thus be interesting to see the results of similar assays in *E. coli* using deletion mutants of known or putative enzymes. However, since the goal here, in contrast to metabolic footprinting, is to pinpoint the intracellular metabolites involved in the specific reaction, making sense of the multiple and often complex changes in intracellular metabolite profiles following even a single gene deletion may not be always as straightforward.

A recent study of *E. coli* protein complexes revealed a large number of heteromeric complexes of proteins which, interestingly, contained proteins whose activity is known to be associated with metabolism. These include enzymes catalyzing related reactions (metabolon) such as Pgm and Zwf (both using glucose-6-phosphate) (Lasserre et al. 2006). Interestingly, even by accounting for the fact that metabolic enzymes represent one of the largest functional category in the *E. coli* proteome, the results highlight and confirm the apparent physical promiscuity of many metabolic enzymes, possibly reflecting a need to maximize pathway activities through substrate channeling, etc.

Together, the different methods that can be used for the discovery of novel metabolic activities and intermediates, on top of providing functional elucidation, may facilitate the production of established or novel, commercially valuable molecules or pharmaceuticals and reveal new antibiotic targets.

6.3 Deorphanizing enzymatic activities and filling-in metabolic pathway holes

In addition to the above strategies for *de novo* enzymatic search using metabolomic tools (Section 6.2) additional insights may come from other sources of information that can lead to more specific experimental confirmation. Many enzymatic activities that have previously been detected in *E. coli* still have not been assigned to a specific gene/protein and are thus referred to as orphan enzymatic activities. Possible reasons are that historically, many enzymatic activities were detected in extracts but the protein was never purified or alternately that while the protein was isolated, no sequence information was obtained. Several methods have been used to identify orphan activities and most are based on genome organization and phylogenetic information (Osterman and Overbeek 2003; Green and Karp 2004; Chen and Vitkup 2006) or gene expression (Kharchenko et al. 2004). A list of orphan activities has been manually compiled and can be found at both EcoCyc and EchoBASE web sites (See Section 11). The information listed comes from different sources including activities with no assigned enzyme from the literature and the iJR904 genome-scale *E. coli* model (Reed et al. 2003).

The metabolic pathways of *E. coli* are also known to contain holes, defined as metabolic enzymes that should be present based on the sequence of reactions in a metabolic pathway but for which there is no protein annotated with that specific function or reaction. The metabolic pathways of multiple organisms, including *E. coli* have been analyzed in this way and a method to identify missing enzymes or pathway holes was developed (Green and Karp 2004).

Together, the missing links in pathways are obvious targets to screen for specific activities using recombinant proteins or other experimental means. Since many of these reactions may not be easily amenable to detection using standard methods, generic metabolomic approaches, as described in Section 6.2, can contribute to fill in the blanks and finally link these orphan activities to a specific protein.

6.4 Phenotype microarrays as reporters of metabolic phenotype

The metabolic phenotypes of cells, including *E. coli* can be interrogated in various analytical ways as discussed earlier. Cells can also be cultured under different conditions and the effects measured at a higher level such as growth phenotype. Such studies have been performed in yeast (Ross-Macdonald et al. 1999; Giaever et al. 2002) and *E. coli* (Serina et al. 2004) and can help to gain insight into gene/protein function. However, these studies looked at only a limited number of conditions. On the other hand, while not a metabolomic platform *per se*, the phenotype microarray technology distributed by Biolog, Inc. can assist efforts in functional genomics and also possibly provide links to intracellular metabolism (Bochner et al. 2001; Bochner 2003). Briefly, the system is used to monitor in real-time and in parallel the growth of cells such as *E. coli* under hundreds of different environmental conditions, grouped into larger categories such as carbon, nitrogen, sulfur or phosphorus sources (especially relevant to metabolism), pH and ionic conditions, and diverse chemicals or drugs (Bochner et al. 2001). Instead of monitoring cellular growth *per se*, the assays measure cellular respiration through a proprietary colorimetric method making use of tetrazolium redox chemistry. While potentially very interesting, the system provides a high-level phenotypic read-out that may not always be easy to interpret. In addition, disruption of almost any gene function, whether it encodes a metabolic enzyme or not, is bound to result in changes in array profiles and, therefore, details about the exact function may not be readily apparent. However, from a functional genomics point of view the resulting phenotypes can be a powerful tool to categorize genes of unknown function by comparing their profiles to those of genes of known activity. Moreover, in cases where mutants display highly specific phenotypes for only a few growth conditions, this can lead to specific conclusions about involvement in a specific metabolic pathway. Thus, in concert with metabolite profiling data, the system can provide powerful functional information.

In a limited scale study, the system was used to analyze knockouts of more than 100 genes including all two-component systems in *E. coli* under nearly 2000 growth conditions (Zhou et al. 2003a). In contrast, more recently, the technology has been successfully used to analyze a large number (1440) of different *E. coli* gene disruptants on a more limited number of conditions, namely 95 different carbon sources (Ito et al. 2005). While details about the actual function of many y-genes included in the dataset remain to be resolved, it was shown that functional classification of many genes of unknown function with those of well-characterized activity is possible. In addition, as can probably be expected, there appears to be some significant link between the deletion of metabolic enzymes and the extent of changes observed on carbon source arrays. Interestingly, a large number of single gene deletion mutants displayed growth defects on acetate, mannose, and alpha-ketoglutarate highlighting the dense connectivity of the pathways utilizing these carbon sources. Finally, the potential of such screenings as a gateway to the discovery of whole new metabolic pathways and intermediates was recently reported (Loh et al. 2006).

The phenotype microarray technology can thus allow functional inference from higher-level analysis similarly to metabolic footprinting (Section 6.1). Linking information obtained from such phenotype screenings with the lower-level intracellular metabolite concentration data and profiles (Section 6.2) can thus significantly increase the chances of making functional discoveries that would not be possible otherwise.

7 Metabolomics to facilitate metabolic engineering of *E. coli*

Because of its rapid growth, its ability to grow on simple and well-characterized culture medium, and the relative ease with which it can be genetically manipulated, *E. coli* remains an essential resource for the production of commercially or pharmaceutically valuable metabolites. Metabolic engineering aims at controlling and improving intracellular fluxes to optimize the yield of desirable biomolecules. Since central carbon metabolism pathways (glycolysis, pentose phosphate shunt, and TCA cycle) are the source of multiple intermediates which act as precursors to a large variety of other biological molecules, they have been the focus of much work in metabolic engineering (Buchholz et al. 2002). Metabolomics can play an important role in the evaluation of metabolic fluxes and metabolite concentrations can be used to derive *in vivo* kinetic parameters for modeling and engineering purposes (Buchholz et al. 2002).

Numerous studies have aimed at improving the yield of specific *E. coli* metabolites (acetate, lactate, formate, succinate, amino acids) (Berry 1996; Chang et al. 1999; Kramer et al. 2003; Zhu and Shimizu 2004) or at their *de novo* synthesis using exogenous sources of enzymes, such as for optically pure L-lactate, a form not normally produced by *E. coli* (Zhou et al. 2003b). In addition, there are examples of reconstituted plant biosynthetic pathways in *E. coli* (Watts et al. 2006). Most such studies focused on the measurements of only a few target metabolites, usually the desired end-product or one of its close metabolic neighbors. However, in order to better understand the effect of deleting metabolic genes and the resulting global rerouting of metabolic activities and fluxes in *E. coli*, metabolomics is a promising tool for metabolic engineering applications. Some of the methods and results of quantitative measurements of metabolites for metabolic engineering purposes were introduced in Section 3. In addition, metabolic engineering is intrinsically linked with metabolic flux analysis (next section). As was proposed for genomic data (Gill 2003), it may be possible to use metabolomic data to evaluate the metabolic phenotype of specific natural or engineered mutations in industrial *E. coli* strains and facilitate the strain improvement process. Metabolomics can also be used detect cross-contamination during industrial fermentation, and facilitate strain identification (Wang et al. 2006).

The effect of single or combined gene mutations on the global profile of metabolites (both intracellular and extracellular) can be evaluated and used to pinpoint potential sources of difficulties or bottle necks and thus further drive improvements. Careful interpretation of metabolomic data, ideally in combination with transcript and protein expression level data, may allow to predict which specific genes should be additionally targeted to further increase the yield and productivity of fermentation strains. Therefore, together with flux analysis and proteomic measurements (Shimizu 2004) global metabolomic analyses promise even better yields and open the way to the successful production of until now difficult to synthesize compounds. As an important component of the systems biology approach in engineering, metabolomic data will thus likely play a significant role for bacterial strain improvement (Stephanopoulos et al. 2004; Lee et al. 2005).

For a more exhaustive description of engineering applications the reader is referred to other publications (Vaidyanathan 2005; van der Werf et al. 2005; Wang et al. 2006).

8 Metabolomics in flux analysis

Complementary to metabolite concentration measurements, the actual survey of the flow of metabolites across pathways, metabolic flux analysis, can provide higher level information that cannot be easily derived from concentration measurements. Flux analysis can reveal the extent of use of reactions in a pathway, bottlenecks, and the traffic through multiple alternative routes. It is one of the major tools used by metabolic engineers to globally evaluate cellular response to gene manipulation. Metabolic flux analysis combines the use of stoichiometric reaction models with measurements of extracellular metabolite consumption and secretion rates. In addition, it is most often combined with metabolic labeling methods to derive intracellular pathway utilization. This is typically performed by analyzing the extent and position of carbon labeling in proteogenic amino acids following metabolic labeling with a stable-isotope labeled precursor, usually ^{13}C-labeled glucose, using a combination of GC-MS and NMR analysis (Szyperski 1998; Wittmann 2002). Such methods have been used to quantitatively analyze flux changes in *E. coli* in response to the mutation of several enzyme genes in central carbon metabolism as well as environmental changes (Fischer and Sauer 2003) and in other studies described in Section 3. Chapter 7 describes in more details the basics of metabolic flux analysis. In addition to the standard analysis of amino acids, the use of quantitative metabolomic data for evaluating fluxes by direct concentration measurements of the pathway intermediates is attractive, though simple and reliable analytical methods have yet to emerge. While flux analysis is traditionally somewhat difficult and time-consuming to perform, computational methods and software that can be applied on a large-scale are appearing (Lee et al. 2003; Sauer 2004; Zamboni and Sauer 2004).

By globally analyzing metabolic fluxes, *E. coli* was shown to respond to environmental perturbations by manipulating only a few key metabolic reactions that display high fluxes, without apparent changes in other parts of the metabolic network where a multitude of reactions seem to support relatively little metabolic flux (Almaas et al. 2004). These findings highlight an uneven use of metabolic network topology that can have important repercussions for the optimization of metabolite production for metabolic engineering.

9 Adaptive evolution in *E. coli*, metabolomics, and metabolic phenotype

Like other organisms, upon environmental or genetic perturbation, *E. coli* will usually respond by manipulating its metabolic network in an attempt to optimize growth. This phenomenon, termed adaptive evolution (AE) has been well studied and B. Palsson and colleagues have made a considerable contribution to the field using *E. coli* as a model for AE (Fong et al. 2003; Fong and Palsson 2004). One of the main findings is that the *E. coli* metabolic network has evolved to principally optimize growth (Ibarra et al. 2002). Using this as an objective function, the power of the constraint-based *in silico* model of metabolic network (see Section 10) was demonstrated through successful predictions of the end-point of adaptive evolution (Fong et al. 2003, 2005; Fong and Palsson 2004). While gene expression was used to make sense of some of the genetic and regulatory changes taking place during AE, the large-scale measurement of metabolite concentrations using metabolomics has the potential to provide further insights into the mechanisms of AE. It will be interesting to see how the measurement of metabolic intermediate concentrations in *E. coli* strains that have undergone adaptive evolution can provide a clearer picture of the rewiring of the metabolic network necessary for *E. coli* to optimize growth rate and biomass production. Since different patterns of gene expression can lead to convergent growth phenotypes in adapted strains (Fong et al. 2005), observing the impact on the overall metabolite levels should provide new insights into the phenomenon.

Recently, combining gene expression and flux analysis data, both pathway redundancy resulting from the expression of cryptic activities and increased metabolic pathway capacity were shown to be the main strategies used by *E. coli* to readily adapt to the loss of key metabolic enzymes (Fong et al. 2006). Interestingly, such redundancy was also recently suggested to form an important obstacle to the development of new antibiotics targeting metabolic enzymes in Salmonella (Becker et al. 2006). This apparent robustness in the *E. coli* metabolic network is a function of its gene network structure and regulation which is ultimately expressed as changes in metabolite concentrations. In addition to gene and protein expression data presented in these studies, it will therefore be interesting to see how metabolomic datasets can reveal additional downstream mechanistic principles and provide a more detailed signature of adaptation to environmental or genetic change.

10 Metabolic models of *E. coli*: the role of metabolomics

Many types of mathematical models and simulation platforms have been developed and can play a role in the functional understanding of an organism (Ishii et al. 2004). Because of its intrinsic value as an integrator of cellular phenotype, metabolome analysis can play an important part in dynamic models (Arita et al. 2005; Ishii et al. 2005).

The development of a large-scale dynamic model of *E. coli*, making use of kinetic parameters, is one of the long-term objectives at the Institute for Advanced Biosciences, Keio University (Ishii et al. 2004, 2005). Obviously, parameter requirements and complexity grow rapidly with model size. However, the current lack of available parameters collected in a consistent way make this objective, at least on a large scale, appear still somewhat distant (Edwards and Palsson 2000a). Moreover, deriving system parameters from metabolite concentrations (system variables) is an inverse and difficult problem (Kell 2004, 2006).

On a limited scale, a classic kinetic model of *E. coli* central carbon metabolism was originally developed by Chassagnole and co-workers (Chassagnole et al.

2002). This pioneering study used metabolite concentrations to obtain the required kinetic parameters for dynamic simulation. Their model could capture and explain the previously observed oscillations in glucose intermediates (Schaefer et al. 1999), thus demonstrating the usefulness of such an approach. A similar study also showed the power of metabolomic data to build dynamic models based on metabolite concentration measurements (Buchholz et al. 2002).

To bypass the difficulties in obtaining kinetic parameters on a large-scale, others have successfully, over several years, used various types of constraints to generate genome-wide metabolic network reconstructions (Edwards and Palsson 2000a; Covert and Palsson 2003; Reed et al. 2003). The models are based on the construction of a stoichiometric matrix of *E. coli* components that represents its known metabolic activities. An early version of such a model was used to predict the minimal set of reactions required to sustain *E. coli* growth in glucose or acetate medium (Burgard et al. 2001). The latest version, iJR904, can be used to analyze and integrate transcriptomic, proteomic, metabolomic and flux data (Reed et al. 2003). Versions of the model that encompass both gene regulatory and metabolic networks (Covert and Palsson 2002, 2003; Covert et al. 2004) were used to successfully predict the effects of gene knockouts or environmental changes on growth phenotypes (Ibarra et al. 2002; Covert et al. 2004; Fong and Palsson 2004). The method is based on the use of physico-chemical constraints to evaluate and limit the possible phenotypic states that an enzyme system can reach (Edwards and Palsson 2000b; Palsson 2000). Commonly used constraints include the steady-state mass balance, the irreversibility of some reactions due to thermodynamic limitations and the flux capacity that can be handled by enzymes or transporters. In addition to gene expression information, metabolic flux data has been used to put additional constraints on the solution space of such models (Wiback et al. 2004). Overall, the models have been used to successfully predict growth phenotypes and can also be used to develop novel minimal media formulations necessary to support the growth of *E. coli* (Imielinski et al. 2006).

A recent method that aims to estimate flux values, makes use of only a limited number of parameters that do not include the classical kinetic constants and as such seems to bridge the detailed kinetic modeling approach with constraint-based models (Fievet et al. 2006).

In contrast to the carefully curated iJR904 model described above (Reed et al. 2003), the GEM system can automatically generate metabolic models based on genome information and has been applied to *E. coli* and multiple other organisms. The resulting model was found to encompass, almost in its entirety, the available *E. coli* data from KEGG and EcoCyc as well the reactions in the iJR904 model (Arakawa et al. 2006). This model accounts for the presence of at least 1195 metabolites. The system can considerably simplify the generation of complex whole genome models and can also be used to estimate the size and composition of an organism's metabolome (see Section 4).

Based on the reconstructed *E. coli* metabolic network model iJR904, an interesting thermodynamics-based approach was recently reported (Kummel et al. 2006). It was used to make sense of metabolite concentrations measurements as well as to predict the concentration of difficult to measure intermediates and to identify potential regulatory sites. This network-embedded thermodynamic analysis (NET analysis) uses the Gibbs energies of formation of components of a reaction to derive the Gibbs energies of the reaction. Constraints are applied according to the interdependencies of reactions in a pathway. In this way, the authors presented evidence that suggests that enzyme activity regulatory sites can be predicted even when metabolite concentrations are missing and information about compartmentalization of reactions can also be inferred from data originating from disrupted cell extracts (average measurements). The authors achieved this using relatively small datasets obtained from targeted analysis of metabolites. We can thus expect that much larger datasets originating from global profiling of intracellular metabolites will yield an even larger number of interesting findings concerning the regulation of *E. coli* metabolism.

Overall, mathematical models are important for biochemical engineering and can also provide greater understanding of cellular functions without having to perform an unreasonably large number of experiments. Functional discovery, evaluation of data inconsistency, prediction of difficult to test results, and the generation of new hypotheses are all tasks that can benefit from the use of biochemical network models (Edwards and Palsson 2000b; Edwards et al. 2001; Covert et al. 2004). The results of modeling can be verified and the resulting data used to further modify the model in an iterative process. For a more extensive discussion of metabolic pathway models, refer to Chapter 5.

11 Databases and resources

A large amount of information about *E. coli* metabolites is available from highly popular sites such as KEGG, EcoCyc, and BRENDA. These major resources are very useful to get details about known metabolites and to facilitate their identification in metabolomic analysis through physico-chemical properties. They can also assist in the reconstitution of whole metabolic pathways maps and models. Several of the major databases and on-line resources that are useful to the *E. coli* metabolomic community are listed in Table 2.

The ARM database is a unique resource for tracing the route linking two metabolites in metabolic pathways through actual atomic position information (Arita 2003). It is a tool that readily allows to follow, *in silico*, the fate or particular atoms in individual reactions of a pathway just as one could do experimentally using radiolabels, to identify the links between any two metabolites. The pathways included in the linked database are not specific to *E. coli* but much of the information originates from this organism.

While not specific to *E. coli*, the BRENDA database contains a wealth of information about metabolites and enzymes including kinetic parameters and physico-chemical properties, co-factors, inhibitors etc. that can be very useful for researchers in the field of metabolomics (Pharkya et al. 2003). Its breadth and exhaustiveness, however, sometimes come at the cost of usability.

Table 2. Some on-line resources relevant to *E. coli* metabolomic research.

Resource	URL	Notes and reference(s)
ARM	http://www.metabolome.jp/	Metabolic map viewer. (Arita 2003)
BRENDA	http://www.brenda.uni-koeln.de/index.php4	Comprehensive enzyme information system. (Pharkya et al. 2003)
Project CyberCell	http://redpoll.pharmacy.ualberta.ca/CCDB/index.html	Basic info about *E. coli* metabolites. (Sundararaj et al. 2004)
EchoBASE	http://www.ecoli-york.org/	Integrated post-genomic database for *E. coli*. (Misra et al. 2005)
EcoCyc	http://www.ecocyc.org/	Encyclopedia of *E. coli* genes and metabolism. (Keseler et al. 2005)
EcoliHub	http://www.ecolihub.org/	Community page. Currently under development.
Escherichia coli and Salmonella	http://www.ecosal.org/ecosal/toc/index.jsp (subscription)	The classic textbook about *E. coli* physiology.
Genobase	http://ecoli.naist.jp/GB6/search.jsp	*E. coli* genome and functional genomic database.
GenProtEC	http://genprotec.mbl.edu/	*E. coli* genome and proteome database. (Serres et al. 2004)
KEGG	http://www.genome.ad.jp/dbget-bin/www_bfind?e.coli	Subset of KEGG database for *E. coli*. (Kanehisa et al. 2004)
OUBCF	http://chase.ou.edu/oubcf/	Oklahoma University *E. coli* gene expression database.
PubChem	http://pubchem.ncbi.nlm.nih.gov/	NCBI's database of biological information about small molecules. (Wheeler et al. 2006)
RegulonDB	http://regulondb.ccg.unam.mx/index.html	A database about gene regulation in *E. coli* and operon structure. (Salgado et al. 2006)
UCSD Systems Biology Research Group	http://systemsbiology.ucsd.edu/organisms/ecoli.html	Constraints-based *E. coli* models. (Reed et al. 2003)
The University of Minnesota Biocatalysis/Biodegradation Database (UM-BBD)	http://umbbd.msi.umn.edu/	Microbial biocatalytic reactions and biodegradation pathways. (Ellis et al. 2003, 2006)
University of Wisconsin	http://www.genome.wisc.edu/functional.htm	*E. coli* DNA and phenotype microarray data.

The CyberCell database and web site form an integrated resource of qualitative and quantitative data about *E. coli* (Sundararaj et al. 2004). Among the subcomponents, the CyberCell metabolite (CCMD) browser contains useful information about the metabolome and basic information about metabolites can be browsed or searched using various user-selected criteria. Detailed information can be obtained by looking at the corresponding Metabocard, where physico-chemical parameters are indicated and structural data can be downloaded. Details of associated metabolic enzymes are also provided.

EchoBASE provides curated functional information (both computational and experimental) about *E. coli* genes and their products (Misra et al. 2005). It is another large and integrated post-genomic database system that is *E. coli*-centric but that is not yet populated with metabolomic data since such datasets are still only scarcely available. An updated list of orphan enzymes is also available and integrates information from three different sources.

EcoCyc is one of the most exhaustive resources concerning biological and post-genomic information about *E. coli* (Keseler et al. 2005). It forms an integrated informatic resource for *E. coli*-specific post-genomic research. Most of the information originates and is curated from the primary literature. Of particular relevance for metabolomics is the information about metabolic pathways. Data about reactions, compounds, and complete metabolic maps are easily searchable and quantitative information about metabolite levels can be plotted on the maps using the Omics Viewer. It is definitely, one of the most useful resources.

The EcoliHub site, a NIH-funded initiative led by B. Wanner at Purdue University is still in development and currently mostly consists of links to other databases. However, it aims to become a central gateway for the community and will also eventually likely include information about metabolomic research.

The ecosal.org site, the on-line and continually updated version of the classic textbook "*Escherichia coli* and Salmonella" by Frederick Neidhardt remains the authority for single-source, encyclopedic, and integrated knowledge about the physiology and metabolism of *E. coli*, and several chapters are dedicated to metabolism. However, access is limited to subscribers.

GenoBase is an *E. coli*-centric functional genomic database that can be of interest to the metabolomics community since it contains information about expression (both gene and protein), protein-protein interactions, protein localization, growth phenotype, etc. that can all assist in the interpretation of metabolomic datasets. Numerous links to external databases are available. It is also the gateway to the complete library of single gene deletion mutants of all non-essential *E. coli* genes, the KEIO collection (Baba et al. 2006), and ASKA, the complete his-tagged open reading frame (ORF) expression library of *E. coli* W3110 (Kitagawa et al. 2005) with or without fusion to green fluorescent protein, two extremely valuable resources for experimentation.

Similarly, GenProtEC, containing information about protein modules and their classification by biochemical mechanisms is also useful for *E. coli* metabolomics (Serres et al. 2004).

The Kyoto Encyclopedia of Genes and Genomes (KEGG), probably the best known, all-around database of metabolic information, remains a central reference

(Ogata et al. 1999; Kanehisa et al. 2002, 2004). A subset for *E. coli* is searchable and the PATHWAY, COMPOUND, and LIGAND components of the system are especially relevant to metabolomics. Recent updates to the database include KEGG BRITE, an ontology for pathway reconstruction based on hierarchical classification and KEGG DRUG, a collection of drug structure maps that aim to include information about exogenous molecules (Kanehisa et al. 2006).

Among the numerous NCBI resources, PubChem is of particular interest for researchers in metabolomics. Information about small molecules can be searched in multiple ways including the possibility to perform structure based-searches. As for KEGG, the system is composed of multiple databases including PubChem Substance, Compound, and Bioassay and links to multiple external resources.

The RegulonDB (Salgado et al. 2006) can also assist the *E. coli* metabolomic community by providing information on *E. coli* gene regulatory network and operon structure, both of which are relevant to research on enzymes and metabolic pathways.

The bioinformatics core facility site at the University of Oklahoma (OUBCF) is rich in gene expression data relevant to glucose-lactose diauxie and other growth condition changes that can facilitate the analysis of metabolomic data. Similarly, the University of Wisconsin's *E. coli* MG1655 genome project includes gene expression data as well as searchable results of Biolog's phenotype microarray (PM) analysis for a subset of gene disruptants.

The University of Minnesota Biocatalysis/Biodegradation Database (UM-BBD) aims to provide information about existing metabolic reactions in microorganisms to facilitate the production of specialty chemicals and for bioremediation.

For complete metabolic models based on genome reconstructions (see Section 10) the web site of the Systems Biology Research Group at UCSD contains useful information and is the actual repository where *E. coli* metabolic network data files can be downloaded.

Finally, significant repositories of large-scale *E. coli* metabolomic data have yet to emerge and currently appear to exist mostly in the form of publications and/or the associated supplementary information. Similarly to efforts in transcriptomics and proteomics, it will be important to promote and follow standards for the distribution of data such as those promoted in the Minimum Information About a METabolomics experiment (MIAMET) (Bino et al. 2004; Jenkins et al. 2004) and the Metabolomics Standards Initiative (http://msi-workgroups.sourceforge.net/) and to create central and public repositories. The newly proposed MeMo system provides a data structure to facilitate annotation and data management that can also be useful for *E. coli* data dissemination (Spasic et al. 2006).

12 Data integration and visualization

The study of the metabolome represents in itself a way of observing the integrated response of the cell beyond the transcriptome and proteome levels. However, the complexity of changes that can occur in the metabolome in response to even single gene inactivation, as well as the inherent biological and technical variability in its analysis mean that the ability to integrate metabolomic datasets with those obtained at other levels of biological information (transcriptomics, proteomics, etc), if not a prerequisite, is highly desirable to make sense of the data. Numerous studies have already highlighted the advantages of integrating different levels of molecular information (ter Kuile and Westerhoff 2001; Hirai et al. 2004; Gibon et al. 2006; Kresnowati et al. 2006). In addition, Chapter 12 also discusses in more details the integration of metabolomic and proteomic data.

One of the most effective ways to make sense of metabolomic information is to visualize the results onto maps of metabolic pathways. This is now possible using various tools that were either designed specifically for this purpose or designed for more general transcriptomic and proteomic information. Some of the better known include MapMan (Thimm et al. 2004; Usadel et al. 2005) and the Omics viewer (Paley and Karp 2006). Another easy-to-use tool utilizes KEGG maps to generate Flash vector images of metabolic pathways (Arakawa et al. 2005). By specifically selecting or importing *E. coli* pathways, users can visualize absolute or relative concentration of metabolites or metabolic enzymes on the relevant pathway maps. In addition, the general purpose and powerful interaction network visualization software Cytoscape (Shannon et al. 2003) can also be used for representing and integrating large-scale metabolomic data. Together, these tools provide more intuitive interfaces and a level of insight into the data that would not be possible otherwise.

13 Future prospects and developments

The success of *E. coli* metabolomics will require further significant improvements in analytical technologies. This means both easier and more effective ways of preparing bacterial samples with faster and more efficient quenching of metabolism and improvements in extraction efficiency. Promising developments in this direction are already emerging (Brauer et al. 2006; Schaub et al. 2006). As described in Section 2.1, the use of cultures growing on filters and dry media will likely have some impact. In addition, coupling of a bioreactor to a sampling device maintained at high temperature (<95°C) can allow to rapidly quench intracellular metabolism and quantitatively extract metabolites while also permitting very high frequency sampling ($5\ s^{-1}$) and short sample processing time (< 30 s). Further improvements in the sensitivity and selectivity of analytical devices will also be important. This may include an increasing use of multi-dimensional separation methods making use of CE, LC, and GC or combinations of these (Liu et al. 2006) and further incremental improvements in MS and NMR instruments. For the foreseeable future, MS- and NMR-based methods are bound to grab the bulk of the attention. However, there is a pressing need to also develop less invasive methods to observe and quantify metabolites dynamically in single cells (Dovichi and Hu 2003; Fehr et al. 2004; Valet 2005). In this regard, the rapid developments in microfluidics are bound to have an impact (Di Carlo et al. 2006; Liu et al. 2006). In addition, a larger number of experimental conditions will need to be used, and this may become possible with the use of miniaturized devices providing increased throughput for culture sampling and analysis of cells (Balaban et al. 2004). Only then, can the metabolome of *E. coli*, being a dynamic entity, be revealed and analyzed in its full-scale. Finally, data analysis, which currently remains an important bottleneck, is bound to benefit from new algorithms and statistical approaches that are customized to the specific needs and nature of metabolomic data.

Overall, most of these expected improvements are not specific to *E. coli* and are bound to affect all applications in metabolomics, whether it is bacterially-, tissue-, or organism-oriented.

14 Concluding remarks

As one of its traditional roles, the legacy of knowledge obtained from experiments in *E. coli* holds tremendous value far beyond the borders of the microbial world, and this is especially true for metabolomic data due to the conserved nature of metabolites and pathways. This unicellular organism will thus continue to provide fundamental discoveries from increasingly large amounts of metabolomic data that are likely to pave the road for related findings in other organisms. The full complexity of *E. coli* metabolic components and their interaction networks still remains to be grasped. Because of its intrinsic character that integrates the responses of the transcriptome, proteome and environment, metabolomics will stay at the forefront of the integrative concept behind systems biology. Of course, many challenges remain, mostly technical, analytical, and computational that new solutions will help overcome. Questions such as "What are the grand rules governing the regulation of metabolism?" or "What controls heterogeneous cell population dynamics?" and many others remain and will shape our future efforts. The more light is shed on the *E. coli* metabolome complexity, the more other fields will also benefit. Clearly, the central role of *E. coli* is here to stay, simply because of its numerous advantages and broad applicability. This simple model, both ideal biological factory and important human symbiont deserves all the respect it can get in the metabolomic era. Models of its metabolism will find numerous applications for the improvement of biochemical engineering processes and also provide a helping hand toward the design of new pathways in this unique organism and beyond.

Acknowledgement

The authors are grateful to Masanori Arita, Richard Baran, and Kenji Nakahigashi for advice and suggestions on the manuscript. The authors also acknowledge the existence of many other relevant publications that could not be cited in this chapter because of length consideration and wish to apologize to their authors.

References

Allen J, Davey HM, Broadhurst D, Heald JK, Rowland JJ, Oliver SG, Kell DB (2003) High-throughput classification of yeast mutants for functional genomics using metabolic footprinting. Nat Biotechnol 6:692-696

Almaas E, Kovacs B, Vicsek T, Oltvai ZN, Barabasi AL (2004) Global organization of metabolic fluxes in the bacterium *Escherichia coli*. Nature 427:839-843

Arakawa K, Kono N, Yamada Y, Mori H, Tomita M (2005) KEGG-based pathway visualization tool for complex omics data. In Silico Biol 5:419-423

Arakawa K, Yamada Y, Shinoda K, Nakayama Y, Tomita M (2006) GEM System: automatic prototyping of cell-wide metabolic pathway models from genomes. BMC Bioinformatics 7:168

Arita M (2003) *In silico* atomic tracing by substrate-product relationships in *Escherichia coli* intermediary metabolism. Genome Res 13:2455-2466

Arita M (2004) The metabolic world of *Escherichia coli* is not small. Proc Natl Acad Sci USA 101:1543-1547

Arita M, Robert M, Tomita M (2005) All systems go: launching cell simulation fueled by integrated experimental biology data. Curr Opin Biotechnol 16:344-349

Baba T, Ara T, Hasegawa M, Takai Y, Okumura Y, Baba M, Datsenko KA, Tomita M, Wanner BL, Mori H (2006) Construction of *Escherichia coli* K-12 in-frame, single-gene knockout mutants: the Keio collection. Mol Syst Biol 2:2006.0008

Bajad SU, Lu W, Kimball EH, Yuan J, Peterson C, Rabinowitz JD (2006) Separation and quantitation of water soluble cellular metabolites by hydrophilic interaction chromatography-tandem mass spectrometry. J Chromatogr A 1125:76-88

Balaban NQ, Merrin J, Chait R, Kowalik L, Leibler S (2004) Bacterial persistence as a phenotypic switch. Science 305:1622-1625

Baran R, Kochi H, Saito N, Suematsu M, Soga T, Nishioka T, Robert M, Tomita M (2006) MathDAMP: a package for differential analysis of metabolite profiles. BMC Bioinformatics 7:530

Becker D, Selbach M, Rollenhagen C, Ballmaier M, Meyer TF, Mann M, Bumann D (2006) Robust Salmonella metabolism limits possibilities for new antimicrobials. Nature 440:303-307

Berry A (1996) Improving production of aromatic compounds in *Escherichia coli* by metabolic engineering. Trends Biotechnol 14:250-256

Bhattacharya M, Fuhrman L, Ingram A, Nickerson KW, Conway T (1995) Single-run separation and detection of multiple metabolic intermediates by anion-exchange high-performance liquid chromatography and application to cell pool extracts prepared from *Escherichia coli*. Anal Biochem 232:98-106

Bino RJ, Hall RD, Fiehn O, Kopka J, Saito K, Draper J, Nikolau BJ, Mendes P, Roessner-Tunali U, Beale MH, Trethewey RN, Lange BM, Wurtele ES, Sumner LW (2004) Potential of metabolomics as a functional genomics tool. Trends Plant Sci 9:418-425

Birkemeyer C, Luedemann A, Wagner C, Erban A, Kopka J (2005) Metabolome analysis: the potential of *in vivo* labeling with stable isotopes for metabolite profiling. Trends Biotechnol 23:28-33

Bochner BR (2003) New technologies to assess genotype-phenotype relationships. Nat Rev Genet 4:309-314

Bochner BR, Gadzinski P, Panomitros E (2001) Phenotype microarrays for high-throughput phenotypic testing and assay of gene function. Genome Res 11:1246-1255

Brauer MJ, Yuan J, Bennett BD, Lu W, Kimball E, Botstein D, Rabinowitz JD (2006) Conservation of the metabolomic response to starvation across two divergent microbes. PNAS 103:19302-19307

Buchholz A, Hurlebaus J, Wandrey C, Takors R (2002) Metabolomics: quantification of intracellular metabolite dynamics. Biomol Eng 19:5-15

Buchholz A, Takors R, Wandrey C (2001) Quantification of intracellular metabolites in *Escherichia coli* K12 using liquid chromatographic-electrospray ionization tandem mass spectrometric techniques. Anal Biochem 295:129-137

Burgard AP, Vaidyaraman S, Maranas CD (2001) Minimal reaction sets for *Escherichia coli* metabolism under different growth requirements and uptake environments. Biotechnol Prog 17:791-797

Castrillo JI, Hayes A, Mohammed S, Gaskell SJ, Oliver SG (2003) An optimized protocol for metabolome analysis in yeast using direct infusion electrospray mass spectrometry. Phytochemistry 62:929-937

Chang DE, Jung HC, Rhee JS, Pan JG (1999) Homofermentative production of D- or L-lactate in metabolically engineered *Escherichia coli* RR1. Appl Environ Microbiol 65:1384-1389

Chassagnole C, Noisommit-Rizzi N, Schmid JW, Mauch K, Reuss M (2002) Dynamic modeling of the central carbon metabolism of *Escherichia coli*. Biotechnol Bioeng 79:53-73

Chen H, Pan Z, Talaty N, Raftery D, Cooks RG (2006) Combining desorption electrospray ionization mass spectrometry and nuclear magnetic resonance for differential metabolomics without sample preparation. Rapid Commun Mass Spectrom 20:1577-1584

Chen L, Vitkup D (2006) Predicting genes for orphan metabolic activities using phylogenetic profiles. Genome Biol 7:R17

Clayton TA, Lindon JC, Cloarec O, Antti H, Charuel C, Hanton G, Provost JP, Le Net JL, Baker D, Walley RJ, Everett JR, Nicholson JK (2006) Pharmaco-metabonomic phenotyping and personalized drug treatment. Nature 440:1073-1077

Cooks RG, Ouyang Z, Takats Z, Wiseman JM (2006) Detection Technologies. Ambient mass spectrometry. Science 311:1566-1570

Coulier L, Bas R, Jespersen S, Verheij E, van der Werf MJ, Hankemeier T (2006) Simultaneous quantitative analysis of metabolites using ion-pair liquid chromatography-electrospray ionization mass spectrometry. Anal Chem 78:6573-6582

Covert MW, Knight EM, Reed JL, Herrgard MJ, Palsson BO (2004) Integrating high-throughput and computational data elucidates bacterial networks. Nature 429:92-96

Covert MW, Palsson BO (2002) Transcriptional regulation in constraints-based metabolic models of *Escherichia coli*. J Biol Chem 277:28058-28064

Covert MW, Palsson BO (2003) Constraints-based models: regulation of gene expression reduces the steady-state solution space. J Theor Biol 221:309-325

Crockford DJ, Holmes E, Lindon JC, Plumb RS, Zirah S, Bruce SJ, Rainville P, Stumpf CL, Nicholson JK (2006) Statistical heterospectroscopy, an approach to the integrated analysis of NMR and UPLC-MS data sets: application in metabonomic toxicology studies. Anal Chem 78:363-371

de Koning W, van Dam K (1992) A method for the determination of changes of glycolytic metabolites in yeast on a subsecond time scale using extraction at neutral pH. Anal Biochem 204:118-123

Di Carlo D, Aghdam N, Lee LP (2006) Single-cell enzyme concentrations, kinetics, and inhibition analysis using high-density hydrodynamic cell isolation arrays. Anal Chem 78:4925-4930

Dovichi NJ, Hu S (2003) Chemical cytometry. Curr Opin Chem Biol 7:603-608

Edwards JS, Ibarra RU, Palsson BO (2001) *In silico* predictions of *Escherichia coli* metabolic capabilities are consistent with experimental data. Nat Biotechnol 19:125-130

Edwards JS, Palsson BO (2000a) The *Escherichia coli* MG1655 *in silico* metabolic genotype: its definition, characteristics, and capabilities. Proc Natl Acad Sci USA 97:5528-5533

Edwards JS, Palsson BO (2000b) Metabolic flux balance analysis and the *in silico* analysis of *Escherichia coli* K-12 gene deletions. BMC Bioinformatics 1:1

Ellis LB, Hou BK, Kang W, Wackett LP (2003) The University of Minnesota Biocatalysis/Biodegradation Database: post-genomic data mining. Nucleic Acids Res 31:262-265

Ellis LB, Roe D, Wackett LP (2006) The University of Minnesota Biocatalysis/Biodegradation Database: the first decade. Nucleic Acids Res 34:D517-521

Emmerling M, Bailey JE, Sauer U (1999) Glucose catabolism of *Escherichia coli* strains with increased activity and altered regulation of key glycolytic enzymes. Metab Eng 1:117-127

Fehr M, Ehrhardt DW, Lalonde S, Frommer WB (2004) Minimally invasive dynamic imaging of ions and metabolites in living cells. Curr Opin Plant Biol 7:345-351

Fiehn O (2002) Metabolomics - the link between genotypes and phenotypes. Plant Mol Biol 48:155-171

Fievet JB, Dillmann C, Curien G, de Vienne D (2006) Simplified modelling of metabolic pathways for flux prediction and optimization: lessons from an *in vitro* reconstruction of the upper part of glycolysis. Biochem J 396:317-326

Fischer E, Sauer U (2003) Metabolic flux profiling of *Escherichia coli* mutants in central carbon metabolism using GC-MS. Eur J Biochem 270:880-891

Fong SS, Joyce AR, Palsson BO (2005) Parallel adaptive evolution cultures of *Escherichia coli* lead to convergent growth phenotypes with different gene expression states. Genome Res 15:1365-1372

Fong SS, Marciniak JY, Palsson BO (2003) Description and interpretation of adaptive evolution of *Escherichia coli* K-12 MG1655 by using a genome-scale *in silico* metabolic model. J Bacteriol 185:6400-6408

Fong SS, Nanchen A, Palsson BO, Sauer U (2006) Latent pathway activation and increased pathway capacity enable *Escherichia coli* adaptation to loss of key metabolic enzymes. J Biol Chem 281:8024-8033

Fong SS, Palsson BO (2004) Metabolic gene-deletion strains of *Escherichia coli* evolve to computationally predicted growth phenotypes. Nat Genet 36:1056-1058

Gerdes S, Edwards R, Kubal M, Fonstein M, Stevens R, Osterman A (2006) Essential genes on metabolic maps. Curr Opin Biotechnol 17:448-456

Giaever G, Chu AM, Ni L, Connelly C, Riles L, Veronneau S, Dow S, Lucau-Danila A, Anderson K, Andre B, Arkin AP, Astromoff A, El-Bakkoury M, Bangham R, Benito R, Brachat S, Campanaro S, Curtiss M, Davis K, Deutschbauer A, Entian KD, Flaherty P, Foury F, Garfinkel DJ, Gerstein M, Gotte D, Guldener U, Hegemann JH, Hempel S, Herman Z, Jaramillo DF, Kelly DE, Kelly SL, Kotter P, LaBonte D, Lamb DC, Lan N, Liang H, Liao H, Liu L, Luo C, Lussier M, Mao R, Menard P, Ooi SL, Revuelta JL, Roberts CJ, Rose M, Ross-Macdonald P, Scherens B, Schimmack G, Shafer B, Shoemaker DD, Sookhai-Mahadeo S, Storms RK, Strathern JN, Valle G, Voet M, Volckaert G, Wang CY, Ward TR, Wilhelmy J, Winzeler EA, Yang Y, Yen G, Youngman E, Yu K, Bussey H, Boeke JD, Snyder M, Philippsen P, Davis RW, Johnston M (2002) Functional profiling of the *Saccharomyces cerevisiae* genome. Nature 418:387-391

Gibon Y, Usadel B, Blaesing OE, Kamlage B, Hoehne M, Trethewey R, Stitt M (2006) Integration of metabolite with transcript and enzyme activity profiling during diurnal cycles in *Arabidopsis* rosettes. Genome Biol 7:R76

Gill RT (2003) Enabling inverse metabolic engineering through genomics. Curr Opin Biotechnol 14:484-490

Green ML, Karp PD (2004) A Bayesian method for identifying missing enzymes in predicted metabolic pathway databases. BMC Bioinformatics 5:76

Harada K, Fukusaki E, Kobayashi A (2006) Pressure-assisted capillary electrophoresis mass spectrometry using combination of polarity reversion and electroosmotic flow for metabolomics anion analysis. J Biosci Bioeng 101:403-409

Hattori M, Okuno Y, Goto S, Kanehisa M (2003) Development of a chemical structure comparison method for integrated analysis of chemical and genomic information in the metabolic pathways. J Am Chem Soc 125:11853-11865

Hirai MY, Yano M, Goodenowe DB, Kanaya S, Kimura T, Awazuhara M, Arita M, Fujiwara T, Saito K (2004) Integration of transcriptomics and metabolomics for understanding of global responses to nutritional stresses in *Arabidopsis thaliana*. Proc Natl Acad Sci USA 101:10205-10210

Hollywood K, Brison DR, Goodacre R (2006) Metabolomics: Current technologies and future trends. Proteomics 6:4716-4723

Hoque MA, Ushiyama H, Tomita M, Shimizu K (2005) Dynamic responses of the intracellular metabolite concentrations of the wild type and pykA mutant *Escherichia coli* against pulse addition of glucose or NH3 under those limiting continuous cultures. Biochem Eng J 26:38-49

Ibarra RU, Edwards JS, Palsson BO (2002) *Escherichia coli* K-12 undergoes adaptive evolution to achieve *in silico* predicted optimal growth. Nature 420:186-189

Imielinski M, Belta C, Halasz A, Rubin H (2005) Investigating metabolite essentiality through genome-scale analysis of *Escherichia coli* production capabilities. Bioinformatics 21:2008-2016

Imielinski M, Belta C, Rubin H, Halasz A (2006) Systematic analysis of conservation relations in *E. coli* genome-scale metabolic network reveals novel growth media. Biophys J 90:2659-2672

Ishii N, Robert M, Nakayama Y, Kanai A, Tomita M (2004) Toward large-scale modeling of the microbial cell for computer simulation. J Biotechnol 113:281-294

Ishii N, Soga T, Nishioka T, Tomita M (2005) Metabolome analysis and metabolic simulation. Metabolomics 1:29-37

Ito M, Baba T, Mori H (2005) Functional analysis of 1440 *Escherichia coli* genes using the combination of knock-out library and phenotype microarrays. Metab Eng 7:318-327

Jankowski J, Stephan N, Knobloch M, Fischer S, Schmaltz D, Zidek W, Schluter H (2001) Mass-spectrometry-linked screening of protein fractions for enzymatic activities - a tool for functional genomics. Anal Biochem 290:324-329

Jenkins H, Hardy N, Beckmann M, Draper J, Smith AR, Taylor J, Fiehn O, Goodacre R, Bino RJ, Hall R, Kopka J, Lane GA, Lange BM, Liu JR, Mendes P, Nikolau BJ, Oliver SG, Paton NW, Rhee S, Roessner-Tunali U, Saito K, Smedsgaard J, Sumner LW,

Wang T, Walsh S, Wurtele ES, Kell DB (2004) A proposed framework for the description of plant metabolomics experiments and their results. Nat Biotechnol 22:1601-1606

Kaderbhai NN, David I, Broadhurst, Ellis DI, Goodacre R, Kell DB (2003) Functional genomics via metabolic footprinting: monitoring metabolite secretion by *Escherichia coli* tryptophan metabolism mutants using FT-IR and direct injection electrospray mass spectrometry. Comp Funct Genomics 4:376-391

Kanehisa M, Goto S, Hattori M, Aoki-Kinoshita KF, Itoh M, Kawashima S, Katayama T, Araki M, Hirakawa M (2006) From genomics to chemical genomics: new developments in KEGG. Nucleic Acids Res 34:D354-357

Kanehisa M, Goto S, Kawashima S, Nakaya A (2002) The KEGG databases at GenomeNet. Nucleic Acids Res 30:42-46

Kanehisa M, Goto S, Kawashima S, Okuno Y, Hattori M (2004) The KEGG resource for deciphering the genome. Nucleic Acids Res 32:D277-280

Kell DB (2004) Metabolomics and systems biology: making sense of the soup. Curr Opin Microbiol 7:296-307

Kell DB (2006) Theodor Bucher Lecture. Metabolomics, modelling and machine learning in systems biology - towards an understanding of the languages of cells. Delivered on 3 July 2005 at the 30th FEBS Congress and the 9th IUBMB conference in Budapest. FEBS J 273:873-894

Kell DB, Brown M, Davey HM, Dunn WB, Spasic I, Oliver SG (2005) Metabolic footprinting and systems biology: the medium is the message. Nat Rev Microbiol 3:557-565

Keseler IM, Collado-Vides J, Gama-Castro S, Ingraham J, Paley S, Paulsen IT, Peralta-Gil M, Karp PD (2005) EcoCyc: a comprehensive database resource for *Escherichia coli*. Nucleic Acids Res 33:D334-337

Kharchenko P, Vitkup D, Church GM (2004) Filling gaps in a metabolic network using expression information. Bioinformatics 20 Suppl 1:I178-I185

Kitagawa M, Ara T, Arifuzzaman M, Ioka-Nakamichi T, Inamoto E, Toyonaga H, Mori H (2005) Complete set of ORF clones of *Escherichia coli* ASKA library (A Complete Set of *E. coli* K-12 ORF Archive): Unique Resources for Biological Research. DNA Res 12:291-299

Koek MM, Muilwijk B, van der Werf MJ, Hankemeier T (2006) Microbial metabolomics with gas chromatography/mass spectrometry. Anal Chem 78:1272-1281

Koeniger SL, Merenbloom SI, Valentine SJ, Jarrold MF, Udseth HR, Smith RD, Clemmer DE (2006) An IMS-IMS analogue of MS-MS. Anal Chem 78:4161-4174

Kopka J, Schauer N, Krueger S, Birkemeyer C, Usadel B, Bergmuller E, Dormann P, Weckwerth W, Gibon Y, Stitt M, Willmitzer L, Fernie AR, Steinhauser D (2005) GMD@CSB.DB: the Golm Metabolome Database. Bioinformatics 21:1635-1638

Kramer M, Bongaerts J, Bovenberg R, Kremer S, Muller U, Orf S, Wubbolts M, Raeven L (2003) Metabolic engineering for microbial production of shikimic acid. Metab Eng 5:277-283

Kresnowati MTAP, van Winden WA, Almering MJH, Proell A, Ras C, Knijnenburg TA, Daran-Lapujade PAS, Pronk JT, Heijnen JJ, Daran JM (2006) When transcriptome meets metabolome: fast cellular responses of yeast to sudden relief of glucose limitation. Mol Syst Biol 2:49

Kummel A, Panke S, Heinemann M (2006) Putative regulatory sites unraveled by network-embedded thermodynamic analysis of metabolome data. Mol Syst Biol 2:E1-E10

Kuznetsova E, Proudfoot M, Gonzalez CF, Brown G, Omelchenko MV, Borozan I, Carmel L, Wolf YI, Mori H, Savchenko AV, Arrowsmith CH, Koonin EV, Edwards AM, Yakunin AF (2006) Genome-wide analysis of substrate specificities of the *Escherichia coli* haloacid dehalogenase-like phosphatase family. J Biol Chem 281:36149-36161

Kuznetsova E, Proudfoot M, Sanders SA, Reinking J, Savchenko A, Arrowsmith CH, Edwards AM, Yakunin AF (2005) Enzyme genomics: Application of general enzymatic screens to discover new enzymes. FEMS Microbiol Rev 29:263-279

Larsson G, Tornkvist M (1996) Rapid sampling, cell inactivation and evaluation of low extracellular glucose concentrations during fed-batch cultivation. J Biotechnol 49:69-82

Lasserre JP, Beyne E, Pyndiah S, Lapaillerie D, Claverol S, Bonneu M (2006) A complexomic study of *Escherichia coli* using two-dimensional blue native/SDS polyacrylamide gel electrophoresis. Electrophoresis 27:3306-3321

Lee DY, Yun H, Park S, Lee SY (2003) MetaFluxNet: the management of metabolic reaction information and quantitative metabolic flux analysis. Bioinformatics 19:2144-2146

Lee SY, Lee DY, Kim TY (2005) Systems biotechnology for strain improvement. Trends Biotechnol 23:349-358

Lemke N, Heredia F, Barcellos CK, Dos Reis AN, Mombach JC (2004) Essentiality and damage in metabolic networks. Bioinformatics 20:115-119

Li M, Ho PY, Yao S, Shimizu K (2006) Effect of lpdA gene knockout on the metabolism in *Escherichia coli* based on enzyme activities, intracellular metabolite concentrations and metabolic flux analysis by (13)C-labeling experiments. J Biotechnol 122:254-266

Liu BF, Xu B, Zhang G, Du W, Luo Q (2006) Micro-separation toward systems biology. J Chromatogr A 1106:19-28

Liu X, Ng C, Ferenci T (2000) Global adaptations resulting from high population densities in *Escherichia coli* cultures. J Bacteriol 182:4158-4164

Loh KD, Gyaneshwar P, Markenscoff Papadimitriou E, Fong R, Kim KS, Parales R, Zhou Z, Inwood W, Kustu S (2006) A previously undescribed pathway for pyrimidine catabolism. Proc Natl Acad Sci USA 103:5114-5119

Lowry OH, Carter J, Ward JB, Glaser L (1971) The effect of carbon and nitrogen sources on the level of metabolic intermediates in *Escherichia coli*. J Biol Chem 246:6511-6521

Maharjan RP, Ferenci T (2003) Global metabolite analysis: the influence of extraction methodology on metabolome profiles of *Escherichia coli*. Anal Biochem 313:145-154

Misra RV, Horler RS, Reindl W, Goryanin, II, Thomas GH (2005) EchoBASE: an integrated post-genomic database for *Escherichia coli*. Nucleic Acids Res 33:D329-333

Neidhardt FC, Ingraham JL, Schaechter M (1990) Physiology of the bacterial cell: A molecular approach. Sinauer Associates, Sunderland, MA

Nobeli I, Ponstingl H, Krissinel EB, Thornton JM (2003) A structure-based anatomy of the *E. coli* metabolome. J Mol Biol 334:697-719

Nobeli I, Thornton JM (2006) A bioinformatician's view of the metabolome. Bioessays 28:534-545

Noteborn HP, Lommen A, van der Jagt RC, Weseman JM (2000) Chemical fingerprinting for the evaluation of unintended secondary metabolic changes in transgenic food crops. J Biotechnol 77:103-114

Ochoa ML, Harrington PB (2005) Chemometric studies for the characterization and differentiation of microorganisms using *in situ* derivatization and thermal desorption ion mobility spectrometry. Anal Chem 77:854-863

Ogata H, Goto S, Sato K, Fujibuchi W, Bono H, Kanehisa M (1999) KEGG: Kyoto Encyclopedia of Genes and Genomes. Nucleic Acids Res 27:29-34

Osterman A (2006) A hidden metabolic pathway exposed. Proc Natl Acad Sci USA 103:5637-5638

Osterman A, Overbeek R (2003) Missing genes in metabolic pathways: a comparative genomics approach. Curr Opin Chem Biol 7:238-251

Paley SM, Karp PD (2006) The pathway tools cellular overview diagram and omics viewer. Nucleic Acids Res 34:3771-3778

Palsson B (2000) The challenges of *in silico* biology. Nat Biotechnol 18:1147-1150

Pan Z, Raftery D (2007) Comparing and combining NMR spectroscopy and mass spectrometry in metabolomics. Anal Bioanal Chem 387:525–527

Peng L, Arauzo-Bravo MJ, Shimizu K (2004) Metabolic flux analysis for a ppc mutant *Escherichia coli* based on (13)C-labelling experiments together with enzyme activity assays and intracellular metabolite measurements. FEMS Microbiol Lett 235:17-23

Pharkya P, Nikolaev EV, Maranas CD (2003) Review of the BRENDA Database. Metab Eng 5:71-73

Pohl NL (2005) Functional proteomics for the discovery of carbohydrate-related enzyme activities. Curr Opin Chem Biol 9:76-81

Proudfoot M, Kuznetsova E, Brown G, Rao NN, Kitagawa M, Mori H, Savchenko A, Yakunin AF (2004) General enzymatic screens identify three new nucleotidases in *Escherichia coli*: Biochemical characterization of SurE, YfbR, and YjjG. J Biol Chem 279:54687-54694

Raamsdonk LM, Teusink B, Broadhurst D, Zhang N, Hayes A, Walsh MC, Berden JA, Brindle KM, Kell DB, Rowland JJ, Westerhoff HV, van Dam K, Oliver SG (2001) A functional genomics strategy that uses metabolome data to reveal the phenotype of silent mutations. Nat Biotechnol 19:45-50

Rahman M, Hasan MR, Oba T, Shimizu K (2006) Effect of rpoS gene knockout on the metabolism of *Escherichia coli* during exponential growth phase and early stationary phase based on gene expressions, enzyme activities and intracellular metabolite concentrations. Biotechnol Bioeng 94:585-595

Reed JL, Vo TD, Schilling CH, Palsson BO (2003) An expanded genome-scale model of *Escherichia coli* K-12 (iJR904 GSM/GPR). Genome Biol 4:R54

Ross-Macdonald P, Coelho PS, Roemer T, Agarwal S, Kumar A, Jansen R, Cheung KH, Sheehan A, Symoniatis D, Umansky L, Heidtman M, Nelson FK, Iwasaki H, Hager K, Gerstein M, Miller P, Roeder GS, Snyder M (1999) Large-scale analysis of the yeast genome by transposon tagging and gene disruption. Nature 402:413-418

Saghatelian A, Cravatt BF (2005) Discovery metabolite profiling - forging functional connections between the proteome and metabolome. Life Sci 77:1759-1766

Saghatelian A, Trauger SA, Want EJ, Hawkins EG, Siuzdak G, Cravatt BF (2004) Assignment of endogenous substrates to enzymes by global metabolite profiling. Biochemistry 43:14332-14339

Saito N, Robert M, Kitamura S, Baran R, Soga T, Mori H, Nishioka T, Tomita M (2006) Metabolomics approach for enzyme discovery. J Proteome Res 5:1979-1987

Salgado H, Gama-Castro S, Peralta-Gil M, Diaz-Peredo E, Sanchez-Solano F, Santos-Zavaleta A, Martinez-Flores I, Jimenez-Jacinto V, Bonavides-Martinez C, Segura-Salazar J, Martinez-Antonio A, Collado-Vides J (2006) RegulonDB (version 5.0): *Escherichia coli* K-12 transcriptional regulatory network, operon organization, and growth conditions. Nucleic Acids Res 34:D394-397

Sauer U (2004) High-throughput phenomics: experimental methods for mapping fluxomes. Curr Opin Biotechnol 15:58-63

Schaefer U, Boos W, Takors R, Weuster-Botz D (1999) Automated sampling device for monitoring intracellular metabolite dynamics. Anal Biochem 270:88-96

Schaub J, Schiesling C, Reuss M, Dauner M (2006) Integrated sampling procedure for metabolome analysis. Biotechnol Prog 22:1434-1442

Schwab W (2003) Metabolome diversity: too few genes, too many metabolites? Phytochemistry 62:837-849

Serina S, Nozza F, Nicastro G, Faggioni F, Mottl H, Deho G, Polissi A (2004) Scanning the *Escherichia coli* chromosome by random transposon mutagenesis and multiple phenotypic screening. Res Microbiol 155:692-701

Serres MH, Goswami S, Riley M (2004) GenProtEC: an updated and improved analysis of functions of *Escherichia coli* K-12 proteins. Nucleic Acids Res 32:D300-302

Shannon P, Markiel A, Ozier O, Baliga NS, Wang JT, Ramage D, Amin N, Schwikowski B, Ideker T (2003) Cytoscape: a software environment for integrated models of biomolecular interaction networks. Genome Res 13:2498-2504

Shimizu K (2004) Metabolic flux analysis based on 13C-labeling experiments and integration of the information with gene and protein expression patterns. Adv Biochem Eng Biotechnol 91:1-49

Siddiquee KA, Arauzo-Bravo MJ, Shimizu K (2004) Effect of a pyruvate kinase (pykF-gene) knockout mutation on the control of gene expression and metabolic fluxes in *Escherichia coli.* FEMS Microbiol Lett 235:25-33

Soga T, Baran R, Suematsu M, Ueno Y, Ikeda S, Sakurakawa T, Kakazu Y, Ishikawa T, Robert M, Nishioka T, Tomita M (2006) Differential metabolomics reveals ophthalmic acid as an oxidative stress biomarker indicating hepatic glutathione consumption. J Biol Chem 281:16768-16776

Soga T, Heiger DN (2000) Amino acid analysis by capillary electrophoresis electrospray ionization mass spectrometry. Anal Chem 72:1236-1241

Soga T, Kakazu Y, Robert M, Tomita M, Nishioka T (2004) Qualitative and quantitative analysis of amino acids by capillary electrophoresis-electrospray ionization-tandem mass spectrometry. Electrophoresis 25:1964-1972

Soga T, Ohashi Y, Ueno Y, Naraoka H, Tomita M, Nishioka T (2003) Quantitative metabolome analysis using capillary electrophoresis mass spectrometry. J Proteome Res 2:488-494

Soga T, Ueno Y, Naraoka H, Matsuda K, Tomita M, Nishioka T (2002a) Pressure-assisted capillary electrophoresis electrospray ionization mass spectrometry for analysis of multivalent anions. Anal Chem 74:6224-6229

Soga T, Ueno Y, Naraoka H, Ohashi Y, Tomita M, Nishioka T (2002b) Simultaneous determination of anionic intermediates for Bacillus subtilis metabolic pathways by capillary electrophoresis electrospray ionization mass spectrometry. Anal Chem 74:2233-2239

Spasic I, Dunn WB, Velarde G, Tseng A, Jenkins H, Hardy N, Oliver SG, Kell DB (2006) MeMo: a hybrid SQL/XML approach to metabolomic data management for functional genomics. BMC Bioinformatics 7:281

Stephanopoulos G, Alper H, Moxley J (2004) Exploiting biological complexity for strain improvement through systems biology. Nat Biotechnol 22:1261-1267

Strelkov S, von Elstermann M, Schomburg D (2004) Comprehensive analysis of metabolites in *Corynebacterium glutamicum* by gas chromatography/mass spectrometry. Biol Chem 385:853-861

Sugimoto M, Kikuchi S, Arita M, Soga T, Nishioka T, Tomita M (2005) Large-scale prediction of cationic metabolite identity and migration time in capillary electrophoresis mass spectrometry using artificial neural networks. Anal Chem 77:78-84

Sundararaj S, Guo A, Habibi-Nazhad B, Rouani M, Stothard P, Ellison M, Wishart DS (2004) The CyberCell Database (CCDB): a comprehensive, self-updating, relational database to coordinate and facilitate *in silico* modeling of *Escherichia coli*. Nucleic Acids Res 32:D293-295

Szyperski T (1998) 13C-NMR, MS and metabolic flux balancing in biotechnology research. Q Rev Biophys 31:41-106

ter Kuile BH, Westerhoff HV (2001) Transcriptome meets metabolome: hierarchical and metabolic regulation of the glycolytic pathway. FEBS Lett 500:169-171

Thimm O, Blasing O, Gibon Y, Nagel A, Meyer S, Kruger P, Selbig J, Muller LA, Rhee SY, Stitt M (2004) MAPMAN: a user-driven tool to display genomics data sets onto diagrams of metabolic pathways and other biological processes. Plant J 37:914-939

Tomita M, Nishioka T (2005) Metabolomics: The frontier of systems biology. Spinger-Verlag, Tokyo

Tweeddale H, Notley-McRobb L, Ferenci T (1998) Effect of slow growth on metabolism of *Escherichia coli*, as revealed by global metabolite pool ("metabolome") analysis. J Bacteriol 180:5109-5116

Tweeddale H, Notley-McRobb L, Ferenci T (1999) Assessing the effect of reactive oxygen species on *Escherichia coli* using a metabolome approach. Redox Rep 4:237-241

Usadel B, Nagel A, Thimm O, Redestig H, Blaesing OE, Palacios-Rojas N, Selbig J, Hannemann J, Piques MC, Steinhauser D, Scheible WR, Gibon Y, Morcuende R, Weicht D, Meyer S, Stitt M (2005) Extension of the visualization tool MapMan to allow statistical analysis of arrays, display of corresponding genes, and comparison with known responses. Plant Physiol 138:1195-1204

Vaidyanathan S (2005) Profiling microbial metabolomes: what do we stand to gain? Metabolomics 1:17-28

Valet G (2005) Cytomics, the human cytome project and systems biology: top-down resolution of the molecular biocomplexity of organisms by single cell analysis. Cell Prolif 38:171-174

van der Werf MJ, Jellema RH, Hankemeier T (2005) Microbial metabolomics: replacing trial-and-error by the unbiased selection and ranking of targets. J Ind Microbiol Biotechnol 32:234-252

Villas-Boas SG, Hojer-Pedersen J, Akesson M, Smedsgaard J, Nielsen J (2005a) Global metabolite analysis of yeast: evaluation of sample preparation methods. Yeast 22:1155-1169

Villas-Boas SG, Mas S, Akesson M, Smedsgaard J, Nielsen J (2005b) Mass spectrometry in metabolome analysis. Mass Spectrom Rev 24:613-646

Wagner C, Sefkow M, Kopka J (2003) Construction and application of a mass spectral and retention time index database generated from plant GC/EI-TOF-MS metabolite profiles. Phytochemistry 62:887-900

Wang QZ, Wu CY, Chen T, Chen X, Zhao XM (2006) Integrating metabolomics into a systems biology framework to exploit metabolic complexity: strategies and applications in microorganisms. Appl Microbiol Biotechnol 70:151-161

Watts KT, Lee PC, Schmidt-Dannert C (2006) Biosynthesis of plant-specific stilbene polyketides in metabolically engineered *Escherichia coli*. BMC Biotechnol 6:22

Wheeler DL, Barrett T, Benson DA, Bryant SH, Canese K, Chetvernin V, Church DM, DiCuccio M, Edgar R, Federhen S, Geer LY, Helmberg W, Kapustin Y, Kenton DL, Khovayko O, Lipman DJ, Madden TL, Maglott DR, Ostell J, Pruitt KD, Schuler GD, Schriml LM, Sequeira E, Sherry ST, Sirotkin K, Souvorov A, Starchenko G, Suzek TO, Tatusov R, Tatusova TA, Wagner L, Yaschenko E (2006) Database resources of the National Center for Biotechnology Information. Nucleic Acids Res 34:D173-180

Wiback SJ, Mahadevan R, Palsson BO (2004) Using metabolic flux data to further constrain the metabolic solution space and predict internal flux patterns: the *Escherichia coli* spectrum. Biotechnol Bioeng 86:317-331

Wittmann C (2002) Metabolic flux analysis using mass spectrometry. Adv Biochem Eng Biotechnol 74:39-64

Wittmann C, Kromer JO, Kiefer P, Binz T, Heinzle E (2004) Impact of the cold shock phenomenon on quantification of intracellular metabolites in bacteria. Anal Biochem 327:135-139

Wu L, Mashego MR, van Dam JC, Proell AM, Vinke JL, Ras C, van Winden WA, van Gulik WM, Heijnen JJ (2005) Quantitative analysis of the microbial metabolome by isotope dilution mass spectrometry using uniformly 13C-labeled cell extracts as internal standards. Anal Biochem 336:164-171

Yu Y, Ko KS, Zea CJ, Pohl NL (2004) Discovery of the chemical function of glycosidases: design, synthesis, and evaluation of mass-differentiated carbohydrate libraries. Org Lett 6:2031-2033

Zamboni N, Sauer U (2004) Model-independent fluxome profiling from 2H and 13C experiments for metabolic variant discrimination. Genome Biol 5:R99

Zea CJ, Pohl NL (2004) Kinetic and substrate binding analysis of phosphorylase b via electrospray ionization mass spectrometry: a model for chemical proteomics of sugar phosphorylases. Anal Biochem 327:107-113

Zhou L, Lei XH, Bochner BR, Wanner BL (2003a) Phenotype microarray analysis of *Escherichia coli* K-12 mutants with deletions of all two-component systems. J Bacteriol 185:4956-4972

Zhou S, Shanmugam KT, Ingram LO (2003b) Functional replacement of the *Escherichia coli* D-(-)-lactate dehydrogenase gene (ldhA) with the L-(+)-lactate dehydrogenase gene (ldhL) from *Pediococcus acidilactici*. Appl Environ Microbiol 69:2237-2244

Zhu J, Shimizu K (2004) The effect of pfl gene knockout on the metabolism for optically pure D-lactate production by *Escherichia coli*. Appl Microbiol Biotechnol 64:367-375

Robert, Martin

Institute for Advanced Biosciences, Keio University, 403-1 Daihoji, Tsuruoka, Yamagata, 997-0017 Japan
mrobert@ttck.keio.ac.jp

Soga, Tomoyoshi

Institute for Advanced Biosciences, Keio University, 403-1 Daihoji, Tsuruoka, Yamagata, 997-0017 Japan

Tomita, Masaru
Institute for Advanced Biosciences, Keio University, 403-1 Daihoji, Tsuruoka, Yamagata, 997-0017 Japan

Abbreviations

E. coli: *Escherichia coli*
MS: mass spectrometry
CE-MS: capillary electrophoresis mass spectrometry
CE-TOF-MS: capillary electrophoresis time-of-flight mass spectrometry
GC-MS: gas chromatography mass spectrometry
LC-MS liquid chromatography mass spectrometry
IP-LC-ESI-MS: ion-pair liquid chromatography coupled to electrospray ionization mass spectrometry
DESI: desorption electrospray ionization mass spectrometry
DART: direct analysis in real time
ASKA: A complete set of *E. coli* K-12 ORF archive
NMR: nuclear magnetic resonance
TLC: thin-layer chromatography
UV: ultra-violet
TCA: tricarboxylic acid
NADPH: reduced nicotinamide adenine dinucleotide phosphate
PEP: phosphoenol pyruvate
cAMP: cyclic adenosine monophosphate
ATP: adenosine triphosphate
IMP: inosine monophosphate
KEGG: Kyoto Encyclopedia of Genes and Genomes
GEM: Genome-based modeling
MW: molecular weight
FANCY: functional analysis by co-responses in yeast
ORF: open reading frame
MIAMET: Minimum Information About a METabolomics experiment

The exo-metabolome in filamentous fungi

Ulf Thrane, Birgitte Andersen, Jens C. Frisvad, Jørn Smedsgaard

Abstract

Filamentous fungi are a diverse group of eukaryotic microorganisms that have a significant impact on human life as spoilers of food and feed by degradation and toxin production. They are also most useful as a source of bulk and fine chemicals and pharmaceuticals. This chapter focuses on the exo-metabolome in filamentous fungi, which comprises more than 30,000 known secondary metabolites. Profiles of this diverse range of secondary metabolites have, for more than 25 years, been central in development of fungal systematics, taxonomy, and ecology, today integrated in a multidisciplinary and polyphasic approach to applied mycology. Lead discovery is an example of the successful integration of metabolite profiling and natural product chemistry in mycology.

1 Introduction

Fungi are eukaryotic organisms belonging to a kingdom of their own, *Fungi*, which is estimated to contain 1.5 million species (Hawksworth 1991) divided into four phyla (divisions): *Ascomycota*, *Zygomycota*, *Basidiomycota*, and *Chytridiomycota*. The filamentous fungi are mainly found in *Ascomycota* and *Zygomycota*. Whereas most mushrooms belong to *Basidiomycota*, some well-known and highly sought-after mushrooms, such as morels and truffles, are ascomycetes. Many fungi have a significant impact on human life. As spoilers, they degrade food and feed stuff and also produce toxins. They can also be most useful for producing bulk and fine chemicals and pharmaceuticals (Adrio and Demain 2003). From a biotechnological perspective, microorganisms from the kingdom of *Fungi* are by far the most important production organisms. As an example, the single-celled fungi, baker's yeast *Saccharomyces cerevisiae* is used in ethanol production etc. Amongst the filamentous fungi, genera such as *Aspergillus*, *Penicillium, Trichoderma*, and *Fusarium,* are used in the production of enzymes, antibiotics, and other pharmaceuticals, and they all belong to the *Ascomycetes.* Within the *Zygomycota* species of *Rhizopus*, *Mucor*, and *Blakeslea* are other examples of important filamentous fungi in the bio-industry producing enzymes and colorants for use in the food industry (Archer 2000; Dufosse et al. 2005; Mapari et al. 2005).

From the beginning, metabolomics and metabolome analysis has aimed at linking the various omics's (genomics – proteomics – metabolomics) to describe the dynamics of the cell or organism (Kell 2004; Chapter 1). In this context, much fo-

Topics in Current Genetics, Vol. 18
J. Nielsen, M. C. Jewett (Eds.): Metabolomics
DOI 10.1007/4735_2007_0230 / Published online: 13 June 2007

cus has been given to central metabolism (earlier known as the primary metabolism), carbon flux, and energy turnover. From a holistic point of view, metabolomics is much more, and could be extended to cover all the metabolites originating from the entire metabolic machinery in an organism. A special segment of the metabolic activity of a fungus is called the exo-metabolome (earlier known as the secondary metabolism), which consists of all the metabolites produced by the organism intended for interaction with the environment. These metabolites function as chemical signals during interaction between individual organisms, such as fungus-fungus attractants, insect repellents, biological active metabolites directed against bacteria (antibiotics), against plants (phytoalexins) or against vertebrates (mycotoxins) (Larsen et al. 2005). The compounds responsible for these interactions are often called secondary metabolites to distinguish them from the central or primary metabolites. Fungi have most of their central metabolism in common with other eukaryotes and the yeast, *S. cerevisiae*, has been used as a model system (as discussed in previous chapters), and metabolic models have been published (Hohmann 2005).

Whereas central metabolism is common to all filamentous fungi, the production of secondary metabolites, or exo-metabolites, is more or less genus or species specific (Frisvad et al. 1998). Secondary metabolites reflect a slow adaptation to the environment and are often a result of co-evolution between fungi and other organisms. In contrast, central metabolism reflects a much more rapid adaptation to the current situation. The production of secondary metabolites requires a significant amount of carbon (sometimes also nitrogen) and consumes a lot of energy. More than 12,000-15,000 genes (ORFs) in filamentous fungi give these organisms the capability to produce a vast diversity of secondary metabolites (Keller et al. 2005; Yu and Keller 2005). Furthermore, the complexity of many of these metabolites is encoded in several genes often assembled in large gene clusters. Today more than 30,000 secondary metabolites with a molecular weight below ca. 2000 Da are known and they can be classified by their biosynthetic origin as polyketides and terpenes, derived from amino acids or the tricarboxylic acid cycle, as well as mixed biosynthetic routes (Frisvad et al. 2004). As mentioned, the general feature of a secondary metabolite is that it is limited in its distribution throughout the fungal kingdom and cannot be found in every species within a given fungal family. Therefore, secondary metabolites are very powerful descriptors to us for making systems and keys to identify and describe the numerous genera and species in the fungal kingdom.

2 Exo-metabolome and taxonomy

For centuries secondary metabolites have been used - indirectly - to differentiate between fungal species. Characteristics, such as pigments or odours that were specific to one fungal species, have been used in the formal species descriptions and in the identification keys. These characteristics were seen as part of the morphological features, and not as chemical compounds or metabolites in their own right.

In many descriptions and keys, fungal species are differentiated by different shades of same colour, i.e., bluish green, dark green, greyish green etc. Such terms are highly subjective and very difficult to communicate to users. To compensate for this, references have attempted to use standardized colour schemes, such as the popular "Methuen Handbook of Colour" (Kornerup and Wanscher 1978). Despite this, colour is still a perceptual phenomenon and does convey many practical problems for the users. Subjectivity of colour, and the sensitivity of pigment formation to growth conditions, has often been used as arguments against the use of any secondary metabolites in fungal systematics. A decade ago, as the use of molecular sequence data in fungal systematics increased, the criticism of the use of chemical characters increased as well. Although they are still being used as the fundamental characteristics in fungal descriptions, all phenotypic (descriptive) characteristics, including micro-morphology, were not objective enough compared to a DNA sequence (Hibbett and Donoghue 1998; Prillinger et al. 2002; Taylor et al. 2000). However, no fungal species has yet been described based on sequence data alone or strict phylogenetic based fungal taxonomies. Species discovered by sequence analyses have always been linked up with distinctive phenotypic features in order to be formally described (Aoki et al. 2005; Nirenberg and O'Donnell 1998). Taking an ecological view-point, fungal taxonomy and systematics should be based on traits of importance in an ecological system – functional characteristics – such as micro-morphology, which covers the physical and mechanical properties and also the exo-metabolome, which covers chemical interactions (Frisvad and Samson 2004). Having said this, it is important to stress that the key features in an ecological context is not the individual chemical compound, but the profile of metabolites as synergism among metabolites is commonly observed. Hence, the exo-metabolome, the full spectrum of secondary metabolites, is the key feature for a living fungal culture, the most important functional character of a fungal species and thereby also the important feature in fungal systematics (Frisvad et al. 1998).

In a wide ecological context the exo-metabolome may also cover extra-cellular enzymes used for degradation of complex polymers into digestible smaller units such as sugars, amino acids, as well as necessary metal ions and other trace compounds. However, this chapter will only focus on determination of secondary metabolites profiles in important filamentous fungi. The development of extraction methods, separation, and detection techniques during the last three decades of mycological and chemotaxonomical research will be highlighted.

3 Exo-metabolome and fungal growth

As with all phenotypic characteristics, the production of secondary metabolites is susceptible to changes in environmental conditions. From an ecological point of view, all fungi are able to respond with different secondary metabolites as part of their survival and adaptation strategy. The challenge is to trigger this production under laboratory conditions as a given fungal species or a fungal isolate may need specific stimuli to produce and accumulate secondary metabolites (Dombrink-

Kurtzman and Blackburn 2005; Filtenborg et al. 1990). There are no set rules when it comes to growth conditions to ensure a maximum production of all metabolites that a fungal genome encodes for. In general most filamentous fungi express most of their secondary metabolites in detectable amounts when growing on solid surface substrates like an agar medium (Hölker et al. 2004). In some cases shake cultures or still liquid cultures may be beneficial for production of one or few specific metabolites, but there will be significant differences in overall metabolite production between different fungal species. The fundamental growth condition can be varied in multiple ways and should be taken into consideration when determining the exo-metabolic potential of a fungus. For example, mineral-clay pellets (LECA nuts) coated with a semisolid agar substrate constitutes a larger surface area for fungal growth than a simple agar surface in a Petri dish (Nielsen et al. 2004a). The result in yield in terms of chemical diversity is much higher on these pellets than in both shake flasks and classical agar substrates in Petri dishes.

An important part of the growth conditions or cultivation of fungi is the choice of substrate ingredients. Much information can be found in the literature dealing with the mycotoxin potential of various fungal species, because toxicity of fungi towards humans and livestock has been evaluated in a high number of studies. On the other hand, many reports are quite conservative in their choice of growth substrate, as often they use in-house substrates based on natural products, like maize, rice, and wheat grains. These ingredients may be used directly as the substrate or they may become an integrated part of the agar substrates. When it comes to mycotoxin production, a substrate mimicking the real raw material is relevant as a model for laboratory experiments; however, if the task is to illustrate the entire exo-metabolome of a fungus, a broader view is needed. In general, a substrate containing many simple nutrients that can be taken up readily usual gives the most diverse metabolite production. Good results have been obtained by using semi-synthetic substrates where a crude yeast extract serves as the primary nitrogen source, such as Yeast Extract Sucrose agar (YES) and Czapek-Dox Yeast Autolysate agar (CYA), though the latter does also contain nitrate. Additional media, semi-natural substrates like Potato Dextrose Agar (PDA), or Oat meal Agar (OA) may be used and combined with YES and CYA depending on the fungal genus under examination (Nielsen et al. 2004b). Many substrates are available as pre-mixed powders, which can ensure a good reproducibility from batch to batch of substrate; however, it should be noted that different brands of the same substrate may be quite different. To compensate for this variation, it is recommended always to add magnesium sulphate (in case of YES) and trace metals to all substrates (Filtenborg et al. 1990). In any case, it is advantageous to apply a broad view and ensure to use a palette of different substrates for exploration of the exo-metabolome of fungi.

4 Visualisation of the exo-metabolome

Thin layer chromatography (TLC) is a well-known, classic method for separation of compounds in a mixture. In addition, it is by far the cheapest when compared to other chromatographic methods. The value of TLC patterns of fungal metabolites in mycological systematics was introduced through a study of common food-borne *Penicillium* species and their production of mycotoxins (Filtenborg and Frisvad 1980). This method does not depend on an extraction step, as the agar plugs are directly applied to the TLC plate. Plugs (6 mm in diameter) are taken from seven day old fungal cultures on the Petri dish and placed, agar-side down, on the application line of the TLC plate for a couple of seconds. This will allow the exo-metabolites in the agar to be adsorbed by the silica gel. Exo-metabolites that are bound to the mycelium can be released by an *in situ* extraction, adding a drop of extraction solvent on the mycelium side of the plug (Filtenborg et al. 1983). After a few seconds the plug is turned up-side-down (mycelium side down) and briefly allowed to touch the TLC plate, either in the same lane/track as the agar plug or in a new separate track. When the TLC plate is dry, regular TLC procedures are followed, such as different elution systems to promote different migration patterns for the exo-metabolome and various chemical spray reagents to enhance colour development of individual metabolites (Frisvad and Thrane 2000).

To perform chemotaxonomic studies, it is important to be able to compare a high number of TLC tracks. This may involve a high number of TLC plates, which might be analysed on different days and in different batches of eluents. Since TLC is sensitive to the quantitative composition of the eluent resulting in variations in migration distance of a compound between different TLC plates, calibration standards on each TLC plate have to be used. Griseofulvin, which is commercially available, cheap and easy to recognize, migrates approximately 65% compared to the liquid front and is therefore suitable as an external TLC standard. The migration of all metabolites in the exo-metabolome is then calculated relative to the griseofulvin standard, indicated as the R_{fg} value of each metabolite (Thrane 1986). By this agar-plug TLC technique it is very easy to analyse a high number of experiments, such as comparison of different growth conditions or different fungal isolates, in a short time, because there are no time-consuming extraction and purification steps. Identification of individual metabolites in the exo-metabolome requires standards and preferably specific detection methods, such as development of coloured spots after spraying and heating the TLC plate (Andersen 1991; Andersen et al. 1995, 2004; Frisvad and Thrane 2000). For many years TLC patterns have been documented by photographic slides, which were useful in those days, but somewhat outdated now as most images are digitalised and much easier to access and compare. In addition to the recording of coloured spots on the TLC plates it was possible to evaluate a developed TLC plate by a UV-VIS scanner that were able to record a reflectance spectrum in the UV-VIS spectrum. The idea of qualifying the individual spots by a unique reflectance spectrum has been used in publications on metabolites from *Penicillium brevicompactum* (Andersen 1991) and *Stemphylium* species (Andersen et al. 1995).

The agar-plug TLC technique has been the cornerstone in the mycological research area at Technical University of Denmark (DTU). This approach has been used in many research and student projects in the 1980's and 1990's. These projects covered many aspects of chemosystematics within *Penicillium* and other common filamentous fungi, such as *Aspergillus*, *Fusarium*, and *Alternaria*. The agar-plug technique has revolutionised mycological research by introducing chemical profiling in systematics and in identification procedures. To the best of our knowledge, the method has only been applied at CABI BioScience (Paterson and Bridge 1994) and in Norway (Stenwig and Liven 1988) for profiling of fungi. Research projects at DTU have successfully used an integrated approach originating from TLC patterns to clarify taxonomic problems within the genera *Penicillium* (Frisvad and Filtenborg 1983), *Talaromyces* (Frisvad et al. 1990), *Stemphylium* (Andersen et al. 1995), and *Fusarium* (Thrane and Hansen 1995). In addition, TLC patterns have been used for identification purposes and as of today are available in a widely distributed manual for food and airborne fungi (Samson et al. 2004). Current research projects at DTU still use the TLC methods for screening and classification purposes as support to other chromatographic and chemical methods, exemplified by a revision of *Penicillium* species associated to flower bulbs and onions (Overy and Frisvad 2003).

Figure 1 shows a TLC plate (agar plug method) of the exo-metabolome of *Penicillium verrucosum* (tracks 1-3), *P. persicinum* (4), *Fusarium culmorum* (5-6), *F. venenatum* (7-8), *Cladosporium cladosporioides* (9), *Aspergillus flavus* (10), *Asp. oryzae* (11), external standard (12), *Asp. niger* (13-14), *Asp. japonicus* (15), *Asp. lanosus* (16-17), *Alternaria tenuissima* (18), *Alt. infectoria* (19) and *Alt. arborescens* (20). The TLC plate shows that different species in some genera have many metabolites in common (e.g. *Penicillium* and *Fusarium*), while different species in other genera have very few metabolites in common (e.g. *Aspergillus* and *Alternaria*). Research has shown that the production of metabolites in a fungus' exo-metabolome is consistent, as can be seen for *Asp. niger* and *Asp. lanosus,* and independent of origin (Andersen 1991, 2004; Thrane et al. 2004). The TLC method is not recommended when the exo-metabolome contains a large number of metabolites or when comparisons are made between closely related species (e.g. *F. culmorum* and *F. venenatum*). In such cases HPLC-UV or HPLC-MS should be used, due to HPLC's superior resolution and sensitivity compared to that of TLC.

5 Extraction of the exo-metabolome

To achieve a more precise identification of individual metabolites in the exo-metabolome and to be able to compare different exo-metabolomes from different species, it is necessary to extract the metabolites from the fungal cultures. The challenge of extracting the complete exo-metabolome equals the challenge encountered in making the fungus produce it in the first place. Much information on extraction procedures can again be found in the mycotoxin literature; however, in most cases the recommended procedures are focussed on the optimal extraction of

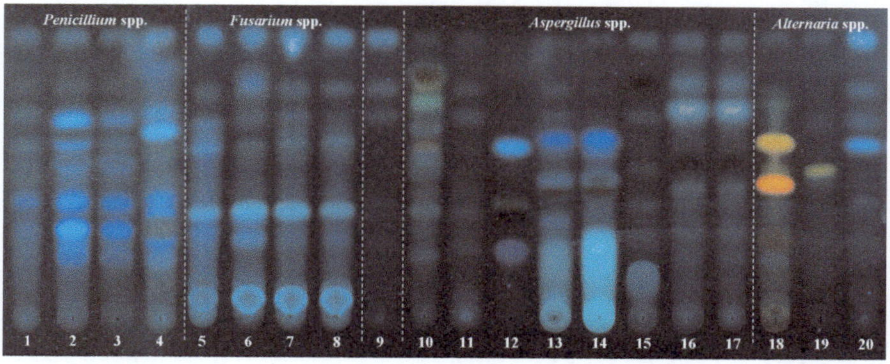

Fig. 1. Thin layer chromatography (TLC) of 19 fungal cultures. Silica gel 60 TLC plate in UV light (366 nm) after spraying with H_2SO_4. The identity of the cultures is as follows: *Penicillium verrucosum* (tracks 1-3), *P. persicinum* (4), *Fusarium culmorum* (5-6), *F. venenatum* (7-8), *Cladosporium cladosporioides* (9), *Aspergillus flavus* (10), *Asp. oryzae* (11), external standard (12), *Asp. niger* (13-14), *Asp. japonicus* (15), *Asp. lanosus* (16-17), *Alternaria tenuissima* (18), *Alt. infectoria* (19), and *Alt. arborescens* (20).

one or a few metabolites, often of chemically related structure. For profiling purposes, all metabolites need be extracted, and as with growth conditions no universal procedure exists, that extracts all the chemically diverse compounds of filamentous fungi. Using a single extraction solvent will in most cases favour some metabolites, whereas others may not be extracted at all, or with very low efficiency. As an alternative, a mixture of solvents may be used in a one-step extraction and this has been done successfully. A further improvement on metabolite diversity of the extracts is to use sequential extractions of the same fungal culture. Two or three different extraction solvents with different chemical affinities are used and the resulting extracts are subsequently pooled into one extract for final analysis. Alternatively, the different extracts may be kept and analysed separately to avoid unwanted chemical reactions, such as precipitation, complex formation, degradation etc. upon mixing the individual fractions.

In the 1980's, fungal profiling by high performance liquid chromatography (HPLC) in the authors' laboratory was a laborious task. The entire content of eight Petri dishes, approximately 150 gram of biomass and agar, was extracted twice by a total of 300 ml organic solvents in a homogeniser. The organic phase was evaporated to dryness, dissolved, transferred, de-fatted, transferred again, and filtered before HPLC analysis (Frisvad and Thrane 1987). Bearing in mind that the aim for exo-metabolome profiling and chemotaxonomic studies was to analyse a high number of cultures from each fungal species, this procedure was expensive and extremely time consuming yielding no more than eight extracts per day. As 21st century analytical equipment has become much more sensitive, it is now only necessary to extract 0.5 grams of biomass for qualitative profiling of a fungal culture. In line with the availability of better equipment, the extraction procedure has been modified and scaled-down (Smedsgaard 1997). Instead of extracting the entire culture, this new procedure extracts 3-10 agar plugs (0.5-2.5 cm^2 area of culture) in a 2-ml vial using no more than 0.5-1 ml of solvent. The extraction itself takes place in an ultrasonic bath. Depending on the analytical equipment and methods, the solvent may be analysed directly after filtration or after drying and dissolving steps. This procedure allows for up to 100 extracts to be prepared during one working day in a labour saving way and at significantly reduced cost. Another advantage is that from a single fungal culture, several samples can be collected, for example, to use different extraction procedures, or as multiple samples of same origin.

6 Analysis of the exo-metabolome by high performance liquid chromatography

HPLC is one of the most common separation techniques used for analysis of secondary metabolites and mycotoxins. During the last 25 years, there have been a number of significant technical developments in HPLC technology. Today, HPLC equipment is easy to operate and maintain, analyses can be automated and costs for purchase and operation have dropped. In addition, column material has also been through a continuous development so a variety of columns with different properties are available. For fungal secondary metabolites, it is very common to use reversed phase system where the column material typically is 3 µm beads coated with a C_8 or C_{18} phase and a polar liquid phase, such as water-acetonitrile or water-methanol with a gradient elution system. In most cases, an HPLC system is an integrated unit consisting of a separation system and a detection system. Modern columns are very robust and very similar from batch to batch but still the separation efficiency of technique is sensitive to variation in chromatographic parameters, especially the retention time may differ between batches of column or solvents. To minimise this variation the use of external standards or a retention time index has been introduced (Frisvad and Thrane 1987). During sequential analysis of fungal extracts a sample of seven alkylphenones is analysed and the retention times for these compounds can be used a set of seven fixed points within this series of analyses allowing the retention times for all other chromatographic peaks to be calculated relative to these fix points. These index values (RI) are relatively constant among different analytical runs as long as the type of column, solvent composition and the solvent gradient are kept constant. For profiling the RI values are of high importance for facilitated comparison of data from different analyses as both known and unknown compounds can be assigned by their RI value under the given chromatographic conditions.

For analysis and profiling of fungal metabolites, it has been successful to use HPLC with gradient elution on reversed phase columns followed by detection of a UV detector. In the early days of HPLC profiling, simple UV detectors with one fixed wavelength were used; however, despite the simplicity species specific peak patterns were recorded by inspection of the printouts from the detector. An informative qualification of the separated compounds was obtained with the use of photodiode array detectors that allow a UV-VIS spectrum to be recorded on-line,

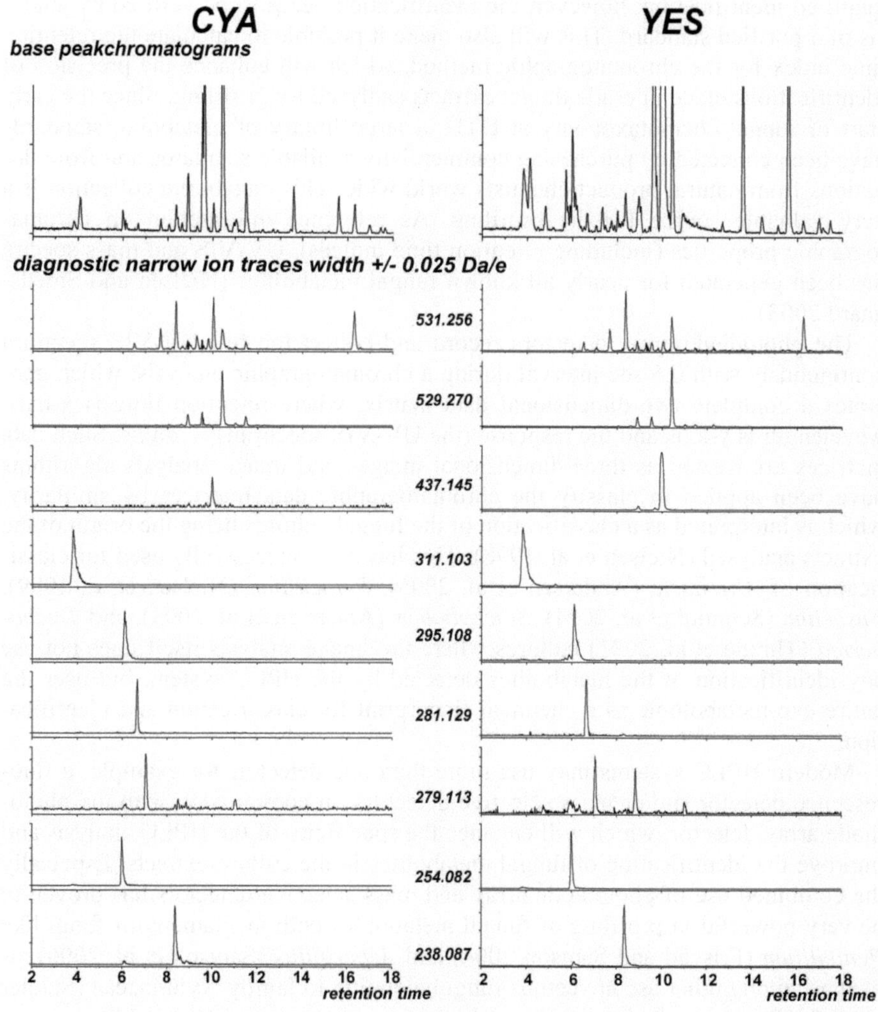

Fig. 2. Narrow ion traces from LC-MS analysis confirm the presence of all the metabolites listed in Table 1 in the cultural extracts of *Penicillium discolor*. Note that the metabolite daldinin D with the protonated mass 437.1369 Da/e is primarily seen on Yeast Extract Sucrose agar and only in very small amounts on Czapek-Dox Yeast Autolysate agar. Several known isomers of the chaetoglobosins are seen as multiple chromatographic peaks in the diagnostic traces from these metabolites.

which was a major step forward for chemical profiling of fungi. By this technique it is possible to identify or partial characterise the fungal metabolites by their UV spectrum linked with their retention time index. Many fungal metabolites have a characteristic UV spectrum within the wavelength range 200-600 nm that allow a qualified identification; however, the identification needs to be verified by analysis of a purified standard. This will also make it possible to calculate the retention time index for the chromatographic method, which will enhance the precision of identifications made in crude fungal extracts analysed for profiling. Since the early start of fungal chemotaxonomy at DTU, a large library of metabolite standards have been collected by purchasing commercially available standards and from donations from natural product chemists world-wide. This metabolite collection is a very valuable source for the profiling. As reference information on chromatographic properties (including retention time indices), UV-VIS and mass spectra has been generated for nearly all known fungal metabolites (Nielsen and Smedsgaard 2003).

The photodiode array detectors record and collect the full UV-VIS spectrum continuously with 0.5 sec interval during a chromatographic analysis, which generates a complete two-dimensional data matrix, where retention time is x-axis, wavelength is y-axis and the response (the UV-VIS spectrum) is z-axis. Such data matrices are treated as three-dimensional images and image analysis algorithms have been applied to classify the chromatographic data matrices by similarity, which is interpreted as a classification of the fungal cultures being the origin of the extracts analysed (Nielsen et al. 1998). This has been successfully used for classification of *Alternaria* (Andersen et al. 2005), *Penicillium* (Nielsen et al. 1999), *Fusarium* (Schmidt et al. 2004), *Stachybotrys* (Andersen et al. 2003), and *Trichoderma* (Thrane et al. 2001) cultures where the image analysis itself does not use any identification of the metabolites detected by the HPLC system, but uses the entire exo-metabolome as a chemical fingerprint for classification and identification.

Modern HPLC systems may use more than one detector, for example, a fluorescence detector and/or mass selective detectors, in combination with the photodiode array detector, which will enhance the specificity of the HPLC analysis and improve the identification of fungal metabolites in the culture extracts. Especially the combined use of photodiode array and mass selective detectors has proven to be very powerful in profiling of fungal metabolites both in filamentous fungi like *Penicillium* (Frisvad and Samson 2004) and *Aspergillus* (Samson et al. 2006), as well as among other ascomycetous fungi belonging to family Xylariaceae (Stadler et al. 2007). Figure 2 shows two total ion chromatograms of HPLC-MS analyses of fungal extracts of *Penicillium discolor* grown on two agar substrates, CYA and YES, and it is obvious that the two different substrates yield two different metabolite profiles as the overall peak patterns are different. However, by extraction of narrow ion traces, here +/- 0.025 Da around the protonated mass of known metabolites, it is possible to confirm the presence of metabolites known from this species (Table 1, Fig. 3). It should be noted that the metabolite daldinin D with the protonated mass 437.1369 Da/e is primarily on YES, and that several isomers of the chaetoglobosins can be seen as multiple chromatographic peaks in the diagnostic traces from these metabolites. Additional extraction of all possible ion traces from the total ion chromatogram gives an exhaustive picture of the chemistry, demonstrating more than 150 metabolites in these samples. Exploitation of the entire information of the data may be laborious and does require advanced skills in

Table 1. Metabolites known to be produced by *Penicillium discolor* with the mass of their protonated and sodiated pseudo-molecular ion as normally seen in positive ESI-MS. The actual mass values found in Figure 4 are listed to illustrate the accuracy (mostly < 6 ppm) except those marked with an asterisk (mass error up to 110 ppm). The ion corresponding to cyclopenol at 311.1032 Da/e was used as internal mass reference. For the structures see Figure 3.

	no		$M+H^+$		$M+Na^+$	
		formula	calculated	found	calculated	found
Viridicatin	1	$C_{16}H_{13}O_2N$	238.0868	238.0865	260.0687	260.0686
Viridicatol	2	$C_{15}H_{11}O_3N$	254.0817	254.0809	276.0637	276.0540
Dehydro-viridicatol	3	$C_{17}H_{14}O_2N_2$	279.1133	279.1164	302.0953	302.1201*
Cyclopeptin	4	$C_{17}H_{16}O_2N_2$	281.1290	281.1305	303.1109	Not detected
Cyclopenin	5	$C_{17}H_{14}O_3N_2$	295.1082	295.1087	317.0902	317.0998*
Cyclopenol	6	$C_{17}H_{14}O_4N_2$	311.1032	Reference	333.0851	333.0857
Daldinin D	7	$C_{21}H_{24}O_{10}$	437.1369	Not detected	459.1267	Not detected
Chaetoglobosin A-D	8	$C_{32}H_{36}O_5N_2$	529.2702	529.2653	551.2522	551.2504
Chaetoglobosin E-G	9	$C_{32}H_{38}O_5N_2$	531.2781	531.2838	553.2678	553.2601
Chaetoglobosin A-D after loss of water		$C_{32}H_{34}O_4N_2$	511.2597	511.2595	533.2416	533.3008*
Chaetoglobosin E-G after loss of water		$C_{32}H_{36}O_4N_2$	513.2753	513.2739	535.2573	Not detected

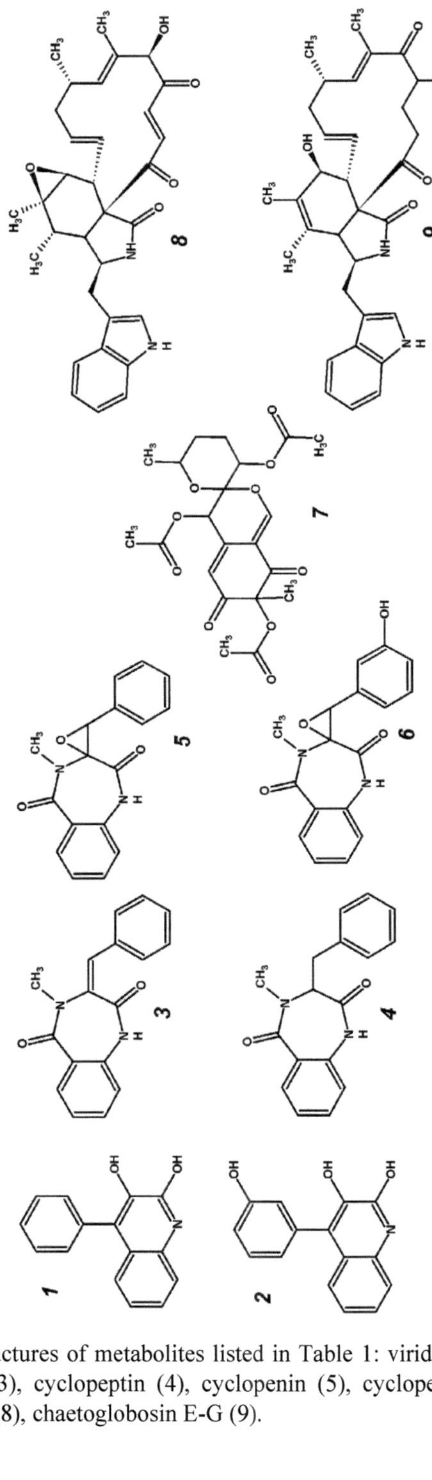

Fig. 3. Chemical structures of metabolites listed in Table 1: viridicatin (1), viridicatol (2), dehydro-viridicatol (3), cyclopeptin (4), cyclopenin (5), cyclopenol (6), daldinin D (7), chaetoglobosin A-D (8), chaetoglobosin E-G (9).

mass spectroscopy and analytical chemistry; however, HPLC-MS is very efficient for identification of fungal metabolites in crude extracts. By using powerful computer technology, it is possible to scan for a high number of known fungal metabolites (as long as a list of expected ions can be made).

7 Direct infusion electrospray mass spectrometry for profiling

The use of an electrospray interface (ESI) to the mass spectrometer makes it possible to limit the fragmentation of sample compounds by optimizing the analytical parameters. In addition, it is also possible to obtain only protonated molecules from each compound, which will yield a profile of masses upon injecting a sample directly into the mass spectrometer without any prior separation on an HPLC. Such analyses will only take a few minutes per sample and mass profiles can easily be stored in a database and used as a library for identification of mass profiles from unknown fungal cultures. This ultimate profiling tool was initially applied onto a group of cereal borne *Penicillium* species (Smedsgaard and Frisvad 1996) and subsequently expanded to the majority of the species within *Penicillium* subgenus *Penicillium* (Smedsgaard and Frisvad 1997). As an example, Figure 4 shows a mass profile from direct infusion nano-electrospray analysis of a crude extract of *Penicillium discolor* cultivated for seven days on YES medium. Ions corresponding to most metabolites listed in Table 1 can be found with accuracy better than 15 ppm (most < 6 ppm) except those marked with an asterisk in Table 1 (mass error up to 110 ppm). The studies on *Penicillium* at DTU have shown that the mass profiles are species specific but also depending on standardised culture conditions as not all metabolites may be produced in a single culture. An extensive study on *Penicillium* species using this technique showed both extracts from YES and CYA could be used, but with a higher chemical diversity from the CYA cultures (Smedsgaard et al. 2004). However, from analyses of independent extracts from independent cultures using different batches of substrates it was possible to identify an unknown extract of a fungal culture to species level including a suggestion of which substrate had been used. For other less explored fungal genera it might be other agar substrates that would be superior in terms of chemical diversity, thus a screening for optimal substrate and cultivation conditions is highly recommended before a more complete set of extracts are prepared and analysed.

An advantage of direct infusion electrospray mass spectrometry is that the metabolite profile can be determined within a few minutes even without any *a priori* knowledge of the metabolites in the extract, and this technique has been used in studies on bacteria (Vaidyanathan et al. 2002) and actinomycetes (Higgs et al. 2001) and should serve as an efficient base for an automated identification system.

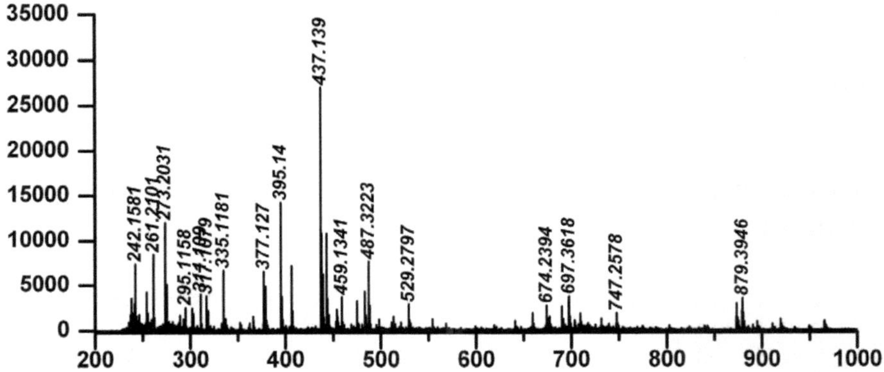

Fig. 4. Mass profile from direct infusion nano-electrospray analysis of a crude extract of *Penicillium discolor* cultivated on Yeast Extract Sucrose agar. Ions corresponding to most metabolites listed in Table 1 can be found with accuracy better than 15 ppm (most < 6 ppm).

8 Outlook – a polyphasic approach

Natural classifications in mycology are based on ecological functional traits that include all disciplines of importance for interactions with the environment, such as physiological, morphological, and chemical characteristics. For a deeper understanding of the actual interactions as well as regulation and expression of the genes coding for the metabolites involved, the recent developments in molecular biology, genetics, and molecular bioinformatics are of high importance and are integrated into true multidisciplinary systematics of fungi. In this polyphasic approach, the exo-metabolome are very important as they reflect the major signals of importance for interactions with the environment. Furthermore, the exo-metabolome is a powerful feature for fungal characterisation and can be used for identification of fungal cultures. The ongoing development in analytical chemistry and in mass spectrometry will make metabolite profiling even more powerful in the future. This is because the techniques are constantly being improved, which improves the quality of the generated data, and also because there is an improvement in data management, data handling, and data exploitation by advanced mathematical algorithms. Continued improvement in computer power is naturally an important factor as well. However, it is important to remember the importance of using a polyphasic approach by incorporation of information on the origin of the fungal cultures, their cultural and physiological characteristics as well as the genotypic information. This underscores the most important functional traits of fungi and is of major interest for the biotechnological industry in their search for new products or as an efficient way to avoid contamination of their production strains and products. All together, a polyphasic approach in fungal systematics with focus on the exo-metabolome gives a complete picture of the organisms with the very best opportunities to explore and exploit the fungi as safe cell factories of the future.

Acknowledgements

The authors would like to thank Centre for Advanced Food Studies (LMC) for long-term support to their mycological research.

References

Adrio JL, Demain AL (2003) Fungal biotechnology. Int Microbiol 6:191-199

Andersen B (1991) Consistent production of phenolic compounds by *Penicillium brevicompactum*. Ant van Leeuwenhoek 60:115-123

Andersen B, Hansen ME, Smedsgaard J (2005) Automated and unbiased image analyses as tools in phenotypic classification of small-spored *Alternaria* spp. Phytopathology 95:1021-1029

Andersen B, Nielsen KF, Thrane U, Szaro T, Taylor JW, Jarvis BB (2003) Molecular and phenotypic descriptions of *Stachybotrys chlorohalonata* sp. nov. and two chemotypes of *Stachybotrys chartarum* found in water-damaged buildings. Mycologia 95:1227-1238

Andersen B, Smedsgaard J, Frisvad JC (2004) *Penicillium expansum*: consistent production of patulin, chaetoglobosins and other secondary metabolites in culture and their natural occurrence in fruit products. J Agric Fd Chem 52:2421-2428

Andersen B, Solfrizzo M, Visconti A (1995) Metabolite profiles of common *Stemphylium* species. Mycol Res 99:672-676

Aoki T, O'Donnell K, Scandiani MM (2005) Sudden death syndrome of soybean in South America is caused by four species of *Fusarium*: *Fusarium brasiliense* sp. nov., *F. cuneirostrum* sp. nov., *F. tucumaniae*, and *F. virguliforme*. Mycoscience 46:162-183

Archer DB (2000) Filamentous fungi as microbial cell factories for food use. Curr Opin Biotechnol 11:478-483

Dombrink-Kurtzman MA, Blackburn JA (2005) Evaluation of several culture media for production of patulin by *Penicillium* species. Int J Food Microbiol 98:241-248

Dufosse L, Galaup P, Yaron A, Arad SM, Blanc P, Murthy KNC, Ravishankar GA (2005) Microorganisms and microalgae as sources of pigments for food use: a scientific oddity or an industrial reality? Trends Food Sci Technol 16:389-406

Filtenborg O, Frisvad JC (1980) A simple screening method for toxigenic fungi in pure cultures. Lebensm Wiss Technol 13:128-130

Filtenborg O, Frisvad JC, Svendsen JA (1983) Simple screening method for moulds producing intracellular mycotoxins in pure cultures. Appl Environ Microbiol 45:581-585

Filtenborg O, Frisvad JC, Thrane U (1990) The significance of yeast extract composition on metabolite production in *Penicillium*. In: Samson RA, Pitt JI (eds) Modern concepts in *Penicillium* and *Aspergillus* classification. New York: Plenum Press, pp 433-441

Frisvad JC, Filtenborg O (1983) Classification of terverticillate penicillia based on profiles of mycotoxins and other secondary metabolites. Appl Environ Microbiol 46:1301-1310

Frisvad JC, Filtenborg O, Samson RA, Stolk AC (1990) Chemotaxonomy of the genus *Talaromyces*. Ant van Leeuwenhoek 57:179-189

Frisvad JC, Samson RA (2004) Polyphasic taxonomy of *Penicillium* subgenus *Penicillium*. A guide to identification of food and air-borne terverticillate penicillia and their mycotoxins. Stud Mycol 49:1-173

Frisvad JC, Smedsgaard J, Larsen TO, Samson RA (2004) Mycotoxins, drugs and other extrolites produced by species in *Penicillium* subgenus *Penicillium*. Stud Mycol 49:201-242

Frisvad JC, Thrane U (1987) Standardized high-performance liquid chromatography of 182 mycotoxins and other fungal metabolites based on alkylphenone indices and UV-VIS spectra (diode-array detection). J Chromatogr 404:195-214

Frisvad JC, Thrane U (2000) Mycotoxin production by common filamentous fungi. In: Samson RA, Hoekstra ES, Frisvad JC, Filtenborg O (eds) Introduction to food- and airborne fungi. 6th edition. Utrecht: Centraalbureau voor Schimmelcultures, pp 321-331

Frisvad JC, Thrane U, Filtenborg O (1998) Role and use of secondary metabolites in fungal taxonomy. In: Frisvad JC, Bridge PD, Arora DK (eds) Chemical fungal taxonomy. New York: Marcel Dekker, pp 289-319

Hawksworth DL (1991) The fungal dimension of biodiversity: Magnitude, significance, and conservation. Mycol Res 95:641-655

Hibbett DS, Donoghue MJ (1998) Integrating phylogenetic analysis and classification in fungi. Mycologia 90:347-356

Higgs RE, Zahn JA, Gygu JD, Hilton MD (2001) Rapid method to estimate the presence of secondary metabolites in microbial extracts. Appl Environ Microbiol 67:371-376

Hohmann S (2005) The yeast systems biology network: mating communities. Curr Opin Biotechnol 16:356-360

Hölker U, Höfer M, Lenz J (2004) Biotechnological advantages of laboratory-scale solid-state fermentation with fungi. Appl Microbiol Biotechnol 64:175-186

Kell DB (2004) Metabolomics and systems biology: making sense of the soup. Curr Opin Microbiol 7:296-307

Keller NP, Turner G, Bennett JW (2005) Fungal secondary metabolism - from biochemistry to genomics. Nat Rev Microbiol 3:937-947

Kornerup A, Wanscher JH (1978) Methuen handbook of colour. London: Eyre Methuen

Larsen TO, Smedsgaard J, Nielsen KF, Hansen ME, Frisvad JC (2005) Phenotypic taxonomy and metabolite profiling in microbial drug discovery. Nat Prod Rep 22:672-695

Mapari SAS, Nielsen KF, Larsen TO, Frisvad JC, Meyer AS, Thrane U (2005) Exploring fungal biodiversity for water-soluble pigments as potential natural food colorants. Curr Opin Biotechnol 16:231-238

Nielsen KF, Larsen TO, Frisvad JC (2004a) Lightweight expanded clay aggregates (LECA), a new up-scaleable matrix for production of microfungal metabolites. J Antibiot 57:29-36

Nielsen KF, Smedsgaard J (2003) Fungal metabolite screening: database of 474 mycotoxins and fungal metabolites for dereplication by standardised liquid chromatography-UV-mass spectrometry methodology. J Chromatogr A 1002:111-136

Nielsen KF, Smedsgaard J, Larsen TO, Lund F, Thrane U, Frisvad JC (2004b) Chemical identification of fungi: Metabolite profiling and metabolomics. In: Arora DK (ed) Fungal biotechnology in agricultural, food and environmental applications. Marcel New York: Dekker, Inc., pp 19-35

Nielsen N-PV, Carstensen JM, Smedsgaard J (1998) Aligning of single and multiple wavelength chromatographic profiles for chemometric data analysis using correlation optimised warping. J Chromatogr A 805:17-35

Nielsen N-PV, Smedsgaard J, Frisvad JC (1999) Full second-order chromatographic/spectrometric data matrices for automated sample identification and component analysis by non data reducing image analysis. Anal Chem 71:727-735

Nirenberg HI, O'Donnell K (1998) New *Fusarium* species and combinations within the *Gibberella fujikuroi* species complex. Mycologia 90:434-458

Overy DP, Frisvad JC (2003) New *Penicillium* species associated with bulbs and root vegetables. Syst Appl Microbiol 26:631-639

Paterson RRM, Bridge PD (1994) Biochemical techniques for filamentous fungi. Wallingford: CAB International

Prillinger H, Lopandic K, Schweigkofler W, Deak R, Aarts HJM, Bauer R, Sterflinger K, Kraus GF, Maraz A (2002) Phylogeny and systematics of the fungi with special reference to the *Ascomycota* and *Basidiomycota*. Chem Immunol 81:207-295

Samson RA, Hoekstra ES, Frisvad JC (eds) (2004) Introduction to Food- and Airborne Fungi. 7th Edition. Utrecht: Centraalbureau voor Schimmelcultures

Samson RA, Hong SB, Frisvad JC (2006) Old and new concepts of species differentiation in *Aspergillus*. Med Mycol 44 Suppl:133-148

Schmidt H, Adler A, Holst-Jensen A, Klemsdal SS, Logrieco A, Mach RL, Nirenberg HI, Thrane U, Torp M, Vogel RF, Yli-Mattila T, Niessen L (2004) An integrated taxonomic study of *Fusarium langsethiae, Fusarium poae* and *Fusarium sporotrichioides* based on the use of composite datasets. Int J Food Microbiol 95:341-349

Smedsgaard J (1997) Micro-scale extraction procedure for standardized screening of fungal metabolite production in cultures. J Chromatogr A 760:264-270

Smedsgaard J, Frisvad JC (1996) Using direct electrospray mass spectrometry in taxonomy and secondary metabolite profiling of crude fungal extracts. J Microbiol Meth 25:5-17

Smedsgaard J, Frisvad JC (1997) Terverticillate penicillia studied by direct electrospray mass spectrometric profiling of crude extracts. I. Chemosystematics. Biochem Syst Ecol 25:51-64

Smedsgaard J, Hansen ME, Frisvad JC (2004) Classification of terverticillate penicillia by electrospray mass spectrometric profiling. Stud Mycol 49:243-251

Stadler M, Fournier J, Quang DN, Akulov AY (2007) Metabolomic studies on the chemical ecology of the Xylariaceae (Ascomycota). Nat Prod Comm 2:287-304

Stenwig H, Liven E (1988) Mycological examination of improperly stored grains. Acta Agr Scand 38:199-205

Taylor JW, Jacobson DJ, Kroken S, Kasuga T, Geiser DM, Hibbett DS, Fisher MC (2000) Phylogenetic species recognition and species concepts in fungi. Fungal Genet Biol 31:21-32

Thrane U (1986) Detection of toxigenic *Fusarium* isolates by thin layer chromatography. Lett Appl Microbiol 3:93-96

Thrane U, Adler A, Clasen P-E, Galvano F, Langseth W, Lew H, Logrieco A, Nielsen KF, Ritieni A (2004) Diversity in metabolite production by *Fusarium langsethiae, Fusarium poae,* and *Fusarium sporotrichioides*. Int J Food Microbiol 95:257-266

Thrane U, Hansen U (1995) Chemical and physiological characterization of taxa in the *Fusarium sambucinum* complex. Mycopathologia 129:183-190

Thrane U, Poulsen SB, Nirenberg HI, Lieckfeldt E (2001) Identification of *Trichoderma* strains by image analysis of HPLC chromatograms. FEMS Microbiol Lett 203:249-255

Vaidyanathan S, Kell DB, Goodacre R (2002) Flow-injection electrospray ionization mass spectrometry of crude cell extracts for high-throughput bacterial identification. J Am Soc Mass Spectrom 13:118-128

Yu JH, Keller N (2005) Regulation of secondary metabolism in filamentous fungi. Annu Rev Phytopathol 43:437-458

Andersen, Birgitte
Center for Microbial Biotechnology, BioCentrum-DTU, Technical University of Denmark, Søltofts Plads 221, DK-2800 Kgs. Lyngby, Denmark

Frisvad, Jens C.
Center for Microbial Biotechnology, BioCentrum-DTU, Technical University of Denmark, Søltofts Plads 221, DK-2800 Kgs. Lyngby, Denmark

Smedsgaard, Jørn
Center for Microbial Biotechnology, BioCentrum-DTU, Technical University of Denmark, Søltofts Plads 221, DK-2800 Kgs. Lyngby, Denmark

Thrane, Ulf
Center for Microbial Biotechnology, BioCentrum-DTU, Technical University of Denmark, Søltofts Plads 221, DK-2800 Kgs. Lyngby, Denmark
ut@biocentrum.dtu.dk

The importance of anatomy and physiology in plant metabolomics

Ute Roessner and Filomena Pettolino

Abstract

Plant metabolomics offers some unique opportunities in the assignment of biochemical pathways. The genetics of model plants is well-characterized which enables functional genomic approaches, qualitative trait loci identification and genetic engineering. Metabolomics has successfully supported the identification of gene function. As a specialized system, a number of key features of plants create challenges in sample preparation and interpretation of metabolomic data. Significantly, most plant tissues are composed of multiple cell types which are difficult to isolate, often resulting in limited numbers per cell type. This hinders spatial resolution of the analysis of metabolites. Secondly, cells are surrounded by a dynamic cell wall which is in constant turnover, interfering with the metabolome. Thirdly, green plant cells are capable of fixing carbon through photosynthesis producing metabolite-captured energy. This also implies a strong light-dependency in plant metabolism. Finally, plants are characterized by a diversity of secondary metabolites produced in response to environmental stimuli.

1 Introduction

1.1 Importance of plants

What is it about plants that make them so important to us? Apart from their visually pleasing qualities and contribution to some of the most famous landscapes in the world, plants provide the earth and its inhabitants with a large and varied set of irreplaceable resources of biological and economic importance. Plants account for 90 % of the biomass on Earth and contribute to the world's rich diversity with an estimated 350,000 species (Prance 2001). Plant importance begins with its position in the food chain as a primary producer, where energy is harvested from light. The processes of photosynthesis and respiration in plants are crucial in maintaining the life-essential balance of oxygen, carbon dioxide and water in the Earth's atmosphere. Plants provide food (either directly or indirectly), shelter and protection for animals, insects, fungi and even other plants.

A large number of industries revolve around plants and plant products. The most obvious are the agricultural, timber and paper industries which supply crops for food and textiles, building materials, and paper and packaging products. The

Topics in Current Genetics, Vol. 18
J. Nielsen, M. C. Jewett (Eds.): Metabolomics
DOI 10.1007/4735_2007_0218 / Published online: 14 April 2007

food and beverage industries use plant products to manufacture food and drink, and to modify their texture, flavor and/or color. The mining and manufacturing industries use plant products such as gums and resins (including latex for rubber) as binders, adhesives, emulsifiers and processing aids. Many of the medicines used today, including traditional herbal remedies, are either plant derived, based on natural plant products or contain plant extracts. Fossilized plants (coal) have been used as a source of energy for centuries but of increasing interest, and importance, is the use of plant biomass as a renewable energy source.

Plant metabolism is essential for the production of all of these plant products. The current focus of plant research worldwide is primarily the improvement of plants for food use. In addition, researchers are examining novel ways of generating plant products for the timber, pharmaceutical, green energy and textile industries. An understanding of plant metabolism at all levels is vital to the continued success of these research programs.

1.2 Plant metabolomics

Plant metabolism has been the target of research for a long time. Around 100 years ago the first concept of separation for plant specific compounds based on column chromatography was developed by Michael Tswett (1872-1920). The beauty of this technology was that he was able to separate chlorophyll, xanthophyll and carotene, based on their different colors, into clearly separated bands. A major step in plant research was achieved when about 50 years later Melvin Calvin and Andrew Benson discovered the carbon fixing dark reaction of photosynthesis, today commonly called the 'Calvin cycle'. Although of immense importance, the photosynthetic process has not been the only plant feature of interest as other plant specific pathways have been studied in great detail. These include the starch synthetic pathway, cell wall synthesis, vitamin production, protein and lipid metabolism. In the last century, an endless number of analytical methodologies have been developed for the extraction, detection and quantification of plant metabolites, always with the emphasis on increasing our understanding of plant metabolism, improving plant products or increasing crop yield. The exciting development of possibilities to specifically alter plant genomes by either mutations or by introduction of additional genes has opened a new opportunity in plant sciences. The release of the complete sequence of the flowering plant *Arabidopsis thaliana* at the end of the nineties has provided a great improvement in understanding not only plant biology, but also evolution and development. In parallel, novel multi-parallel and/or highly sensitive analytical tools have been developed for a comprehensive analysis of the different cell products. Most prominent amongst these new technologies has been the establishment of protocols for the determination of the expression levels of many thousands of genes in parallel (transcriptomics), the detection, identification and quantification of the protein complement (proteomics) and the determination and the simultaneous identification of a large number of metabolic compounds in a high-throughput manner (metabolomics). Metabolomics today can be considered as the accumulation and combination of knowledge of analytical biochemistry from the last 50 years and its application towards developments of new technologies with greater sensitivity, comprehensiveness, robustness and higher throughput. Currently in the field of metabolomics, both gas and liquid chromatography coupled to various mass spectrometric detection technologies (GC- and LC-MS) are applied to analyze complex metabolite mixtures. In addition, nuclear magnetic resonance spectroscopy (NMR) has been successfully used to fingerprint plant systems. Very recently, the power of capillary electrophoresis coupled either to laser induced fluorescence detection or mass spectrometry has been discovered. The advantage of this technology is its great sensitivity allowing the analysis of a large range of metabolites in very small sample sizes. The principles, advantages and disadvantages of each of the available technologies have been described in great detail in a large number of published reviews and books (e.g. Sumner at al. 2003; Hall 2006; Saito et al. 2006; Villas-Boas et al. 2007) and will therefore not be discussed in this chapter. In addition, endless numbers of publications are available with exciting and impressive applications of metabolomic technologies in many different scientific fields. The future of research will be driven by the exponential growth of metabolomics as its own entity in the 'omics' sciences. It is important to note, that metabolomics has attracted increasing interest, not only from biologists but also from the public and politicians. Concurrent with the evolution of metabolomics is the assured confidence in the validity of the data obtained and in the way it is applied.

In the following we want to present another perspective of plant metabolomics. As plants are unique and essential members amongst all living organisms we would like to place special emphasis on the distinctiveness of plant systems and relate these back to important factors to consider when conducting metabolomics experiments in plant research.

2 Plant anatomy

2.1 Whole plant anatomy

Most plants are immobile and therefore have to quickly and efficiently adapt to changing environments. In general, plants are built of three basic organs: leaves, stems and roots, which are made of four types of tissue including the vascular, the dermal, the ground and the meristematic tissues. The roots anchor the plant in the soil and are required to absorb and transport water and nutrients from the soil to the other parts of the plant. There are 13 minerals essential to all plants, including macronutrients, such as N and P, and micronutrients, such as Na, K, B, Mn, Fe, Ca. If a plant grows in mineral deficient conditions it affects the plant growth dramatically and in the worst case can kill the plant. On the other hand, excess amounts of most of these minerals may be harmful and thus result in the presentation of toxicity symptoms which again affect growth and reproduction. In both cases, the plant has to develop mechanisms to withstand these conditions for survival. Since plant metabolism is dramatically affected by both mineral deficiency and toxicity, metabolomics approaches are currently used to monitor metabolic changes following inadequate mineral supply to gain an increased understanding of plant mechanisms for adaptation or even the development of tolerance.

The stems have two major functions, firstly, to hold up the leaves for optimal exposure to sunlight and secondly, to transport water and nutrients via the xylem and soluble carbon sources and hormones via the phloem within all parts of the plant. In contrast, the main function of leaves is to 'host' the process of photosynthesis. Photosynthesis occurs in chloroplasts, specialized green cellular compartments where light energy is captured for the production of glucose from CO_2 and water.

When analyzing plant metabolism, the anatomic complexity of plants has to be considered. Each plant organ, tissue or cell type is characterized by a specific set of metabolites in a certain distribution/concentration and is often differentially affected by external stimuli. Currently, due to the low sensitivity of analytical technologies used in metabolomics, metabolites from a sufficient amount of tissue have to be extracted for comprehensive coverage of many metabolites simultaneously. Therefore, often many different cell types and tissues may be combined and only the 'average' of the metabolite content determined. Successful attempts at single cell metabolite analysis have already been reported. Schad et al. (2005) collected enough material composed of specific cell types from cryo-preserved and laser micro-dissected tissue to analyze about 68 major metabolites by GC-MS. Unfortunately, often it is very difficult or even impossible to separate and isolate specific cell types from plant tissues. Another exciting approach for cellular as well as subcellular specific determination of metabolite abundance has been presented by Fehr et al. (2004). The authors describe the development of protein-based fluorescent-tagged nanosensors for imaging specific metabolites. One important feature of this technique is that it is almost non-invasive and can be applied to monitor dynamic changes of metabolites and also ion levels in the cells, tissues or organs of interest (Fehr et al. 2004).

2.2 Cell anatomy

All eukaryotic cells share anatomical features. They are surrounded by a plasma membrane, have a nucleus containing the cell's genetic information along with a nucleolus for processing and assembly of ribonucleoprotein subunits, an endoplasmic reticulum and Golgi apparatus, mitochondria, ribosomes, peroxisomes and vacuoles. In addition, plant cells contain plastids and are surrounded by a cell wall (Fig. 1).

The plant cell wall is a rigid semi-permeable structure surrounding all plant cells. The principal component of the plant cell wall is the cellulose microfibril framework which is embedded in a matrix of non-cellulosic polysaccharides. The nature of the polysaccharide matrix is very much dependent on the plant species and the developmental stage of the cell. For higher plants, Gibeaut and Carpita (1998) have defined two types of primary walls. Type I is typical of most monocot

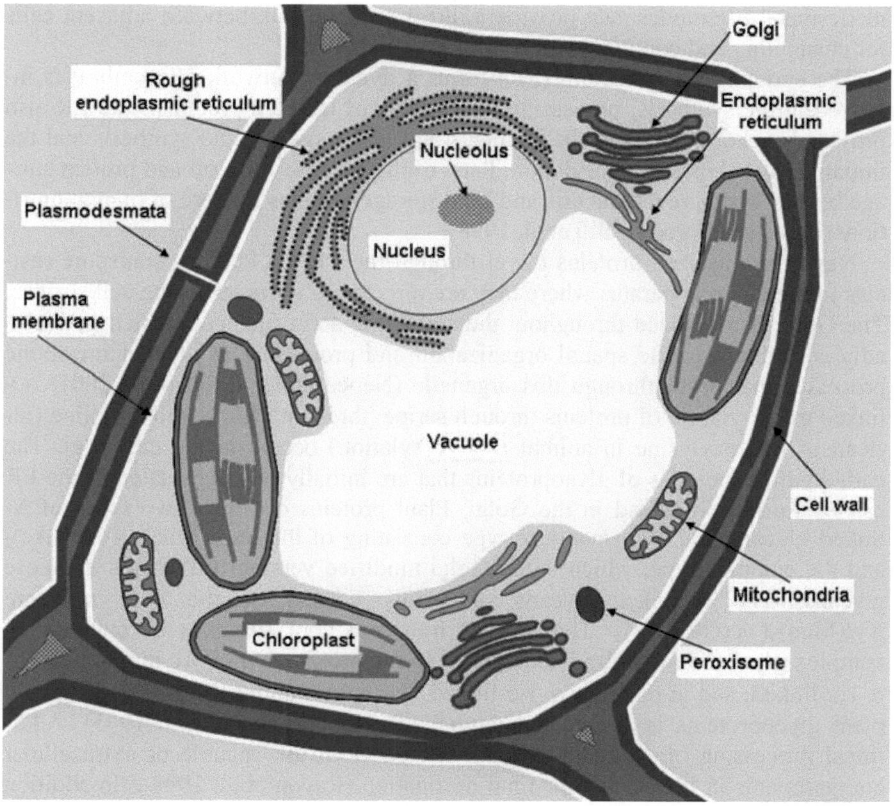

Fig. 1. A typical plant cell. Plant cells are distinguishable from animal cells by the presence of the cell wall, plastids (chloroplast), and large vacuole. Plant cells also possess plasmodesmata that allow for cell-to-cell molecular interactions.

and dicot species where xyloglucan, and/or glucomannan, associates with the cellulose microfibrils to form a framework embedded in a gel-like matrix of pectins. The Type II walls are specific to the commelinoid monocots (e.g. grasses) and contain glucuronoarabinoxylans in place of xyloglucan, and depending on the cell type and stage, also mixed-linkage β-glucans (Gibeaut and Carpita 1998). Proteins serve structural and catalytic roles in the cell wall and are involved in the strengthening and manipulation of the various components of the wall during growth and development. Secondary walls develop internal to the primary walls where further modification of the polymers is evident, including the deposition of lignin and suberin.

Living plant cells are enclosed within a plasma membrane that is restricted against the cell wall due to turgor pressure. The plasma membrane is involved in signal transduction and assists in the regulation of molecular transport into and out of the cell. In plants, particular areas of the plasma membrane combine with elements of the endoplasmic reticulum to form membranous tubes called plasmodesmata. Plasmodesmata provide a direct physical link between adjacent cells for channeling and communication (McLean et al. 1997).

The endoplasmic reticulum (ER) forms a dynamic network of membranes involved in the synthesis, processing and sorting of targeted proteins. The ER also provides anchor sites for actin filaments and is the site of lipid synthesis and the initiation of *N*-linked glycosylation. Parts of the ER also form oil and protein storage bodies where vegetable oils and seed storage proteins that are of human nutritional value are stored (Galili et al. 1998).

Newly synthesized proteins travel through the ER via budding transport vesicles to the Golgi apparatus where they are directed to vacuoles or the cell surface. The Golgi is mobilized throughout the cell along actin filaments which undoubtedly contributes to the spatial organization and processing of cellular metabolic processes that occur through this organelle (Nebenführ and Staehelin 2001). *O*-linked glycosylation of proteins through serine, threonine and hydroxyproline (instead of hydroxylysine in animal *O*-glycosylation) occurs in the *cis*-Golgi. The carbohydrate moieties of glycoproteins that are initially *N*-glycosylated in the ER can be further processed in the Golgi. Plant proteins can have two types of *N*-linked glycans; the high mannose type consisting of the unit $(Man)_{6-9}(GlcNAc)_2$ and the complex type, which is the Golgi modified version of the high mannose glycans. The complex glycans of plants consist of the core structure $Xyl(Man)_3Fuc(GlcNAc)_2$. This differs from the core structure of mammalian complex glycans in that the Fuc attached to the proximal GlcNAc in mammals is α-1,3-linked, and in plants is α-1,6-linked. Furthermore, the Xyl that is present in plant glycoproteins is absent in mammalian glycoproteins (Sturm 1995). Additional processing of *N*-linked glycans can occur in the vacuole or extracellular compartments in transit to their final destination (Rayon et al. 1998.). In addition to containing the enzymes involved in protein and lipid glycosylation, the plant Golgi is also the site of synthesis of pectic and non-cellulosic cell wall polysaccharides.

A defining feature of plant vacuoles is their size, capable of occupying over 30 % of the cell volume. The turgor pressure of the cell is maintained by the osmotic uptake of water as solutes accumulate in the vacuole. Turgor pressure, along with cell wall extensibility, drives plant cell enlargement and expansion. Vacuoles store inorganic ions, important for pH and ionic homeostasis; organic acids, including amino acids; sugars; enzymes such as proteases, nucleases, glycosidases and lipases important for digestion; proteins; and secondary metabolites such as pigments and defensive molecules (phenolics, alkaloids, cyanogenic glycosides, saponins) (Marty 1999). Due to their chemical nature not all of these molecules are likely to be found in any one vacuole. In plant cells at least two types of vacuole have been identified; the neutral, protein-storing vacuoles and the acidic, lytic vacuoles (Staehelin and Newcomb 2000).

The role of peroxisomes in plant cells is organ or tissue specific. Peroxisomes are involved in the conversion of fixed N_2 into nitrogen-rich organic compounds in legume root nodules. Glyoxsomes are specific peroxisomes involved in lipid metabolism in germinating seeds that store fats. In leaves, peroxisomes, in conjunction with mitochondria and chloroplasts, participate in photorespiration. Peroxisomes serve a protective function in that the hydrogen peroxide that is liberated in each of these metabolic processes is destroyed by their resident catalases.

The mitochondria of plant cells are typical of the eukaryotic organelle responsible for the generation of ATP via the citric acid cycle and associated electron transfer chain. Plant mitochondria have a much larger genome than in other organisms, ranging in size from 200,000 to 2,600,000 nucleotides (compared with 15,000 – 18,000 in mammals). The plant mitochondrial genome, which codes for only 16 of the 20 tRNA genes required for protein synthesis, also contains some chloroplast DNA, most of which is non-functional in mitochondria (Staehelin and Newcomb 2000).

Amongst the eukaryotes, plastids are found only in plant and algal cells. There are a number of different plastids that can exist in a plant cell, each of which serves different functions. All plastids begin as proplastids that develop into, or convert from one type of plastid to another. Amyloplasts and leucoplasts are non-pigmented plastids that store starch and are involved in monoterpene synthesis, respectively. Etioplasts, which arise when chloroplast development is arrested due to the lack of light, store tubular membranes as semicrystalline structures called prolamellar bodies. The lipid membranes transform into thylakoids when the etioplast is illuminated and progresses in development to a chloroplast. Chromoplasts synthesize and store carotenes and xanthophylls giving them the yellow, orange or red coloring seen in many fruits, flowers and vegetables. Chloroplasts are the green chlorophyll containing plastids responsible for energy capture from sunlight. The photosynthetic machinery of chloroplasts resides within the thylakoid membrane system composed of stacked grana that are interconnected via the unstacked stroma.

Each of the above described compartments is characterized by their own suite of metabolites as well as concentration patterns. This is especially important to be considered as most metabolomics approaches cover metabolites extracted from whole cells, tissues, organs or even plants. Therefore no information is obtained about metabolite levels and changes within and between compartments, e.g. following environmental stimuli or genetic alteration. Recently, there have been efforts to develop metabolite analysis tools at the subcellular level. Farré et al. (2001) applied a non-aqueous fractionation technique to separate plastids, vacuoles and cytoplasm/mitochondria based on their individual density from potato tubers. The resulting fractions were characterized using compartment-specific enzyme marker assays to determine the distribution of compartments in each fraction. The percentage distribution was further correlated with levels of about 60 metabolites, analyzed using GC-MS, to give an estimation of metabolite concentrations in the different compartments (Farré et al. 2001). As mentioned above, more recent and extremely promising developments for subcellular metabolite imaging are based on fluorescent-tagged nanosensors (Fehr et al. 2004).

3 Plant physiology – Challenges for plant metabolomics

3.1 Photosynthesis

There is essentially a lot of similarity between plant primary metabolism and those of all other organisms. But, the ability of green plants to capture energy from light for the production of high-energy containing molecules, has equipped plants with a number of unique reactions. Most well known and studied is photosynthesis which is characterized by two major processes. The first is the capture of light energy for the production of ATP and the reducing equivalent NADPH. Figure 2 represents a schematic overview of the importance of photosynthesis for the supply of energy and carbon molecules for a range of metabolic processes in plant cells. The prerequisite of this so called light-reaction is the presence of chlorophyll. The second step is light-independent and produces glucose from carbon dioxide and water using ATP and NADPH, and releases oxygen. This is a very advanced process where the enzyme ribulose-1,5-bisphosphate carboxylase (Rubisco), actually the most abundant protein in green tissues, binds 6 molecules of carbon dioxide to 6 molecules of ribulose-1,5-bisphosphate producing twelve molecules of 3-phosphoglycerate. The 3-phosphoglycerates are further metabolized to release one molecule of glucose and resynthesize 6 molecules of ribulose-1,5-bisphosphate for the next cycle. The glucose is then the key metabolite for all down-stream metabolic processes, both for biosynthetic pathways or respiration via glycolysis. In most plants, sucrose is the transport form of carbon throughout the whole plant.

Photosynthesis is therefore the key process dictating the great dependency on light availability and intensity for many metabolic processes in plant cells. A large range of metabolic enzymes are regulated either directly by light or by the resulting glucose or sucrose. As a consequence, substantially different metabolite quantities are present during the day compared to the night. This has been shown in an in-depth analysis of leaf metabolites during diurnal rhythm in potato and rice (Sato et al. 2004; Urbanczyk-Wochniak et al. 2005). Therefore, special care has to be taken with the time of the day when leaf samples for metabolomics studies are harvested. Leaves however, are not the only tissue to show a light-dependent metabolite profile. As demonstrated by Roessner-Tunali et al. (2003a), even heterotrophic tissues such as potato tubers which grow in the dark in the soil, show a differential metabolite profile in the course of a day because they are dependent on the 'delivery' of sucrose from the aerial parts for starch production.

3.2 Photorespiration

Plants have to develop a specialized mechanism to survive in situations where the CO_2 levels inside a leaf become very low. This occurs in very hot and/or dry environments which cause a closure of the stomata to avoid undesired water loss, resulting in insufficient CO_2 uptake. Rubisco is a dual functional enzyme, which in

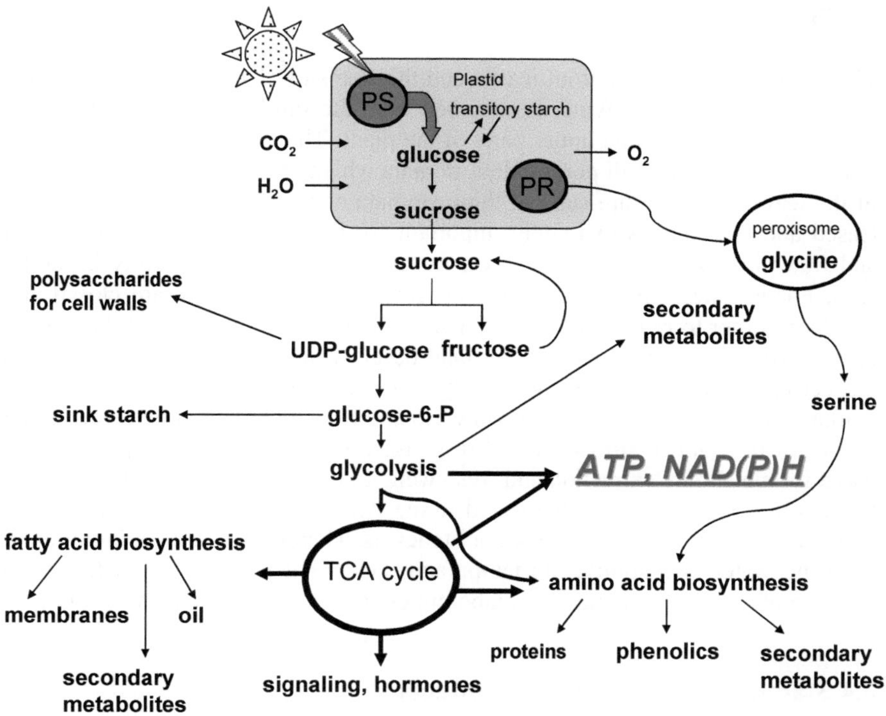

Fig. 2. Simplified scheme of metabolism in green plant tissues. PS – photosynthesis, PR – photorespiration, TCA – tricarbolic acid cycle.

low CO_2 conditions prefers to accept O_2. This leads to the production of 2-phosphoglycolate, which is toxic for the plant cell, and to reduction of ATP production. In this case, the plant uses a series of enzymatic steps to transform 2-phosphoglycolate into non-toxic and even metabolically useful compounds. The first step cleaves the phospho-group to produce glycolate. After transport of this molecule from the plastids to the peroxisomes, it is transformed into glycine, by the release of CO_2, and is then transported to the mitochondria. There, glycine is further converted to serine which can be either channeled into amino acid and protein metabolism or metabolized to form 3-phosphoglycerate, an important intermediate in glycolysis and a useful precursor for other primary metabolites. Unfortunately for the plant, all these conversions result in a net loss of CO_2 and the use of ATP and reducing equivalents. This pathway demonstrates an interesting example of how cells can control carbon flow by separating metabolic reactions into different compartments. The process of photorespiration involves the action of three compartments, the plastid, the peroxisome and the mitochondria. This means that transport proteins specific for each of the compartments and for the respective molecule requiring transport, have to be expressed and activated.

3.3 Transpiration

The evaporation of water from leaves and the stems of plants into the atmosphere are called transpiration. Water is absorbed from the soil by the roots and pumped through the vessels to the upper parts of the plant. The actual process of evaporation occurs through small pores called stomata which are located on the lower side of the leaves. The closure state of these stomata controls the amount of water released and therefore is extremely important for the balance between water gain and water loss of the plant and hence, for the actual water availability in the tissues. For instance, under water limiting conditions, the stomata are closed very rapidly in order to reduce water loss. Therefore, the opening status of the stomata can control the water content of the plant tissue which can have two effects with respect to metabolomic analysis. Firstly, water availability in plant cells has remarkable effects on all metabolic pathways and therefore metabolite levels. In addition it can result in stress-induced responses. Secondly, when samples are prepared for metabolite extraction and fresh weight is used as a way of normalization, the amount of water in the harvested tissue will influence the fresh weight and consequently the evaluation of metabolite levels. Therefore, it is most important to keep the water availability, the temperature and light intensity consistent when growing plants for comparative metabolomics as well as other 'omics' studies.

3.4 Starch and other storage products

Plants are important food components as they store high-energy products, such as carbohydrates, fats and proteins. Carbohydrates can be stored as free sugars, such as hexoses in fruits or sucrose in sugar cane, or polymerized in the form of cell walls and starch. Starch is a plant specific storage product and consists of an endless number of polymerized glucose polymers. In plants, two types of starch are produced, transitory and storage starch. Transitory starch, which is a store for excess glucose made in green leaves during photosynthesis in the light, is degraded during the dark period and distributed throughout the plant for energy production via respiration or for delivery to sink organs for long-term storage. This type of starch mainly occurs in non-photosynthetically active (heterotrophic, non-green) tissues, such as tubers or grains. As mentioned earlier, starch is made of long chains of glucose molecules. The chemistry of these long chains determines the type of starch. On one hand, glucose monomers are linked by α-(1,4)-glucosidic bonds resulting in amylose, a linear, helical polymer that aggregates to form insoluble starch granules. In the other form of starch, amylopectin, these α-(1,4)-glucose chains are further substituted by α-(1,6)-glucosidic linkages forming more complex and branched structures. The biosynthetic pathway of starch starts with the formation of nucleotide-activated glucose by the enzyme ADP-glucose pyrophosphorylase. The ADP-glucose is then used as a substrate by starch synthase enzymes, which add glucose units to the end of a growing polymer chain to build up a starch molecule (releasing the ADP in the process). Branches in the chain are introduced by starch branching enzymes (SBEs), which hydrolyze α-(1,4)-

glycosidic bonds, and in their place, create α-(1,6) bonds with other glucose units. The investigation of the starch synthetic and degradation pathways, and their differences in different species and tissues, has been a target of research for many years particularly with the emphasis on increasing yield of starch-storing crops. The analysis of the intermediates of the starch biosynthetic pathway and also the many other metabolites either directly associated, e.g. from glycolysis or the TCA cycle or metabolites indirectly involved in this pathway will indicate what factors influence the flux of carbon into starch. This will result in improved and more efficient approaches to modify starch with respect to increased yields and altered structure for specific industrial applications.

3.5 Cell wall synthesis

A large proportion of the glucose generated by plant cells is directed towards cell wall synthesis. New cell walls form after nuclear division when a phragmoplast containing actin, myosin and microtubules assembles a cell plate between the nuclei. New wall components carried via Golgi vesicles are deposited at the cell plate which continues to grow from the centre towards the edges of the cell until it fuses with the existing wall. Biosynthesis of the plant cell wall is a highly regulated process due in part to the complex nature of its structure, the location of the biosynthetic machinery and the coordinated changes that take place during growth and development. Cell wall biosynthesis itself involves a number of metabolites and it is the synthesis, conversion and transport of these that are important in cell wall development. The synthesis and shuffle of carbohydrates, nucleotides, proteins, amino acids, lipids (e.g. sterols), phenolics, growth regulators and cofactors (e.g. acetyl-CoA) are critical to the growth and development of the plant cell wall, and in turn, the plant.

Polysaccharides, the major components of plant cell walls are a secondary gene product; the primary gene product being the synthases and transferases responsible for their synthesis. As a secondary gene product no template is available (unlike protein synthesis) and yet the general polysaccharide structures present in a given wall at a specific developmental stage are consistent. The identification of the cellulose synthase genes (Ces A) opened a flood gate for the discovery of a large number of genes encoding putative polysaccharide synthases. These genes have been classified according to their similarity to the CesA genes to give the CSL (cellulose synthase-like) gene families (http://cellwall.stanford.edu/). Interestingly, the genomic approaches to study cell wall biosynthesis have shown that each tissue can have multiple Ces and CSL genes, presumably to account for the different types of polysaccharides synthesized as well as providing the ability to switch on different genes to coincide with a particular developmental stage. Despite the large number of genes identified, only a handful have been unambiguously identified as polysaccharide synthase genes (Scheible and Pauly 2004) and the actual mechanisms involved in cell wall biosynthesis are not clearly defined. Metabolite profiling has the potential to shed light on some of these mechanisms.

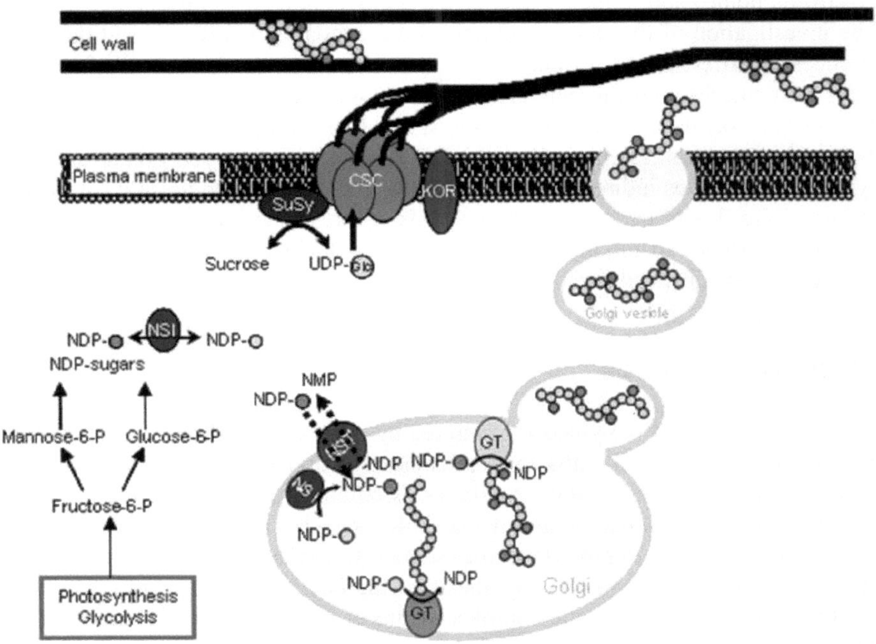

Fig. 3. Primary cell wall synthesis. Cellulose is synthesised at the plasma membrane by CSC, (Cellulose Synthase Complex) and is associated with other enzymes including SuSy (Sucrose Synthase) and KOR (KOR endoglucanase). Nucleotide sugars (NDP) are synthesised and converted in the cytosol (and most likely the Golgi) by Nucleotide Sugar Interconverting enzymes (NSI). Matrix polysaccharides are synthesised by GTs (Glycosyl Transferases) in the Golgi and transported to the wall. NDPs are transported to sites of polysaccharide synthesis including inside the Golgi by Nucleotide Sugar Transporters (NSTs).

In the synthesis of the wall polysaccharides (Fig. 3), nucleotide sugars are the donor substrates that provide the specific monosaccharides to be attached to the growing polysaccharide chain. This is also the case for glycosyltransferases which add specific sugars to a preformed polysaccharide backbone. Cellulose and callose (a developmentally regulated polysaccharide that occurs only in specialized cells and in response to wounding) are synthesized at the plasma membrane. The other cell wall polysaccharides, including xyloglucan, pectins, arabinoxylan and heteromannans are synthesized in the Golgi (Moore and Staehlin 1988) and transported to the cell surface for release into the wall space via vesicular transport mechanisms. The cellulose synthetic machinery that occurs on the plasma membrane forms rosette structures composed of cellulose synthase hexamers. A number of other enzymes appear to be closely associated with the complex, including sucrose synthase (SuSy), which presumably supplies UDP-glucose as the donor substrate to the cellulose synthase (Delmer and Amor 1995), and KOR endoglucanase. There has been some evidence to suggest that the acceptor for initiation of cellulose synthesis is a sitosterol-linked β-glucan and it has been proposed that the KOR endo-glucanase cleaves the lipid-linked glucan chain after initiation of polymerization (Peng et al. 2002).

Nucleotide sugars are synthesized in the cytoplasm and must be transported to the sites of polysaccharide synthesis in both the Golgi and at the plasma membrane. Nucleotide sugar interconverting enzymes have also been associated with the Golgi (Baldwin et al. 2001). The evidence for the presence of nucleotide sugar transporters in Golgi membranes suggests that nucleotide sugars that are synthesized in the cytoplasm can be transported into the Golgi and perhaps converted to raise the population of various sugar nucleotides required for polysaccharide synthesis (Orellana 2005; Reiter and Vanzin 2001). With respect to metabolite profiling, the extraction, fractionation and detection of nucleotide sugars is complicated by their structural similarities and low abundances. Methods currently used are limited to the simultaneous detection of only a few metabolites, but new developments in methodology (e.g. Ramm et al. 2004) are helping to overcome these issues.

The cell wall is a dynamic structure and once deposited will undergo many changes according to developmental stage, tissue type and environmental influences. These include modification of the polysaccharides by hydrolytic, deacetylation and de-methylesterification processes. Acetylation of polysaccharides occurs in the Golgi with acetyl-CoA as the acetyl donor and is thought to protect the polysaccharide from degradation and influence its solubility (Pauly and Scheller 2000). Homogalacturonan is believed to be synthesized in a highly-methylesterified form which is later modified by pectin methylesterases to generate esterified and unesterified regions (Willats et al. 2001), both of which are important in determining the physical properties of pectins and their associations within the wall. In addition, cross-linking of different wall components can occur through polysaccharide-polysaccharide, protein-phenolic, phenolic-polysaccharide and phenolic-phenolic interactions. Cell elongation and expansion therefore requires mechanisms that can disrupt these associations to allow for flexibility. Enzymes such as xyloglucan endotransglycosylase, peroxidases, endoglycanases and esterases and non-enzymic proteins such as expansins are involved in this process and can be influenced by regulators such as auxins and low pH (Cosgrove 2001).

Once the cell has ceased growing, further layers are added to the wall to form the secondary wall thickenings. Secondary walls form three layers inside the primary wall layer (S_1, S_2 and S_3 from outer to inner layers) and are composed of polysaccharides (mostly cellulose) and lignin, although lignin is rarely found in the S_3 layer. Lignin is synthesized by the free radical-driven polymerization of phenylpropanoid monomers including ferulic, coumaric, sinapic, cinnamic and hydoxybenzoic acids.

Biosynthesis of the plant cell wall involves a number of key metabolic processes. Whether or not polysaccharides are strictly defined as metabolites remains to be resolved. However, polysaccharides and oligosaccharides should not be ignored when it comes to plant metabolite profiling since much of plant metabolism revolves around carbohydrate shuffling and incorporation into cell wall and storage polysaccharides. A major challenge facing analysts is the high throughput measurement of complex carbohydrates due to the structural similarity of a vast and diverse range of possible structures. For example, the same hexose units can join together to form 20 different disaccharide structures and 448 trisaccharide structures, while three different hexoses joined together can potentially give rise to 2688 isomers, but in the case of peptides, three different amino acids will give rise to only six different peptides (Oxley et al. 2004).

Recent developments using molecular genetic approaches have contributed significantly to the current understanding of cell wall biosynthesis, but many aspects of its regulation remain a mystery. The use of metabolomics techniques will help to unravel some of these mysteries particularly when developments in spatial resolution of metabolites come to fruition.

3.6 Secondary metabolites

Plant secondary metabolites, organic compounds that are produced by plants but are not directly involved in their growth and development, include an extremely varied and complex array of molecules. In our everyday lives we use these metabolites, either native or chemically modified, in items such as dyes and food colorings, polymers, fiber, adhesives, oils and waxes, flavoring agents, fragrances and drugs. The function of many secondary metabolites *in planta* is not known, but for others, they serve roles in defense against herbivorous and microbial attack, pollination and seed dispersal and act as allelopathic agents. Plant secondary metabolites are classed into three major groups; the terpenoids, the alkaloids and the phenylpropanoids and related phenolic compounds, all of which are produced by extremely complex biosynthetic pathways.

The terpenoid group of compounds are not restricted to plants but are also produced by animals and microorganisms. Plants however possess a much wider variety of terpenoids than other organisms and by developing highly specialized cells (such as glandular epidermal cells) are able to produce and store them in large quantities. Over 22,000 different compounds belong to the plant group of terpenoids, all of which are derived from the 5-carbon (C_5) precursor isopentenyl diphosphate (IPP) (McGarvey and Croteau 1995). IPP is generated from the fusion of 3 acetyl-CoA molecules and gives rise to mevalonic acid which is in turn phosphorylated and decarboxylated to IPP. Repetitive addition of IPP units gives rise to a series of prenyl diphosphate molecules. These are processed by specific terpenoid synthases to yield terpenoid skeletons that are further modified enzymatically to deliver the vast array of terpenoids that exist in nature. The simplest terpenoid is isoprene which is composed of a single C_5 unit (the terpenoid monomer). Examples of the monoterpenes (2 C_5 units) include the components found in the essential oils of flowers, herbs and spices. Sesquiterpenes (3 C_5 units) can also be found in essential oils and are involved in defense against herbivores and microorganisms. The diterpenes (4 C_5 units) include phytol (component of chlorophyll), the giberellin growth regulators, phytoallexins and taxol (an anticancer agent). Brassinosteroids, some wax components and membrane phytosterols (such as β-sitosterol, campesterol and stigmasterol) belong to the triterpene group, which are composed of two C_{15} chains linked together. The tetraterpenes (8 C_5 units) include the carotenoids which are essential for photosynthesis, and examples of polyterpenes (more than 8 C_5 units) include rubber, dolichol (essential for sugar transfer reactions), plastiquinone and ubiquinone (electron carriers).

The alkaloids were historically defined as 'pharmacologically active, nitrogen-containing basic compounds of plant origin' based on the therapeutic use of these compounds in traditional medicines. Alkaloids have since been isolated from animal and insect sources (usually toxins) but we continue to use the plant alkaloids in modern medicine. Over 12,000 alkaloid structures from plants have been described, with approximately 20% of plant species known to accumulate alkaloids (DeLuca and Pierre 2000). These include caffeine, camptothecin, cocaine, codeine, morphine, nicotine, quinine and strychnine. The role of alkaloids in plants is generally a defensive one, with evidence for their involvement in wound responses. Not all the biosynthetic pathways for alkaloid biosynthesis have been elucidated, however, it is known that they are mostly derived from amino acids such as tryptophan, tyrosine, phenylalanine, lysine and ornithine, sometimes in combination with steroid or terpenoid-like groups.

The phenylpropanoids and related phenolic compounds (>2500 compounds) are generated through the shikimic acid or malonite/acetate biochemical pathway. Whilst many of the phenolic compounds serve structural roles in the plant cell wall, others have also been ascribed roles in plant defense, flower color and plant flavors and aromas. Lignins are deposited in secondary cell walls to strengthen and reinforce the wall, while suberin acts to protect tissue from dehydration and pathogen attack. The flavanoid group of compounds includes the anthocyanins that impart color in the way of pigments; condensed tannins; and isoflavanoids that serve as defense and signaling molecules.

Metabolomics of secondary metabolites is complicated by their vast numbers and diverse chemistries. Techniques are continuously developing to incorporate as many secondary metabolites in profiling analysis as possible. Recently, von Roepenack-Lahaye et al. (2004) successfully used capillary liquid chromatography coupled to ESI-QqTOF-MS profile to obtain approximately 2000 mass signals from *Arabidopsis* tissue that covered a large number of secondary metabolites but not mono- and sesquiterpenoids, triterpenoid alcohols, phytosterols, waxes or carotenoids. This is typical of all methods currently being used where only a subset of secondary metabolites is detectable mostly due to extraction procedures and low resolution. Furthermore, this group of compounds is so complex that it is possible that previously unidentified structures exist but are being overlooked.

4 Unique aspects of plant research

4.1 Functional genomics

The ever developing area of functional genomics aims to assign function to the multitude of genes that have been identified by genomic analyses of biological systems. In the "post-genomic" era, the profiling of biological systems at the levels of RNA (transcriptomics), protein (proteomics) and metabolite (metabolomics) is essential to functional genomics. Functional genomics is the ultimate tool for the rational improvement of plants for food, fiber and other commodities that are essential to life as we know it today.

In plants, much of the genetic information gained has been from model systems such as *Arabidopsis thaliana* and commercially important crop plants such as rice, potato and maize. *Arabidopsis* and rice (*Arabidopsis* genome initiative 2000; Yu et al. 2002) have now been fully sequenced. These sequences provide an essential tool for plant functional genomics because of the similarity in gene sequences within the plant kingdom. Based on comparative sequences alone, 54 % of genes in higher plants can be assigned a function (Somerville and Somerville 1999). Useful genetic information has also been gained from other sources where the entire genome is not necessarily available. Expressed sequence tag (EST) libraries have been used to correlate gene expression with developmental processes in plants. For example in potato, ESTs were used to identify genes involved in tuber initiation, dormancy and sprouting (Ronning et al. 2003). Insertion mutant libraries, which are available for *Arabidopsis*, maize, petunia and snapdragon (Somerville and Somerville 1999), and gene silencing by double stranded RNA production, allow for the phenotypic analysis of plants where particular genes have been interrupted or silenced. Developments in gene chip and microarray technologies have also provided essential information by quantitative analysis of gene expression associated with particular treatments or developmental stages (Schena et al. 1997).

Molecular genetic techniques have assisted in identifying entire genomes and transcriptomes because, to a large extent, it is possible to assign gene function based on orthology. However, this does not necessarily help in describing gene function at a cellular level. For example, knowing that a gene codes for a particular enzyme does not provide information on how the enzyme is regulated or what chain of events are triggered by it, nor does it take into account gene duplication. Proteomic technologies are advancing with the development of separation and mass spectrometric platforms for functional genomics. Metabolomic approaches will go one step further in filling the gaps and addressing some of the questions raised by the identification of the vast number of genes discovered from genomic analyses.

4.2 Breeding and QTL analysis

In order to create a novel variety of genotypes and phenotypes, plant genomes can be manipulated in a targeted fashion using breeding. Classical plant breeding deliberately crosses closely or even distantly related species to produce new crops with desired features by introducing genes, and therefore traits, from one species into another genetic background. Basically, plant breeding has been performed since the start of agricultural practices thousands of years ago, but today is approached in a much more sophisticated and organized manner to ensure food security and sustain agriculture. Classical breeding relies on the homologous recombination process between two genomes creating novel genetic diversity. Currently, large breeding programs worldwide for many different plant species are aiming for the development of better crops, e.g. with increased yield and quality of the crop, increased tolerance levels to environmental challenges, resistance to viruses, bacteria, fungi or insects, as well as increased tolerance to certain herbicides. In future, metabolomics technologies may become an important tool as a high-throughput method to screen for desired features of a crossed progeny, e.g. for vitamin, acid and/or sugar contents in fruits.

Another great potential tool for identifying novel genetic variety and new genes involved in plant performance is quantitative trait locus mapping. A quantitative trait locus (QTL) is an interval across a chromosome that is associated with a particular feature of the plant, a trait. QTLs are not necessarily genes themselves but are stretches of DNA that are closely linked to the genes controlling the desired trait. The statistical investigation of the alleles which occur in a locus and the produced trait is called QTL mapping. QTL mapping aims to identify the loci, decipher the genes within these loci and ultimately, to identify the functions of the underlying genes. In the past, most QTL analysis was done on single traits, such as yield, plant height or stress tolerance. More recently, with the development of novel technologies for high-throughput simultaneous analysis of transcripts or metabolites, a great potential for multi-trait analysis has become available. Combining for instance the techniques of QTL analysis with those of metabolomics will offer identification of novel QTLs affecting either the level of a single metabolite or even the levels of many metabolites simultaneously. Two primary and exciting examples of this approach have been presented very recently by Schauer et al. (2006) and Keurentjes et al. (2006). Schauer et al. (2006) utilized a GC-MS based metabolite profiling method to analyze the metabolite profiles of fruits from a cultivated tomato species (*Solanum lycopersicon*) in which marker-defined genome regions were introgressed with homologous regions of a wild and non-ripening tomato species (*Solanum penellii*). The authors describe the identification of a large number of single metabolite QTLs as well as many QTLs affecting whole pathways and/or the metabolic network. The work of Keurentjes et al. (2006) demonstrated the investigation of the variation of metabolite composition in plants by analyzing 14 *Arabidopsis thaliana* accessions using a non-targeted LC-QTOF-MS method for the simultaneous detection of more than 2000 individual mass peaks. In addition, the analysis of the metabolomes of a recombinant inbred line (RIL) population of the two most divergent accessions allowed the detection of respective QTLs for about 75 % of all mass peaks. Both examples can be seen as the pioneer work for future QTL identification and mapping as they demonstrate the potential that metabolomics approaches are offering. Once metabolomics technologies become more robust, faster and easier to automate, it will be one of the most promising and informative methods to study genetic segregation and identify novel genes.

Another, sometimes quicker way of introducing new genetic variety into a species is the use of genetic techniques for the production of genetically altered organisms based on transgenesis. These techniques enable the specific introduction or deletion of targeted genes rather than the random approach used in plant breeding. Several methods are established for doing this but the most common methods include the "gene gun" and the *Agrobacterium* based method. Today, the *Agrobacterium*-mediated genetic transformation is the most commonly used technology for the production of transgenic plants and protocols for a large range of different species have been established. Extensive research in this technology is aiming at improvement of the efficiency of the actual gene transfer. Since the first transgenic plant was created more than 20 years ago, scientists have wanted to analyze the intended as well as unintended effects that the introduced gene has on the plants performance, including the visible phenotype or the abundance of certain cell products. Huge efforts have been made in the development of strategies for risk assessment of genetically modified organisms and metabolomics has been identified as the tool of choice for comprehensive analysis of transgenesis, including effects on plant metabolism as well as on potential interactions with human health and the environment (Rischer and Oksman-Caldentey, 2006).

4.3 Genetic engineering

Plants provide an ideal system for the expression of both foreign and non-foreign genes, either for improved qualities or for the production of selected compounds such as plant secondary metabolites. Worldwide, 90 million hectares of crops are biotech approved across 21 different countries. These include (in order of acreage) USA (49.8 million hectares), Argentina, Brazil, Canada, China, Paraguay, India, South Africa, Uruguay, Australia, Mexico, Romania, the Philippines, Spain, Colombia, Iran, Honduras, Portugal, Germany, France and the Czech Republic (<0.1 million hectares) (James 2005). The industry is worth $5.25 billion, which equates to 18 % of the global commercial seed market. The majority of crops have been modified for pathogen resistance by the introduction of *Bacillus thuringiensis* toxin (BT) (e.g. maize, cotton, canola, rice, potato, tomato) and/or herbicide resistance (e.g. soybean, maize, cotton, canola, rice, sugar beet, tomato). Viral resistance is also available for some plant species (e.g. in squash and papaya). The modified traits that are most common in crops currently focus on farming practices to reduce pesticide use and increase crop yields. The future will see the increase in the introduction of genes into crops to modify nutritional qualities. For example, biotech soybean with high oleic acid content, tomato with increased lycopene levels, and potatoes, maize and wheat with modified starch.

In addition to the modification of crop plants for selected traits, it is possible to use plants as protein or secondary metabolite factories both in the field and in tissue culture. Although potentially expensive, plant tissue culture can offer some advantages over traditional field growing practices. These advantages include the growth of metabolically active cells from rare plants, or plants that are either difficult to cultivate or have long maturation periods. Furthermore, a culture system can be manipulated to occlude the influence of environmental factors such as climate, nutrient availability and disease. Plant-based systems offer a feasible alternative to microbial or mammalian cell culture systems for the production of recombinant proteins and can offer some advantages for the production of medically important proteins. Plants don't carry human pathogens or produce endotoxins and they have the necessary machinery for post-translational modification of proteins. Although glycosylation in plants differs only slightly from mammalian glycosylation, it is different enough to cause potential immunogenic and efficacy concerns. These factors are being overcome by the introduction of mammalian glycosyltransferases into plants to produce proteins with mammalian glycosylation patterns (Palacpac et al. 1999).

A number of medically relevant proteins have been produced in plant cells. These include various immunoglobulins and immunoglobulin fragments, human erythropoietin, interleukins and granulocyte macrophage stimulating factor (see Hellwig et al. 2004 and references therein). Furthermore, the effectiveness of recombinant proteins as oral vaccines has been demonstrated. Hepatitis B surface antigen (HBsAg) expressed in potato and fed to mice elicited an immune response to the antigen (Kong et al. 2001). Similarly, in humans, Norwalk virus capsid protein (NVCP) expressed in potato was shown to stimulate an antibody response against the antigen upon its oral administration (Tacket et al. 2000.)

In addition to protein production, plant cell culture provides a controlled method for the production of secondary metabolites for medicinal purposes. Examples of plant cell culture production of secondary metabolites include codeine and morphine by poppy (*Papaver somniferum*), ginsenosides by *Panax ginseng* and capsaicin by *Capsicum frutescens*. Perhaps the most successful example is the production of taxol by plant tissue culture. Taxol is an alkaloid anticancer agent found in the bark of Pacific yew trees (*Taxus brevifolia*). The species does not grow abundantly and taxol is collected only from trees that are over 50 years old. The natural yields of taxol are low (0.001% by dry weight of bark) and it is difficult and expensive to chemically synthesize. Plant tissue culture techniques have enabled reasonable levels (14 mg/L) of taxol to be produced and accumulated in the medium of *Taxus* cultures (Ketchum and Gibson 1996).

The moss, *Physcomitrella patens*, provides a unique system for the establishment of recombinant technologies relevant to higher plants. The lifecycle of *Physcomitrella* is dominated by the haploid gametophytic stage, which means that there are no dominant/recessive traits that can complicate interpretation of genetic screens through the influence of a second allele. Furthermore, *Physcomitrella* has an extremely efficient homologous recombination system making it a far superior system to any other seed plant and twice as efficient as mouse embryonic stem cells for gene targeting (Reski and Frank 2005). Cultures of *Physcomitrella*, which have the advantage of genetic stability compared to tissue cultured cells of higher plants, have been used to produce a humanized antibody for deep-vein thrombosis prevention (Decker and Reski 2004) and human vascular endothelial growth factor (Baur et al. 2005)

The use of plants as "factories" offers an extremely promising approach to the production of recombinant proteins and secondary metabolites. However, a number of obstacles, particularly low yield, must be overcome before this technology can truly advance. Functional genomic approaches, including metabolomics, will certainly allow this to proceed by describing biosynthetic and regulatory pathways (see Oksman-Caldentey and Inze 2004, and references therein). These approaches will enable rational engineering of biosynthetic pathways to produce metabolites of interest on demand.

5 Recent, current and future of plant metabolomics

5.1 Successful applications

Metabolomics as a technology allows a large number of metabolites of different compound classes to be analyzed and has already been successfully applied to a range of fields in plant sciences. As a tool, metabolomics is applied to answer biological questions that range from the simple to the complex, and to increase our understanding of plant biology and physiology. Metabolomics is also used for the comprehensive phenotyping of genetic varieties or genetically altered plants, for gene identification in functional genomics approaches and to monitor plant behavior and responses to challenging environmental conditions.

The model plant *Arabidopsis thaliana* has been the target of extensive metabolomics studies with different emphases. For example, the metabolome of a large range of mutants has already been analyzed and is the future focus of a large functional genomics program funded by NSF aiming for the identification of the function of all genes in this plant by 2010. In addition, an interesting investigation of the metabolic differences of a range of different accessions of *Arabidopsis* revealed that there exists a large, unexpected diversity between these accessions not only in the amounts of individual metabolites but also in the appearance of certain metabolites (Keutentjes et al. 2006). This study also demonstrated the applicability of metabolomics for high-resolution QTL analysis by untargeted LC-QTOF-MS of the metabolomes of a recombinant inbred line population (RIL) from a cross between two divergent accessions for the identification of QTLs for more than three quarters of the detected mass signals (Keutentjes et al. 2006). As the model plant, the knowledge about *Arabidopsis* genetics and physiology is immense, and large studies have been conducted to investigate cellular responses to a number of different environmentally challenging conditions. For example, the effects on the metabolite profiles have been determined for plants grown in sulphur deficient conditions. Most importantly, these measurements were done in conjunction with transcriptomics analysis to demonstrate the first attempts of integration of both types of datasets (Hirai et al. 2005, Nikiforova et al. 2005). These examples have shown the power of combining metabolomics and transcriptomics analyses for a systems biology approach towards understanding cellular responses and adaptation. Another important stress factor for plants is varying temperature as shown in the detailed characterization of metabolic adaptations to low and high temperatures (Kaplan et al. 2004; Cook et al. 2004). Interestingly, it could be shown that low temperatures have more profound effects than heat, and novel findings of metabolic adaptation to temperature stress were identified (Kaplan et al. 2004). One approach to identify adaptation mechanisms to abiotic stress in plants is to compare the cellular responses of native, stress-tolerant species and ecotypes. Gong et al. (2005) compared transcript and metabolite abundances between a *Arabidopsis* and a highly salt-tolerant related species, *Thellungiella halophila,* in response to salt stress. Some responses were similar in both species but there were also a range of differences identified in how they responded to the increased salt. These differences will lead to the identification of novel mechanisms that are either constitutively or inductively operating in stress tolerance.

Of great importance, from an agricultural point of view, will be the in-depth analysis of economically important crop plants. A number of interesting metabolomics applications have been demonstrated which have resulted in increased understanding of crop development and physiology and deciphered the impacts of certain external factors on crop quality and quantity. Tomato is one of the major crops under investigation and metabolomics methodologies based on GC-MS have been used to analyze fruit metabolites during development, and following transgenic overexpression of an *Arabidopsis*-derived hexokinase (Roessner-Tunali et al. 2003b). Tikunov et al. (2005) have focused their analysis on volatile compounds produced by the fruit resulting in new insights into fruit metabolism. Another pioneer example by Schauer et al. (2006), as previously mentioned, combined metabolomics with conventional QTL analysis to identify metabolic trait loci.

Cereal grains play an important role in nutrition. They are very carbohydrate-rich but also contain high-value proteins. Increasingly, efforts are being undertaken to understand grain development and quality in order to improve yield and nutritional value. Rice is the major primary food for most nations and because of its importance, has been the target for research for many years. To date, rice is chosen as the model plant for cereal and monocot genetics and physiology which has driven the initiative to sequence its whole genome. The application of metabolomics technologies in rice has started only recently with just a few published examples. Tarpley et al. (2005) have monitored metabolite levels in different tissue sections of developing rice seedlings, allowing the identification of biomarker metabolites being influenced by development, environment or genotype. A combination of different metabolomics techniques based on capillary electrophoresis were used to examine diurnal differences in metabolite concentrations in rice leaves (Sato et al. 2004), demonstrating the dependency of a large range of metabolites on light availability that result in changing patterns throughout the day. In addition, wheat, barley and maize are some of the most important cereal crops and a huge amount of genetic information is available and accumulating on these species from previous and ongoing breeding programs aimed at the development of stronger and higher yielding cultivars. Metabolomics as a tool to investigate metabolite levels of e.g. wheat and barley has only just begun and is mainly used to monitor responses to abiotic stress conditions (Roessner et al. 2006, Roessner, unpublished results). Abiotic stress is the major cause of substantial yield losses because tolerance mechanisms are not very well developed in commercial cultivars. The comparison of metabolite responses of these commercial cultivars with those of landraces exhibiting greater tolerance to certain stresses should lead to the determination of the role of both metabolites and genes in stress tolerance, and thus provide new strategies for breeding and genetic engineering of novel stress-resistant crops.

Legumes play a critical role in natural agriculture because of their ability to fix nitrogen in symbiotic interactions which makes them economically and environmentally important crop species. Nodule formation occurs in most legume species once a compatible *Rhizobium* bacteria strain is present in the soil. This process has been investigated in detail using a metabolomics approach by Colabatch et al. (2004) and Desbrosses et al. (2005). These reports provided novel insights into nodule formation processes but are also important examples of studying plant-microbe interactions using metabolomics.

6 Future

In recent years it has become apparent that metabolomics will be one of the most important tools in biological sciences. In the near future, many institutions and laboratories worldwide will have established the physical and intellectual capacities to apply metabolomics in their research programs. In plant research, potential applications for metabolomics are enormous and the outcomes overwhelming. Although the technologies employed in metabolomic analyses are uncovering a huge amount of new knowledge in biology, a range of challenges are still to be faced. One bottleneck in metabolomic analysis is the identification of novel compounds. Additionally, in order to allow greatest spatial resolution, the sensitivity and selectivity of currently available technologies has to be increased. Multi-parallel and high-throughput analyses result in large data sets which need be evaluated, extracted and interpreted. To do this we need to work closely together with computer scientists and bioinformaticians to improve and develop bioinformatics methodologies to extract useful and novel information out of the data flow. One step toward this would be the establishment of an open source database for metabolomics data which will attract bioinformaticians and computer scientists to use the huge data sets for large scale statistical analysis, comparative metabolomics and the development of new methodologies for data analysis, mining, visualization and interpretation. This kind of database will also allow us to compare data between labs which will ultimately lead to a better understanding of our own data. An additional major challenge in the metabolomics field is the integration of metabolic data with genomic and proteomic datasets. The ultimate goal is to comprehensively describe complex biological systems and as such, metabolomics will become an important player in systems biology.

References

Arabidospis genome initiative (2000) Analysis of the genome sequence of the flowering plant *Arabidopsis thaliana*. Nature 408:796-815

Baldwin TC, Handford MG, Yuseff M-I, Orellana A, Dupree P. (2001) Identification and characterization of GONST1, a Golgi-localised GDP-mannose transporter in *Arabidopsis*. Plant Cell 13:2283-2295

Baur A, Reski R, Gorr G (2005) Enhanced recovery of a secreted recombinant human growth factor using stabilizing additives and by co-expression of human serum albumin in the moss *Physcomitrella patens*. Plant Biotech J 3:331-340

Colebatch G, Desbrosses G, Ott T, Krusell L, Montanari O, Kloska S, Kopka J,

Udvardi MK (2004) Global changes in transcription orchestrate metabolic differentiation during symbiotic nitrogen fixation in *Lotus japonicus*. Plant J 39:487-512

Cook D, Fowler S, Fiehn O, Thomashow MF (2004) A prominent role for the CBF cold response pathway in configuring the low-temperature metabolome of *Arabidopsis*. Proc Natl Acad Sci USA 101:15243-15248

Cosgrove DJ (2001) Wall Structure and Wall Loosening. A Look Backwards and Forwards. Plant Physiol 125:131–134

Decker EL, Reski R (2004) The moss bioreactor. Curr Opin Plant Biol 7:166-170

DeLuca V and St Pierre B 2000. The cell and developmental biology of alkaloid biosynthesis. Trends Plant Sci 5: 168-173

Delmer DP, Amor Y (1995) Cellulose biosynthesis. Plant Cell 7:987-100

Desbrosses GG, Kopka J, Udvardi MK. (2005) *Lotus japonicus* metabolic profiling. Development of gas chromatography-mass spectrometry resources for the study of plant-microbe interactions. Plant Physiol. 137:1302-1318

Farré EM, Tiessen A, Roessner U, Geigenberger P, Trethewey RN, Willmitzer L (2001) Analysis of the compartmentation of glycolytic intermediates, nucleotides, sugars, organic acids, amino acids and sugar alcohols in potato tubers using a non-aqueous fractionation method. Plant Physiol. 127:685-700

Fehr M, Ehrhardt DW, Lalonde S, Frommer WB. (2004) Minimally invasive dynamic imaging of ions and metabolites in living cells. Curr Opin Plant Biol. 7:345-351

Galili G, Sengupta-Gopalan C, Ceriotti A (1998) The endoplasmic reticulum of plant cells and its role in protein maturation and biogenesis of oil bodies. Plant Mol Biol 38:1-29

Gibeaut DM, Carpita NC (1994) Biosynthesis of plant cell wall polysaccharides. FASEB J 8:904-915

Gong Q, Li P, Ma S, Indu Rupassara S, Bohnert HJ. (2005) Salinity stress adaptation competence in the extremophile *Thellungiella halophila* in comparison with its relative *Arabidopsis thaliana*. Plant J 44:826-839

Hall RD. (2006) Plant metabolomics: from holistic hope, to hype, to hot topic. New Phytol 169:453-68

Hellwig S, Drossard J, Twyman RM, Fischer R (2004) Plant cell cultures for the production of recombinant proteins. Nature Biotech 22:1415-1422

Hirai MY, Klein M, Fujikawa Y, Yano M, Goodenowe DB, Yamazaki Y, Kanaya S, Nakamura Y, Kitayama M, Suzuki H, Sakurai N, Shibata D, Tokuhisa J, Reichelt M, Gershenzon J, Papenbrock J, Saito K. (2005) Elucidation of gene-to-gene and metabolite-to-gene networks in *Arabidopsis* by integration of metabolomics and transcriptomics. J Biol Chem 280:25590-25595

James C (2005) Executive Summary of Global Status of Commercialized Biotech/GM Crops: 2005. *ISAAA Briefs* No 34 ISAAA: Ithaca, NY

Kaplan F, Kopka J, Haskell DW, Zhao W, Schiller KC, Gatzke N, Sung DY, Guy CL (2004) Exploring the temperature-stress metabolome of *Arabidopsis*. Plant Physiol 136:4159-4168

Ketchum, REB, DM Gibson (1996) Paclitaxel production in suspension cell cultures of *Taxus*. Plant Cell Tiss Org Cult 46:9-16

Keurentjes JJ, Fu J, de Vos CH, Lommen A, Hall RD, Bino RJ, van der Plas LH,

Jansen RC, Vreugdenhil D, Koornneef M (2006) The genetics of plant metabolism. Nature Genet 38:842-849

Kong Q, Richter L, Yang YF, Arntzen CJ, Mason HS, Thanavala Y (2001) Oral immunization with hepatitis B surface antigen expressed in transgenic plants. Proc Natl Acad Sci USA 98: 11539-11544

Marty F (1999) Plant vacuoles. Plant Cell 11:587-599

McGarvey DJ, Croteau R (1995) Terpenoid metabolism. Plant Cell 7:1015-1026

McLean BG, Hempel FD, Zambryski PC (1997) Plant intercellular communication via plasmodesmata. Plant Cell 9:1043-1054

Moore PJ, Staehelin LA (1988) Immunogold localization of the cell-wall-matrix polysaccharides rhamnogalacturonan I and xyloglucan during cell expansion and cytokinesis in *Trifolium prantese* L.; implication for secretory pathways. Planta 174: 433– 445

Nebenführ A, Staehelin LA (2001) Mobile factories: Golgi dynamics in plant cells. Trends Plant Sci 6:160-167

Nikiforova VJ, Kopka J, Tolstikov V, Fiehn O, Hopkins L, Hawkesford MJ, Hesse H, Hoefgen R. (2005) Systems rebalancing of metabolism in response to sulphur deprivation, as revealed by metabolome analysis of *Arabidopsis* plants. Plant Physiol 138:304-318

Oksman-Caldentey K-M, Inze D (2004) Plant cell factories in the post-genomic era: new ways to produce designer secondary metabolites. Trends Plant Sci 9:433-440

Orellana A (2005) Biosynthesis of non-cellulosic polysaccharides in the Golgi apparatus. Topological considerations. Plant Biosystems 139:42-45

Oxley D, Currie G, Bacic A (2004) In: Purifying Proteins for Proteomics. A laboratory manual. Pp 579-637. R.J. Simpson (Ed) Cold Spring Harbour Laboratory Press New York

Palacpac NQ, Yoshida S, Sakai H, Kimura Y, Fujiyama K, Yoshida T, Seki T (1999) Stable expression of human β1,4-galactosyltransferase in plant cells modifies N-linked glycosylation patterns. Proc Natl Acad Sci USA 96:4692-4697

Pauly M, Scheller HV (2000) O-Acetylation of plant cell wall polysaccharides: identification and partial characterization of a rhamnogalacturonan O-acetyl-transferase from potato suspension-cultured cells. Planta 210:659-667

Peng L, Kawagoe Y, Hogan P, Delmer D (2002) Sitosterol-β-glucoside as primer for cellulose synthesis in plants. Sci 295:147-150

Prance GT (2001) Discovering the plant world. Taxon 50:345-359

Ramm M, Wolfender J-L, Queiroz EF, Hostettmann K, Hamburger M (2004) Rapid analysis of nucleotide-activated sugars by high-performance liquid chromatography coupled with diode-array detection, electrospray ionization mass spectrometry and nuclear magnetic resonance. J Chromatogr A 1034:139-148

Rayon C, Lerouge P, Faye L (1998) The protein N-glycosylation in plants. J Exp Bot 49:1463–1472

Reiter W-D, Vanzin GF (2001) Molecular genetics of nucleotide sugar interconversion pathways in plants. Plant Mol Biol 47:95-113

Reski R and Frank W (2005) Moss (*Physcomitrella patens*) functional genomics – Gene discovery and tool development, with implications for crop plants and human health. Briefings in Functional Genomics and Proteomics 4:48-57

Rischer H, Oksman-Caldentey KM. (2006) Unintended effects in genetically modified crops: revealed by metabolomics? Trends Biotech 24:102-104

Roessner-Tunali U, Urbanczyk-Wochniak E, Czechowski T, Kolbe A, Willmitzer ., Fernie AR (2003a) *De novo* amino acid biosynthesis in plant storage tissues is regulated by sucrose levels. Plant Physiol 133:683-692

Roessner-Tunali U, Hegemann B, Lytovchenko A, Carrari F, Bruedigam C, Granot D, Fernie AR (2003b) Metabolic profiling of transgenic tomato plants overexpressing hexokinase reveals that the influence of hexose phosphorylation diminishes during fruit development. Plant Physiol 133:84-99

Roessner U, Patterson J, Forbes MG, Fincher G, Langridge P, Bacic A. (2006) An investigation of boron toxicity in barley using metabolomics. Plant Physiol 142:1087-101

Ronning CM, Stegalkina SS, Ascenzi RA, Bougri O, Hart AL, Utterbach TR, Vanaken SE, Riedmuller SB, White JA, Cho J, Pertea GM, Lee Y, Karamycheva S, Sultana R, Tsai J, Quackenbush J, Griffiths HM, Restrepo S, Smart CD., Fry WE, van der Hoeven R, Tanksley S, Zhang P, Jin H, Yamamoto ML, Baker BJ, Buell CR (2003) Comparative analyses of potato expressed sequence tag libraries. Plant Physiol 131: 419-429

Saito K, Dixon RA, Willmitzer L. (Eds) (2006) Plant Metabolomics. Vol. 57 of the series Biotechnology in agriculture and forestry. Eds. Nagata T., Lörz H. and Widholm JM. Publ. Springer-Verlag Berlin-Heidelberg, Germany

Sato S, Soga T, Nishioka T, Tomita M (2004) Simultaneous determination of the main metabolites in rice leaves using capillary electrophoresis mass spectrometry and capillary electrophoresis diode array detection. Plant J 40:151-163

Schad M, Mungur R, Fiehn O, Kehr J. 2005. Metabolic profiling of laser microdissected vascular bundles of *Arabidopsis thaliana*. Plant Methods 1:2

Schauer N, Semel Y, Roessner U, Gurb A, Balbo I, Carrari F, Pleban T , Perez-Melisa A, Bruedigam C, Kopka J, Willmitzer L, Zamir D, Fernie AR (2006) Quantitative genetics of metabolite accumulation in intraspecific introgressions of tomato. Nature Biotechnol 24: 447-454

Scheible W-R, Pauly M (2004) Glycosyltransferases and cell wall biosynthesis: novel players and insights. Curr Opin Plant Biol 7:285-295

Schena M, Shalon D, Davis RW, Brown PO (1995) Quantitative monitoring of gene expression patterns with a complementary DNA microarray. Science 270:467-470

Somerville C, Somerville S (1999) Plant functional genomics. Science 285:380-383

Staehelin AL, Newcomb EH (2000) Membrane structure and membranous organelles. In Biochemistry and Molecular Biology of Plants, B Buchanan, W Gruissem, R Jones, Eds. American Society of Plant Physiologists, USA

Sturm A. 1995. N-glycosylation of plant proteins. *New Compr. Biochem.* 29a:521–542

Sumner LW, Mendes P, Dixon RA. (2003) Plant metabolomics: large-scale phytochemistry in the functional genomics era. Phytochem 62:817-836

Tacket CO, Mason HS, Losonsky G, Estes MK, Levine MM, Arntzen CJ (2000) Human immune responses to a novel norwalk virus vaccine delivered in transgenic potatoes. J Infect Diseases 182:302–305

Tarpley L, Duran AL, Kebrom TH, Sumner LW (2005) Biomarker metabolites capturing the metabolite variance present in a rice plant developmental period. BMC Plant Biol 5:8

Tikunov Y, Lommen A, de Vos CH, Verhoeven HA, Bino RJ, Hall RD, Bovy AG. (2005) A novel approach for nontargeted data analysis for metabolomics. Large-scale profiling of tomato fruit volatiles. Plant Physiol. 139:1125-1137

Tohge T, Nishiyama Y, Hirai MY, Yano M, Nakajima J, Awazuhara M, Inoue E, Takahashi H, Goodenowe DB, Kitayama M, Noji M, Yamazaki M, Saito K (2005) Functional genomics by integrated analysis of metabolome and transcriptome of *Arabidopsis* plants over-expressing an MYB transcription factor. Plant J 42:218-235

Urbanczyk-Wochniak E, Baxter C, Kolbe A, Kopka J, Sweetlove LJ, Fernie AR. (2005) Profiling of diurnal patterns of metabolite and transcript abundance in potato (*Solanum tuberosum*) leaves. Planta 221:891-903

Willats WGT, Orfila C, Limberg G, Buchholti HC, van Alebeek G-JWM, Voragen AG J, Marcus SE,. Christensen TMIE, Mikkelsen JD, Murray BS, Knox JP (2001) Modulation of the degree and pattern of methyl-esterification of pectic homogalacturonan in plant cell walls. J Biol Chem 276:19404–19413

Villas-Boas S.G., Roessner U., Hansen M., Smedsgaard J., Nielsen J. (2006) Metabolome Analysis. Publ. Wiley & Sons, New Jersey, NJ, USA (in press)

von Roepenack-Lahaye E, Degenkolb T, Zerjeski M, Franz M, Roth U, Wessjohann L, Schmidt J, Scheel D, Clemens S (2004) Profiling of *Arabidopsis* secondary metabolites by capillary liquid chromatography coupled to electrospray ionization quadrupole time-of-flight mass spectrometry. Plant Physiol 134:548–559

Yu J, Hu S, Wang J, Wong GK-S, Li S, Liu B, Deng Y, et al (2002) A draft sequence of the rice genome (*Oryza sativa* L. ssp. *indica*). Sci 296:79-92

Pettolino, Filomena
School of Botany, The University of Melbourne, 3010 Victoria, Australia

Roessner, Ute
Australian Centre for Plant Functional Genomics, School of Botany, The University of Melbourne, 3010 Victoria, Australia
ute.roessner@acpfg.com.au

Index

MIX
Papier aus verantwortungsvollen Quellen
Paper from responsible sources
FSC® C105338

If you have any concerns about our products, you can contact us on
ProductSafety@springernature.com

In case Publisher is established outside the EU, the EU authorized representative is:
Springer Nature Customer Service Center GmbH
Europaplatz 3, 69115 Heidelberg, Germany

Printed by Libri Plureos GmbH
in Hamburg, Germany